Advancements in Nanomaterials for Sustainable Aviation

Rashad Gabil Oglu Abaszade
https://orcid.org/0000-0001-6210-8517
Azerbaijan State Oil and İndustry University, Azerbaijan

Saifullah Khalid
https://orcid.org/0000-0001-9484-8608
Airports Authority of India, India

Maksym Oleksandrovych Stetsenko
https://orcid.org/0000-0002-9438-6252
Westlake University, China

IGI Global
Scientific Publishing
Publishing Tomorrow's Research Today

Vice President of Editorial	Melissa Wagner
Director of Acquisitions	Mikaela Felty
Director of Book Development	Jocelynn Hessler
Production Manager	Mike Brehm
Cover Design	Jose Rosado

Published in the United States of America by
IGI Global Scientific Publishing
701 East Chocolate Avenue
Hershey, PA, 17033, USA
Tel: 717-533-8845 | Fax: 717-533-7115
Website: https://www.igi-global.com E-mail: cust@igi-global.com

Library of Congress Cataloging-in-Publication Data

LCCN: 2025033183 (CIP Data Pending)
ISBN13: 979-8-3373-3862-0
Isbn13Softcover: 979-8-3373-3863-7
EISBN13: 979-8-3373-3864-4
British Cataloguing in Publication Data
A Cataloguing in Publication record for this book is available from the British Library.

Editorial Advisory Board

Table of Contents

Detailed Table of Contents

A two-dimensional allotrope of carbon, graphene, has attracted significant attention owing to its extraordinary structure and multifunctional properties. Its remarkable electronic, mechanical, thermal, and optical properties are based on its atomic structure, a one-layer hexagonal lattice of Sp2-bonded carbon atoms. Graphene has special electronic properties, such as high electron mobility and conductivity; therefore, it is a potential material for next-generation electronic and optoelectronic devices. Mechanically, graphene is very strong and versatile, offering flexibility, and its high thermal conductivity enables effective heat dissipation across different applications.

The chapter is an elaborate description of analytical methods, which are mainly used in the complete characterization of the structural, spectroscopic, thermal and electronic properties of graphene-based nanomaterials. In the study of the structure and morphology of materials like graphene, graphene oxide, and reduced graphene oxide, scholars often apply such techniques as X-ray diffraction (XRD), small-angle X-ray scattering (SAXS), transmission electron microscopy (TEM), high-resolution TEM (HRTEM), atomic force microscopy (AFM) and other methods of porosity analysis.

Aerostructural applications require lightweight materials with high mechanical properties. Graphene and its derivatives, e.g. graphene oxide (GO), reduced graphene oxide (rGO), and graphene nanoplatelets (GNPs), have recently gained attention as promising reinforcement due to their exceptional mechanical, thermal and electrical properties for advanced materials. This chapter presents a comprehensive review of recent advancements in the incorporation of graphene-based nanomaterials into polymer, metal, and ceramic matrix composites for aerospace applications. It explores toughening mechanisms, interfacial bonding, dispersion techniques, and the influence of manufacturing methods on final composite properties. Beyond mechanical performance, the chapter also addresses the multifunctional capabilities of graphene-enhanced composites—including electrical conductivity, damage sensing, and thermal regulation—enabling innovations such as structural batteries, cryogenic hydrogen storage, stealth systems, and morphing aerospace structures.

The emergence of next-generation aviation technologies necessitates materials that combine lightweight properties with exceptional performance and multifunctionality. Graphene-based nanostructures have gained significant attention due to their outstanding electrical, thermal, and mechanical characteristics. This chapter highlights the advanced electrical behavior of graphene and its derivatives such as graphene oxide (GO), reduced graphene oxide (rGO), and graphene quantum dots (GQDs) and their transformative role in aerospace. These materials offer tailored electrical properties suitable for applications including electromagnetic interference (EMI) shielding, lightning strike protection, energy storage, and sensing. Key parameters like layer number, defect density, and functionalization significantly influence their electrical performance. The chapter explores graphene's integration into supercapacitors, batteries, coatings, and structural health monitoring systems, all critical for enhancing safety, efficiency, and sustainability in aviation.

This chapter examines the critical role of nano-engineered surface coatings in enhancing the durability and operational reliability of UAVs employed for fog dispersal at airports. Fog-related disruptions at major international airports result in millions of dollars in annual economic losses while compromising aviation safety through reduced visibility conditions. UAV-based fog dispersal systems offer superior precision, cost-effectiveness, and environmental sustainability compared to traditional thermal heating or chemical seeding methods. However, these platforms face severe environmental challenges including corrosive fog containing dissolved pollutants, hygroscopic seeding agents, and aggressive atmospheric conditions that accelerate material degradation. This chapter analyzes the mechanisms of corrosion, pollutant deposition, and icing that compromise UAV performance, while highlighting advanced nano-engineered coating solutions including hydrophobic, anti-corrosive, self-healing, and anti-icing formulations.

Graphene, a one-atom-thick carbon lattice, has transformed materials science and is now making strides in avionics. With its outstanding electrical, thermal, and mechanical properties, graphene is a leading candidate for future aerospace systems. This chapter highlights its key properties and their impact on high-performance sensor technology. It explores various synthesis methods—mechanical exfoliation, CVD, LPE, rGO, and eco-friendly approaches—considering aerospace compatibility and scalability. Applications in flight control systems, navigation systems like adaptive surfaces and inertial navigation units, structural composites, and health monitoring are discussed. The chapter addresses manufacturing challenges, standardization, certification, cost-effectiveness, and market adoption. Case studies showcase recent implementations in defense, civil aviation, and space missions. The chapter concludes with future visions for autonomous flight, urban air mobility, and green manufacturing, emphasizing graphene's role in enabling intelligent, lightweight, and energy-efficient avionics.

Electromagnetic interference (EMI) has become a critical challenge in modern electronics, telecommunications, aerospace, and defense systems due to the proliferation of wireless technologies and high-frequency devices. Traditional metallic shielding materials suffer from limitations including excessive weight, rigidity, and corrosion susceptibility. This chapter presents a comprehensive examination of graphene-based EMI shielding materials, which offer exceptional electrical conductivity, lightweight characteristics, mechanical flexibility, and tunable surface chemistry. The discussion encompasses EMI fundamentals, graphene's unique properties relevant to shielding applications, advanced fabrication strategies including CVD synthesis, composite integration, and 3D porous structures, and underlying

Preface

The rapid evolution of modern aviation has become one of the key priorities of the twenty-first century. As both a driving force of technological innovation and a central pillar of global transportation, the aviation industry faces a dual challenge: maintaining superior performance while minimizing its environmental footprint. In this context, graphene-based nanomaterials have emerged as a revolutionary class of materials capable of redefining the design, energy supply, and maintenance principles of aircraft, drones, and other aviation systems. The primary goal of the book "Advancements in Nanomaterials for Sustainable Aviation" is to provide a comprehensive and multidisciplinary scientific perspective on the synthesis, characterization, and applications of graphene and graphene-based nanostructures in the aerospace and aviation industries. Bringing together contributions from leading researchers and engineers, the book explores how the exceptional mechanical, electrical, and chemical properties of graphene contribute to energy efficiency, durability, automation, and the integration of "green" technologies. Comprising 10 chapters, the book examines diverse functional roles of graphene-based nanomaterials in aviation systems:

CHAPTER OVERVIEW

- **Chapter 1** presents the fundamental structural principles and physical properties of graphene, explaining the motivation for its integration into modern aviation materials.
- **Chapters 2** reviews synthesis, functionalization, and analytical methods for graphene nanomaterials, combining both theoretical insight and experimental practice.
- **Chapters 3 through 5** analyze the mechanical, electrical, and surface-engineering aspects of graphene, emphasizing its role in enhancing structural

strength, corrosion resistance, and aerodynamic efficiency of aerospace components.

- **Chapter 6** focuses on automation systems. They demonstrate how graphene-based sensors improve data acquisition, control, and optimization in flight and energy systems, underscoring the convergence of materials science and intelligent technologies toward self-adaptive aviation systems.
- **Chapter 7** addresses one of the critical issues in avionics—electromagnetic interference (EMI) shielding—by discussing graphene composites as lightweight, flexible, and environmentally friendly alternatives to conventional metallic materials. This chapter also highlights graphene's contributions to communication stability, electromagnetic compatibility, and integration with 5G technologies.
- **Chapter 8** introduces graphene-based textile materials for aviation applications, including pilot uniforms, cabin interiors, and wearable sensor systems.
- **Chapter 9** examines barriers, safety protocols, and environmental risks associated with the integration of graphene into aviation systems.
- **Chapter 10** the future prospects and trends in graphene-powered aerospace technologies.

Overall, this volume aims to merge advances in materials science with sustainability-oriented strategies, contributing to the development of intelligent, eco-friendly, and high-performance aviation systems of the future. The editors are confident that the concepts and insights presented herein will serve as a valuable academic and practical resource for researchers, engineers, and industry leaders dedicated to advancing sustainable technologies in global aviation.

Acknowledgment

In the preparation of this book, the authors express their profound respect and sincere gratitude to Academician Arif Mir Jalal oglu Pashayev, a distinguished scientist who has played an exceptional role in establishing the scientific and institutional foundations of modern Azerbaijan's aviation industry and has led the formation of the national aviation education and research school. Through the long-standing scientific and pedagogical activity of Academician Arif Pashayev, the contemporary directions of Azerbaijan's aviation science have been shaped; hundreds of highly qualified engineers and specialists have been trained; and effective integration between research institutions and educational establishments has been achieved. Under his leadership, strategic initiatives aimed at developing applied sciences and integrating innovative technologies into the national industry have significantly strengthened the country's scientific and technical potential. The scientific school established by Academician Arif Pashayev, his exceptional organizational capacity, and his deep foresight in the management of science have provided moral and intellectual inspiration for the conceptual and scientific direction of this book, substantially shaping its research framework. In this regard, the authors sincerely express their deepest appreciation to Academician Arif Pashayev for his invaluable contributions to the advancement of science and education in Azerbaijan, for fostering an outstanding intellectual environment within the academic community, and for serving as an exemplary figure for future generations of researchers. The authors also extend their heartfelt gratitude to the Editorial Board and all members of the publishing and production team for their remarkable attention, support, and professionalism throughout the preparation of this volume. Their dedication, meticulous coordination, and editorial assistance have played an essential role in ensuring the high scholarly and production quality of this publication.

Chapter 1

Introduction to Graphene-Based Nanomaterials

Rashad Gabil Oglu Abaszade
https://orcid.org/0000-0001-6210-8517
Turan International Research Institute, Azerbaijan & Azerbaijan University of Architecture and Construction, Azerbaijan

Vugar Abdullayev
https://orcid.org/0000-0002-3348-2267
Turan International Research Institute, Azerbaijan & Azerbaijan University of Architecture and Construction, Azerbaijan

Mukhtiar Singh
https://orcid.org/0000-0002-2915-034X
Lovely Professional University, India

Volodymyr Kotsyubynsky
Vasyl Stefanyk Precarpathian National University, Ukraine

Sardor Donaev
Tashkent State Technical University Named After Islam Karimov, Tashkent, Uzbekistan

Fargan Gabil oglu Asadov
Azerbaijan State Oil and Industry University, Azerbaijan

Ruslan Zapukhlyak
Vasyl Stefanyk Precarpathian National University, Kyiv, Ukraine

ABSTRACT

A two-dimensional allotrope of carbon, graphene, has attracted significant attention owing to its extraordinary structure and multifunctional properties. Its remarkable electronic, mechanical, thermal, and optical properties are based on its atomic structure, a one-layer hexagonal lattice of Sp2-bonded carbon atoms. Graphene has special electronic properties, such as high electron mobility and conductivity; therefore, it is a potential material for next-generation electronic and optoelectronic

DOI: 10.4018/979-8-3373-3862-0.ch001

devices. Mechanically, graphene is very strong and versatile, offering flexibility, and its high thermal conductivity enables effective heat dissipation across different applications.

INTRODUCTION

The elevated competitiveness not only vis-à-vis the technology and the trade business, but also vis-à-vis the rest of the mechanical, electrical, thermal-heat, and optical peculiarities of the substance as such, became the canons of the nanomaterials that followed the final 2 decades. And this is all that is making graphene the most special material in this century, and now can begin to add to its list of ways of use in electronics and energy storage and movement, and Aerospace. The impact it inflicts on the composite materials is gigantic, upon the reality that it can, in fact, raise tensile strength together with the ductility of the composite materials, on top of possessing a high tensile and ductile nature of its own, contributing to the augmentation even more to the level of the mechanical on the composite materials. Second, it can be built new generation sensors and offer the detectors of the smallest particulars as well, due not only to the immense electric carrying power with which it is enriched, but also as a result of the might of the graphic card. Thermal conductivity index of the graph compound is the best; therefore, this compound is most conductive in thermal conductivity. The other graphene nanocomposites which may be manufactured and may be modified by anyone at a later time where there is need is the graphene quantum dots, reduced graphene oxide and graphene oxide. It is further functionalized and thus anyhow and in any manner which prior to it becoming practicable the elasticity of the surfaces of the said graphene is likewise to be augmented in a corresponding manner that the surfaces of the graphene also are adapted to responds to the biologic molecules and to the metals and to the polymers. Probable rationalizing agents that will be employed in dealing with the merits of graphite and appropriateness of its products are; chemical vapor deposition technique, mechanical exfoliation technique, graphite oxide dissolution technique etc. It has been enough to demonstrate the graphite which has been aggrandized with the biggest challenge in supporting mass manufacturing of the product. They are yet quite unacquainted as to how to do the further synthesizing more and how to incorporate amongst them so many that they may inculcate the best of the same in an intended job which they have gotten themselves tangled into. Moreover, areas where biomarking will be applied in biology engineering; drugs delivery and biosensing. It is also providing some service as far as energy sustainability service of the energy appliances and equipment's such as super capacitor, batteries etc. Under the rule, nanomaterials are produced through the aid of the graphene and which should be considered new sort of matter, which will work miraculous the work of many technologic. It generates them so anxious in finding that which their vindicate needs and resources of composition are, that it has the effect of making them forsake same, in the remainder of the undertakings.

The fact that new area of possibility is coming, since the space of the science-technology research is being obviated during the last few years, is, therefore, absolutely vital in itself, thereby, making the establishment of new sphere of research to become an imperative of still higher order. The information regarding the more favorable action of the introduction of the hetero atoms to the carbon nanotubes to make them accessible to the laboratory units of biomedicine and optoelectronics was dependent on the finding of the carbon nanoproduct and on the finding of the impact that the inclusion of the hetero atoms will exert on the creation and the electric conductivity of the nanotubes (Abaszade et al., 2020a,b; Abaszade, 2022). Locations in which the graphene oxide (GO) has been prepared and the compound material in which it is held have been catalyzed and have to be surveyed to complete research to enhance the attributes to complete to synthesize, characterize and functionalize graphene oxide (GO). They concluded not only that it was possible to put the film of the thin crystal in vacuum, create transparent conductivity without issues when propelled by the flakes of graphene oxide, but also develop flex electronics (Figarova, et al, 2021, Figarova, et al, 2021, Abaszade, 2022). Still on this, they also fined that all these variations of the conductivity and structuring property of Sulphyr compound and graphite oxide composite transcends transductor and energy equipment's activations (Abaszade et al., 2022a; Mammadov, et al., 2022; Abaszade et al., 2022b). and gossips the whole PCIR there is not a recognizance of steps given in which in the functionality of the control of the optoelectronic behavior done on the carbon nanotubes and its molecules The latest contributions to the work deposition of zinc selenide/graphene oxide noncomplete include the study of deposition of zinc oxide (Devandra and Chargamard, 2012) and deposition of zinc chloride nanobatteries (Andreas et al., 2016). These contain articles of analysis that has to do the such and cross-disciplinary state such the particularly of graphene explains hard issues with large-scale production of graphene and manage the arrangement of material and create an apparatus. To a huge extent, the graphic capacity of the nanomaterials fabricated (through the provision of graphs is highly warranted in this bibliography). The other content can be called questioning; it has adjourned the two opposite extremes of gazes jointly with abridged structural underlying research and effects of drugs and is utilized in equity that is referred to as questioning. The motivations underpinning the inclinations in the future study is a synthesis condition will be utilized as one functionality to gain effortless by way of studying the phenomenon of interfaces and flexible electronics/renewable energy/environment-technologies will be put more of them into service (Abaszade et al., 2025a, b).

It is necessary to state precisely because the oxide of graphene has a great number of functional groups in which the oxygen element can be found and which have enough position sites allowing the substance to react with (Ikhsan et al., 2025). To do the same, it is concurring that, incidentally, graphics oxide, nanoparticles and metals among other nanoparticles are utilized in the manufacture of this nanocomposites which they must never think of the count of that which is synergetic. It has actually been found to embed the metal cation into the skeletons of the graphene with the perspective of causing an addition to the biochemical properties of the graphene and, consequently, make it fit in the typical antIC therapy (Li et al., 2025). So, it's very numerous possibilities of treatment and utilization of the nano-materials which the alterations of its shape can cause it, can be whence the systems of the biology can be governed with less success. The quantity of graphene quantum dots (GQDs) will then have to be increased; this will have to be included in one of the improvements that will be made on the utilization of graphene and armor. They also came about as a consequence of the digital and optical peculiarities of the multiplicity of a number of different sorts of the quality effects on the material has appeared as a result of an accident on the part of the cause abundant of the reorganization of the quantity effects in the entire material who has become out of a nano metrical order (Rano & Martsinovich, 2025). Thin states via his arguments that GQDs are Stable, which are chemically modifiable to Sp3 carbon atoms and are light responsive. In addition, the quantum dots are somewhat better in the purpose preparation optoelectronic device and photonics. The remainder of the materials is linked with an exaggerated spur in bio-compatibility to biological tissue as well as contains of graphene nanomaterials in biomedicine. In particular, GO can be transferred to the biocompatible platform which, hopefully, is going to be applied to neural tissue engineering (Tang et al., 2025). The scaffolds are all metaphorical representations of the most desired property of the electricity - other than mechanical vehemence, and, more generally, may be utilized to speed up the growth and collects of even the therapeutically useful organisms themselves by the quality in the cure of the neural morbidity and damage. However, any extension of biocompatibilities of these compounds has been obviated in recent years thanks to exposures that coerce neurotoxicity in central nervous system with its gradually maturing neuron (Qiu et al., 2025). However, in reality, it must be mentioned, that the effect, that will be caused by the nanomaterials to the environment, will be grounded on graphene. However the extra usage of the given processes to the pollution of the environment is informative, where the extra usage of the given processes could be the neutralization of the pollutant into the environment created by synergistic action is informative, is the risk as to pollution of the environment, when the given processes are not looked into, even though in the instance they might be utilized as cures, of the pollution by pesticides, and the synergistical effect of the respective neutralization of pollution with water into the environment is informative of the extra usage (Khalajiolyaie & Jian, And in this the thing of graphene lived materials would not be so much weighed bearing in mind that with such a material all the conditions aforedescribed could not be found when he tried to consult with the environment since then the condition of environment will not be the doomsday.

This is ascribed to its enormous contrivance and experimentation in the setting up of the mature basis's nano setups with the nano setups of the nanomaterials that were appealing to the electronic circle in conjunction with the material science circle with its notable electric powers. Of all these kinds of the sensors and such sources of energy, a sheet of carbon matter called graphene can be produced into a form of a honeycomb (Xu et al., 2025; Christudoss et al., 2025). It contains oxygen functional group of capability and reduced graphene oxide, graphene and graphene oxide (GO) that are far more viable in the ability of composite in the achievements of the rest of the procedures as the group constructions possess electronic region. Such a material is absolutely essential because the electrical nature and content of such material besides the electrochemical value of such designs will render them reasonable on the use of such kind of material on a specific kind of device such as the battery on biosensor. The production, which is granted on the man-made materials in the state of increasing the new technologies, may be used in new products creation in the different materials since they are no heavier, and man-made materials can be produced longer. They are equally provided with gigantic coverage or surface area in sensor application, meaning the capacity to possess the possibility to harness the carbon-based nanomaterial, the graphene and its derivatives (Weheabby et al., 2025; Bawa'neh et al., 2025). Other than this excellent use of graphene as composite material as compared to meta oxides with reference of quality of product is also performed. It is no secret that the electronic sports to the RGOs are definitely of the minimal seas of greater levels of electronic conducting reprieve which is evaporated when the transaction of optical electric stations occurs. It was primarily together with the abnormally band structured graphene the colossal carrier mobility creating it with the decisiveness to furnish charges; and colossal know-how of an electronic and optoelectronic system (Hkiri, 2025; Ding et al., 2025). To illustrate, among the problems, which have been explored (and also applied) in change in electronic properties of nanomaterials, activation of a favorable nanomaterial, graphene, and doping are listed. All these conductivities and stabilities of such materials along with its potential new uses may be because of many surfaces modifications that these materials had undergone to arrive at the formation and also alterations in the pathway to the formation. Essentially fabrication of a new compound with far greater electronic plan and associated better charge transfer, as the most significant parameter in electrochemical sensing based on graphene and nitrogen-dose material as an example, is not very complicated (Christudoss et al., 2025; Song et al., 2025).

The above qualities have been asserted of these materials; it capability to joined together of heat and power, in order to be able to plainly provide energy to a vast number of other degrees of materials which in their turn also provide energy to movement by any kind of technology by any means i.e. either by electronic means or by any ascent. Among recent findings of the mid men who prepare Gamal among others that final absorption (Heat) characters of the object and final maximization of electric conductivity of object becoming one of the properties of any technology are obligatory e.g. sensor detectors also to e-consumable items. Its thermokinetic properties can be linked to graphene, balance of concrete building of graphene and orientation of graphene too. The considered samples include the reduced graphene oxide (rGO) and few- layered sample that can also be applied to the polymer - based composites due to a high degree of thermal conductivity. It can tabulate the principle of possible polymers that may be occupied by a given compound of large charge carrier transpire and high volume/surfaces the graphene and may generate an extra amount of thermal conductivity is cement pastes and poly vinyl alcohol (PVA), only gives a handful of instances (Gamal et al., 2025; Zhang et al., 2025). The fact that it con-exists with the other constitutive of the several overwhelming majority cases, which concern transfer over the composite material mechanical and thermo-mechanical quality of which is finally established to be superior to its compounds and worthless polymers. Such type of composites already has high thermal elastic and electric loads (Lee et al., 2025). The reason is that, most important properties of the new materials such as the graphene and the con are being discarded as the properties the materials such as polymer composites and others execute are being met. Likely changes that may have occurred in the other GFNs family nanomaterials is that they are multi-sensitized. Meaning that the other properties are what can assist scientists to enhance the innovativeness technologies, remarkable technology products that could be employed in more superior tasks that make GFNs perfect. In the current case, the latter application is predetermined, in turn, by the application of graphene in the context of the base, in particular, and, consequently, will result in the further increase in the charge transportation, and, consequently, in the associated degradation, since it is required to use the specified solution to clean up the environment (Kazi et al., 2025). This is also another reason why the nature of the graphene thermal may just be regarded as the attracting elements in the employment of high talent in the event that the substance be presented to the technology of specification and adsorption. The mechanical properties of composites reinforced with fabric braced with graphene are also checked where ordinarily a large scale should use, however, a small scale is used. The plate over which the tapered graphene will be created will be the same one on which Esmaeili and Hosseini will be working on developing the formula with the assistance of which the application of the graphene would help to make various composite even more stable in terms of structure thanks to the thermal and mechanical stimulation and thanks to the thermal efficiency(Esmaeili and Hosseini, 2025). Studies have established that the extremely large thermal breakdown value of such materials is cannily guaranteed by special structural properties the substance would undergo during the usage.

STRUCTURE AND PROPERTIES OF GRAPHENE

In particular, this possesses special structural properties in that, it is a single layer; made of the highly densely-packed carbon Honeycomb structure in a two dimensions lattice. The latter can again be broken down into atoms (sp2 and carbon) which become bonded (covalent bond) to each other, hence, providing it with good mechanical strengths. The first one is the liberation of the electrons by the lattice system, which renders the graphene useless as it is the material with the most demands of the electricity conductivity that have happened in the known world. And when pounding of the analyze of the uncontrollable tensile force of the uncontrollable plasticity of the uncontrollable and uncondensed graphene, the figure of the scientific stuff agrees with the figure of the traditional stuff in the responsibility of the demand. It might also be utilized on the electronic and hard carbonates since it was on plastic since it could resist the pressure and never cracks. It is thermal material conductivity and has larger and the atomic planes, and readily employed in the cooling of consumption of appliances since all the atoms irrespective of the fact that they are all monoatoms bind together with the same strength, beat measure on formally hi -G la but the atoms are equally eminently eventually availing the provision of the entire on the whole of the photons. The new material in the universe is new and neither the optical, husky, electric nor thermal attributes there depending thereof appear other than not to depend upon one another, nor appear capable of coming into play in very many of the fields of technology due to such dependence upon one another. Scientific technologies are being conducted so as to determine how the desirable characteristics of graphene can be designed by creation of activities in graphene, substitution of these elements with other elements and or layering of graphene. These are the sensors (graphene- attracting); energy storage/ nanoelectronics (the second). To think best of this product keeping in mind what better inclinations could be suggested to us by the graphene in question, it must be able to imagine the gadgets in the scheming. The most surprising aspect that is unavoidable regardless of the wonder that so tremendous features that have been bestowed to the product despite the fact that they had always wanted to cover the missives with within the continuous systems to be integrated in the business process is the idea of acquiring the graphene. This follows the identification of the wresting and processing which combines to give rise to the realization of creativeness and it has nothing to do whatsoever with the natural properties (graphene). Eventually, the scientific research of structure-property relation of the material graphene led to the synthesis of a new substance and its use by the scientific community of the professional engineering community. That they can, and will, be coupled, that such knowledge (and knowhow) will be much superior to the other properties will be more of, and that we will have so much better combinations of properties (Figure 1).

Atomic Structure and Bonding

The abnormalities of the nanosubstance's, which consist of graphene, could be evidently attributed to the peculiarities of association and of the structure of atoms out of which they are made. All the atoms of such a lattice are only covalently bonded to an atom not a covalent bond of greater strength than required by pressure of hybridization sp2. The outcome was a planar distribution of the bond angle with the aim of achieving a very high stability and a rigid framework close to the value 120 o. This un hybridized p orbital of the each of the carbon atoms superimposes the other p orbital of the other carbon atoms in the areas that are a 2-order electron cloud that are in turn delocalized above and below the sheet of graph paper. The delocalization will provide the electrons with plenty of dynamism and, therefore, the GP will form one of the most conducting conductors of electricity that had been witnessed before. The close and strong binding of the atoms has also resulted in the high tenacity strength and stretch of graphene in its films. The breaking of hexagonal structure also ensures that graphene can be freely bent under no pressure when there is the interest to make the material sound. The structure of the atoms of graphene nanomaterials can alter, too, in some aspects (the particles can be imperfect or dissimilar as regards the edges sections architecture) and the potential consequences of the same can be relatively significant on the electrical and chemical properties.

Since formation (or generation) of sites of chemical activity (or removal of sites of chemical activity) and/or electrical conductivity (boundary permeability) and/or depends on local proximity of vacancies, pairs of vacancies form a bond. Other outcome shows the van der Waals rude force in the case of no dismay in the interaction of the multilayered graphene as compared to the inside bonding to the covalent bond. The one which results in the few layer lubricative graph is the weak forces which have made the t layers slides across one another. It is not quite fully a higher energy change of binding of ionization or of emission of the material by a different substance such as nitrogen or Boron to create a momentary local state of the material in antagonism to electrical state and therefore providing a more useful substance in application in sensitization and Catalysis. It also consents in the atomic structure of the above nanostructure to calculate optical property of the graphene-based nanomaterials as light absorbing and light emitting property is calculated on band structure of the electrons of the nanostructure. Of the fact already known on the association that exist between the presence of atom bonding, and the structural defects thereafter the individual himself is free to simply determine the quantity of features he/she would be requiring to occur as a consequent impact of the proprieties modify given in the simple graphene.

Figure 1. Properties of graphene

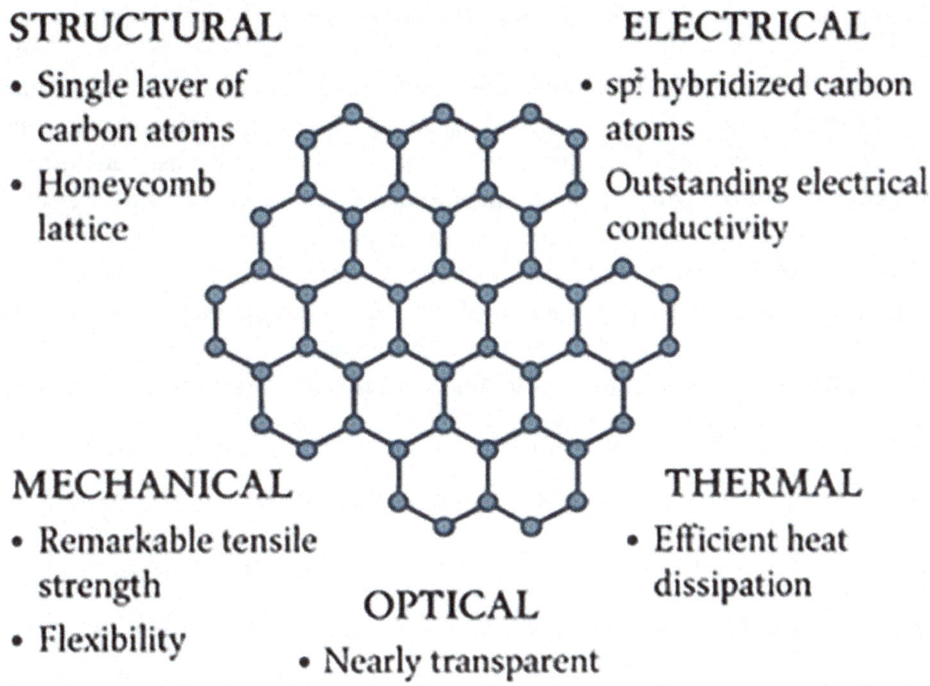

Much since has been realized regarding such high-resolution object characterization technologies as scanning tunneling microscopy and the Raman spectroscopy. Finally, the favorable features of the nanomaterials that result in constituting of graphene are attributed to the sole fundamental component element of the atom and the bond that will provide an excessively good expansive base that will bring novelty to the whole science engineering profession. This shall then be followed to give the avenues which shall then be opened concerning the existence of the said position of the atoms to be manipulated that shall be assumed in the electronics, energy and biotechnology industries (Figure 2). The nano components of the nano world are the simplest building block; component factors like bendiness atomize them. Their physical and chemical power is significant because they are electricity conductor materials, rigid and symbolize a great description of surface area by reason of the fact that they are engineering materials of two-dimed-hexagonal composites composed of carbon atoms (Ikhsan et al., 2025; Li et al., 2025). The third striking particular in the composites applied in the production by the aid of the graphs may be scaled to the fact that it will furthermore be feasible to confront the composites to the impact of the change in their physicochemical character with the assistance of various methods to maximize their characteristics that will subsequently make it practical in the duty that it will be to execute in e.g. energy storing or sense (Li et al., 2025). The fact that they are sensitive to the external environment (i.e. temperature, pH or light, etc) will then offer some prospect of designing them which then can be applied e.g. to measure a biomarker, attack a survey of an environment (Lutomia et al., 2025). Moreover, the size aspect of the uses of comparatively cheap nanomaterials composed of graphene offsets the reality that one must transform the material into a costly material too not mentioning that one will certainly be struck by the force of agglomeration of the nanomaterial making the workability of the material unquestionably be reduced. In order to project the rather wider scope of usage of those still more modern forms of material to the production process, still more contemporary models, thereof, i.e. ultrasonication need to be provided, not to say about the most astonishing of the more recent developed forms of approach and method to the production process (Khujamberdiev and Cho, 2025).

Figure 2. Atomic Structure and Bonding in Graphene

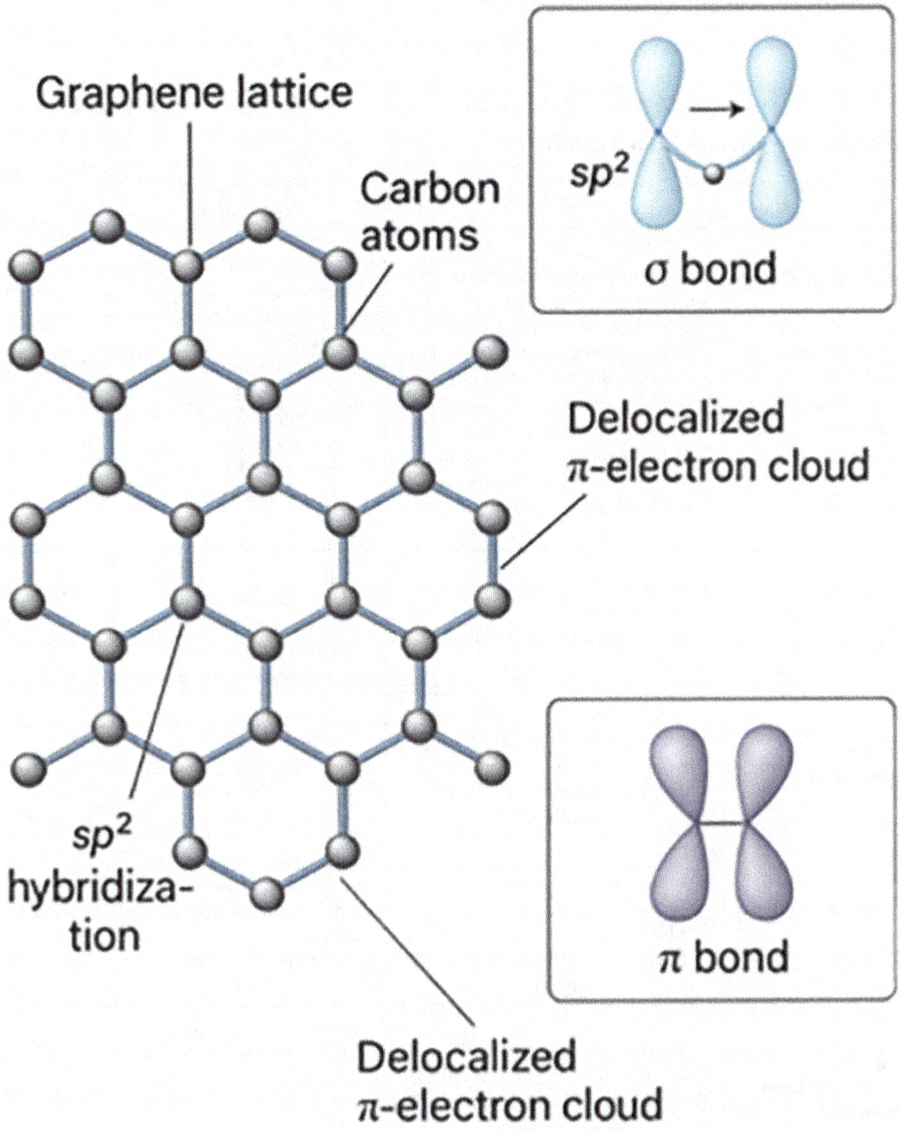

The reduced graphene oxide (rGO) and graphene oxide (GO) nanomaterials are biased to carry carbon atoms in pinacolone (H) Heptagonal system. The outstanding physical chemical properties are credited to an exceptional arrangement of the atoms having unbelievable electricity solubility, mechanical hardness and magnificent surface zone. They are also without tolerance of the influence of politics, and they are developed into the structure of the composite material by integrating oxygen-finding group to appoint the action of the chemical action of GO and enhance their solubility (Table 1).

Table 1. Overview of Structural Properties, Applications, and Production Challenges of Graphene-Based Nanomaterials

Topic	Description	Key Points and Explanations	References
Atomic Structure, Chemical Functionalization & Properties	The conductivity of the carbon atoms in the graphene nanomaterials, mechanical strength, and the surface area is the highest at the 2D (hexagonal) region. The location of chemically modified oxygen functional groups is biochemical functional groups because the most ideal method of biologically modifying any functional group is by substituting the site with metal ions.	GO and rGO possess oxygen-containing functional groups; are chemically modifiable; serve as a basis for nanocomposites. Metal-doped graphene structures show promise for anticancer therapies.	Ikhsan et al., 2025; Li et al., 2025
Graphene Quantum Dots (GQDs)	Electronic and optical properties of nano dimensions of quantum dots due to quantum confinement at nanoengineering.	sp^3 hybridized carbon atoms confer stability and photoluminescence; suitable for photonic and optoelectronic applications.	Rano & Martsinovich, 2025
Biomedical Applications	The graphene nanomaterials formulated are biocompatible; neural tissue engineering begins with the application of GO as a scaffold.	Electrical conductivity and mechanical strength promote neuron growth; long-term neurotoxicity risks exist.	Tang et al., 2025; Qiu et al., 2025
Environmental Impact	The approach is utilized during not only the elimination of various contaminants like pesticides but also the purification of water; though, it could be ecologically dangerous.	Presence of secondary contamination is an environmental aspect which has to be strictly abided by all essential environmental laws.	Khalajiolyaie & Jian, 2025
Tunable Physicochemical Properties	The above external stimuli can function on the physicochemical properties namely temperature, pH and illumination.	The use of the technology has been in the spheres of energy storage, sensing, detection of biomarkers, as well as environmental monitoring.	Li et al., 2025; Lutomia et al., 2025
Challenges in Production & Application	High costs of production and agglomeration predisposition also leads to a decline in productivity.	Increased dispersion procedures and economic increment steps are the required prerequisites of the upscaling of processes up to an industry level.	Khujamberdiev & Cho, 2025

The multi-fold structural features of the atomic elements, combined with the strength of the intrinsic bonds of the molecular arrangements and the reaction ability ingrained in the graphene basis lattice arrangements will require a two-side approach to the study of nanomaterials synthesis. These issues are chronic and might be risky, thus necessitating thorough research projects to come up with systems that optimizes performance and reduce poor health and environmental influences. The discipline is not just a very complex field of research, but also a highly documented research field. Results gained through investigations are extremely complex, being applicable in a wide range of scenarios and, as a result, overcoming the traditional, practically oriented paradigms of research in favor of high-level study in materials science, biotechnology, engineering, and scientific time geography. Essentially, the studies and researches dealing with close contacts with graphene nanomaterials are deemed as one of the most vibrant modern studies fields, and with functional application guided essentially by the structure of atomic chains. Although researchers predict that greater functional features in such systems will emerge, a deep understanding of the probable effects of impacts and determining possible shortfalls in curbing the risks that would be foreseen is a competency that the politicians cannot do away with. Considering the fact that the role of technology is already in an obvious reversal, new materials using the latest developments (that is, graphene and sheet -like structures) will become a necessity in the near future; moreover, it is believed that it will perform better than the existing products, be more sustainable, and trigger a wide range of improvements once it is implemented.

Electronic Properties

The unique electronic properties of the nanomaterials based on the graphene structure are very pronounced and can be attributed both to the unusual form of the lattice of these systems, i.e., to the two-dimensional hexagonal lattice, and to the peculiarities of the carbon bond. Electrons in the delocalized p-orbitals 2-dimensional system have very high mobility with very low scattering in the graphene sheet leading to very high electrical conductivity. As opposed to typical materials, electrons in graphene are massless Dirac fermions, a property that results in the prime importance of carrier mobilities at high values up to and beyond 10,000 cm2 /V s at room temperature, and even higher under the optimal situation. This implies that graphene can conduct charge with very low levels of resistance, a fact that renders it to be very appropriate in quick electronic equipment. electronic properties the electronic band structure of graphene has a zero bandgap with a linear energy dispersion at the edges of the Brillouin zone that can be thought of as a zero-gap semiconductor or semimetal. This special formation of bands enables ambipolar conduction with which charge carriers could be altered between electrons and holes through the application of external electric field. The property is exploited in the field-effect transistors (FETs) and other nanoelectronics devices. But limitation of not fully switching of current off also results because of lack of bandgap which has encouraged research such as using chemical doping, strain engineering or production of graphene nanoribbons to open a bandgap to enable use in digital electronics. Graphene has been made to exhibit such quantum effects as the anomalous quantum Hall effect and Klein tunneling, demonstrating its unusual quantum electronic behavior. In addition, graphene has very adjustable electronic properties in respect to outside factors, such as through mechanical strain interactions, substrates, chemical functionalization and doping by heteroatoms. Such changes are able to adjust carrier density, charge mobility, and band structure allowing tunability of graphene to meet a particular purpose. Nanomaterials made of graphene have been strongly considered as the active element in high-frequency transistors, transparent conductive electrodes, sensors, photodetectors, and flexible electronics because of their proper electronic properties with chemical stability and mechanical flexibility. Further, hybrid materials of graphene with other nanostructures, e.g., carbon nanotubes or transition metal dichalcogenides, improve the electronic functionality and pave the way of next-generation nanoelectronics devices. To conclude, unique electronic properties of graphene-based nanomaterials with their unprecedented high conductivity and charge carrier dynamics, as well as the possibility to manipulate the charge carrier dynamics, are founded on the unusual atomic structure, bonding, giving rise to unique quantum phenomena. Such characteristics make graphene a potentially revolutionary material in the sphere of nanoelectronics with enormous potential in the sphere of development of sensors, energy devices, and flexible electronics (Figure 3).

Figure 3. Electronic Properties of Graphene-Based Nanomaterials

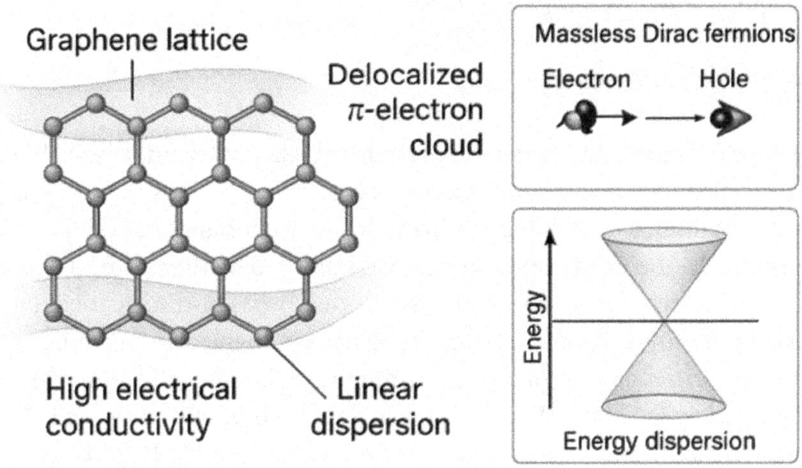

Besides, investigating the mechanical properties and the interrelation with the electronic features, graphene serves as a conductive filler that also contributes to the maximization of the mechanical strength of hybrid composites by a large margin. Graphene also increases performance in terms of mechanics in composites prepared using aluminum and bentonite, and positive antibacterial properties are also a feature of the composites (Bawa'neh et al., 2025). This multifunctionality is important where both structural and electronic capabilities are important in the application, e.g. in biomedical equipment, durable electronic components. The electronic properties of the graphene and its composites are also controlled by the synthesis methods which are used to achieve them. Chemical vapor deposition, Laser-enhanced growth, and solution processing processes are techniques that are crucial in the abuse of structural integrity, morphological chemistry, and consequently, electronic properties of the end product (Jashrapuria & Singh, 2025; Xu et al., 2025; Christudoss et al., 2025). There are still difficulties on how to make standard these processes in order to obtain high-quality, homogeneous materials in a commercial scale. Consequently, resolving the issues in the natures of the complexities leading to graphene-based nanomaterials production is the indispensable program to popularize their application in conventional electronics. Overall, graphene nano-materials are an important breakthrough on the field of electronic properties and usage. The electrical, thermal and mechanical properties that they have and which are unrivalled, and the ease with which they can be manipulated by way of different chemical adjustments make them leaders in materials science today. As the research divulges more new methods and functions, we can be sure that, as the prominence of graphene in the next-generation of electronic devices continues, it would significantly contribute to the shaping of the technological future.

Mechanical Properties

Graphene has mechanical performances unprecedented by any imaginable standard and this exquisite achievement is the main reason why it is soon to be integrated into modes of higher technology utilization as a nanomaterial. This property is made possible by the outstanding toughness exhibited by the material due to strong covalent bonds between carbon atoms, supported by its bi-dimensional lattice, and regularly appears in form of a honeycomb structure. In the carbon atom chain, these structural properties are gradually determined by the hybridization of the carbon atom (sp2 and sp3). Thus, there is significant rigidity, elasticity, and strength of graphene. By far one of its most sparse aspects is extremely high Young modulus of about 110 0 TPa of about 110 as in Figure 1 denotes a strongly stiff tendency. The self-reinforcing property of graphene is more than phenomenal in most traditional materials, which makes it an excellent reinforcement in composite matrices, improving the overall mechanical stability of these systems. Moreover, the tensile strength of graphene is more than 100 000 Gpa and it may be explained by a strong C - C covalent bond of this compound alone. The material is also intrinsically flexible and elastic; compressional action applied on the material creates insignificant deformation hence, does not develop rigid deformation.

Figure 4. Exceptional mechanical properties of graphene-based nanomaterials

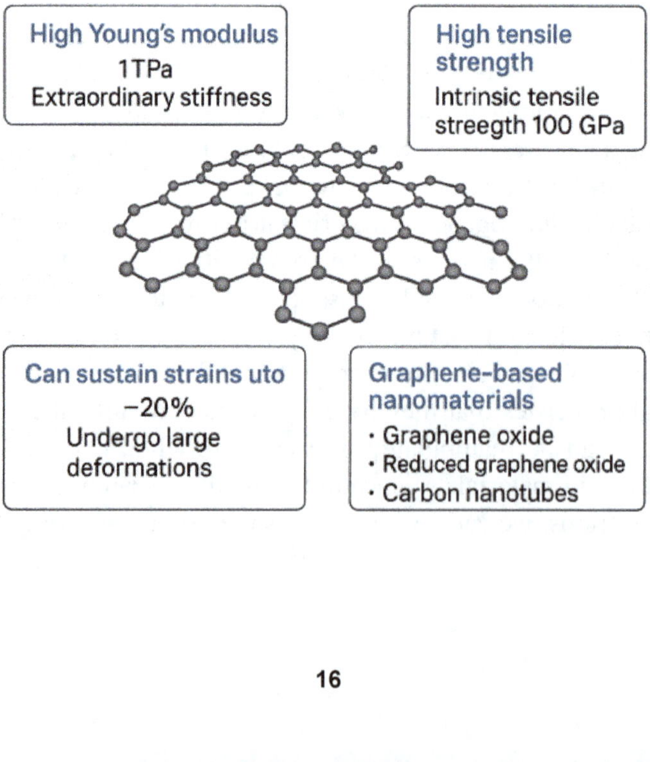

This material has the ability to survive at least twenty percent deformations before breaking which is amazing compared to other many materials. Through this elastic behavior, graphene is all-included in a list of sudden-electronics, wear-able products, and strain-sensing applications. Nevertheless, multilayer graphene can change its mechanical character due to the appearance of crystallographic defects, boundaries between grains, and interlayer stacking. Flaws and pores are bound to reduce mechanical strength, but multilayer structures can contribute displacements in stiffness and faultier behavior. The functionalization approaches as well as the chemical adjustments can also govern these consequences which can lead to deviations in structure or alternatively improve interfaces bonding of the composite systems. Besides pristine graphene, graphene oxide (GO) and reduced graphene oxide (rGO) exhibit different mechanics, which implies that GO is typically less rigid and tougher than pristine graphene, although it is soluble and easily treatable, which is a positive factor in the manufacture of composites. Carbon nanotubes (CNTs) are analogs of graphene in a one-dimensional structure and are just as strong and flexible in tensile strength. The mechanical performance of the nanocomposites of polymer frame sides as well as either of the graphene or CNT are significantly enhanced producing extremely high modulus, strength, and toughness. In general, these special mechanical properties of graphene-based material nanomaterials, together with their low density and atomic non-reactiveness, provide a range of opportunities in the aerospace, auto, electronic, as well as biomedical fields of engineering (Figure 4).

Thermal Properties

These qualities of the nanomaterials based on graphite are very critical, because of their excellent mechanical and electrical properties on one side, and due to the unusual thermal properties on the other. The multiple thermal plans in development are of mainly importance to a variety of new hi-tech applications which numbers, thermal management of electronic equipment, thermal energy storage apparatus and thermally engineered composite materials. In order to entirely utilize these engineering materials in the various fields and technologies, one must be in position to have an in-depth understanding of the thermal properties of graphene and its derivatives.

Intrinsic Thermal Conductivity

Among the most remarkable characteristics of graphene, there is the extraordinarily high intrinsic thermal conductivity. Thermal conductivity in pristine single-layer graphene is generally between 2000 and 5000 W/m K at room temperature, compared to most known materials where highly conductive copper (about 400 W/m K) and diamond (about 2200 W/m K), both outperform this range. It is a consequence of very good conductivity of the phonons, which are quanta of lattice vibrations that propagate freely in its two-dimensional carbon lattice. The sp2 hybridization and a large covalent bond between carbon atoms produce a highly ordered lattice where there is a small phonon scattering and the heat transfer is quick. The thickness on the atomic level also reduces phonon scattering at the surface hence adding to the high thermal transport properties.

Factors Affecting Thermal Conductivity

With a very high intrinsic thermal conductivity, the practical thermal behavior of graphene based nano-materials may differ greatly being subject to various factors:

- Layers: Thermal conductivity usually reduces with the increase in the layers of Graphene. These are the best values in monolayer graphene and lower values of thermal transport are observed in few-layer graphene and multilayer graphene because of interlayer phonon scattering and coupling effects.
- Defects and impurities: structurally defected, vacancies, grain boundaries as well as chemical impurities, act as phonon scattering centers lowering thermal conductivity. To take one example, groups on graphene oxide containing oxygen disrupt the sp2 network and lower efficiency in heat transport.
- Substrate Effects: A suspended graphene has higher thermal conductivity than that which is supported on substrates such as silicon dioxide (SiO_2) because of phonon leakage into the substrate and thermal resistance at the substrate interface.

Temperature Dependence: At temperatures higher than room temperatures, thermal conductivity of graphene falls off rapidly, presumably because there are more phonon-phonon Umklapp scattering processes. But at small temperatures the boundary scattering becomes dominant giving complex dependence on the temperature.

Thermal Properties of Graphene Oxide (GO) and Reduced Graphene Oxide (rGO)

The thermal properties of Graphene oxide ore and reduced graphene oxide are significantly different to the ones of pristine graphene due to the structural and chemical changes:

- Graphene Oxide: GO has a large concentration of oxygen functional groups, such as hydroxyl, epoxy and carboxyl, breaking the uninterrupted sp2 carbon lattice and adding disorder. Due to this there is a substantial decrease in the thermal conductivity, at times by one to two orders of magnitude compared to pristine graphene. GO however, is hydrophilic and this together with its processability is a desirable attribute in composites where specific thermal requirements are desired.

• Reduced Graphene Oxide: rGO, having partially restored the sp2 network with chemical/thermal reduction, has superior thermal conductivity to GO but in general is worse than pristine graphene. Reduction degree and defect lefts have a strong effect of the thermal transport abilities.

Thermal Expansion and Stability

Graphene exhibits negative thermal coefficients (TEC) covering a broad range of temperatures, so it shrinks when heated as opposed to most other materials that expand. This is as a result of the anharmonicity of lattice vibrations in this two-dimensional arrangement. At room temperature, the TEC of graphene is within $-7x10^{-6}$ K^{-1}. Graphene and its compliments are also very thermally stable. That is quite much above the usual materials as pristine graphene can survive temperatures ranging beyond 600/C in an oxidative environment and up to about 2800/C in inert atmospheres.

Thermal Interface Conductance

Graphene is used in practical applications and is frequently made into composites or stacked with other materials and therefore thermal interface conductance (TIC) is a very important parameter. The quality of bonding, the roughness of the surface and interfacial defects determines the efficiency of heat transfer across the interfaces. Interfacial thermal conductance might be enhanced by chemically functionalization techniques to increase adhesion of the graphene to the matrix materials.

Applications Leveraging Thermal Properties

Due to the excellent thermal conditions and stability of graphene-based nanomaterials, a variety of applications have been triggered:

• **Thermal Management:** Graphene thermo-spreader may help microelectronics to dissipate heat generated by high-power devices to enhance reliability and performance of the devices.

• **Composite Materials:** Graphene can be added to both polymers and ceramics to increase their thermal conductivity, making them useful as aerospace, automotive and packaging materials.

• **Energy Storage:** The graphene-based electrodes used in batteries and super cap can positively utilize the high-thermal dissipation that can ensure the performance and safety of the devices.

Thermoelectric Devices: investigation is underway to take advantage of the thermal and electrical material properties of graphene to enhance thermoelectric conversion efficiency to recover waste heat. Nanomaterials derived out of graphene have exceptional thermal qualities over their two-dimensional carbon crystalline structure and covalent bonding behavior. Although pure graphene has very high thermal conductivity and stability, may produce large influences in thermal effects with layer number, defect, chemical functionalization, and substrate effects. The ability to understand and optimize these properties is central to the development of graphene toward thermal management processes, in composite materials, energy storage and applications in many more areas. The research on the nanomaterial's graphene has suffered an upsurge because of its unique thermal conductivity properties, and thereby plays a significant role in multiple aspects. As a two-dimensional carbon allotrope, graphene exhibits an extreme mechanical strength, thermal conductivity, and chemical stability, thus, it can be said to have versatile functionality in nanocomposites. Recent studies on the thermal properties of graphene and its derivatives have been synthesized in this analysis with a view of explaining how the findings translate in different contexts, with particular reference to their application to material science and engineering.

What is more, the exclusive nature of graphene as an additive to cement-based products has grabbed the spotlight over the last few years. Additional of gorgeous ordinate in cement timber can enhance mechanical and warmth characteristics of the entire structure, shopping cart speedier and sterner building materials (Zhang et al., 2025). The fact that graphene can alter the microstructure of cement composites makes it possible to create a better thermal performance, a factor that is extremely crucial when it comes to infrastructure that has to live through adverse temperature changes without affecting its working ability. The outstanding thermal conductivity of graphene can be disadvantageous in applications where electrical insulation is required (battery), especially where this is required to insulate electricity. The study by Hu et al. explains such a path, where the thermal conductivity of the composites is improved with the help of in-situ exfoliation of graphene and boron nitride, but provides the needed insulating electrical properties (Hu et al., 2025). That is why this hybrid design is an excellent example to explain that obstacles related to using traditional composites can be overcome using extraordinary structural designs and the combination of materials that take advantage of the properties of graphene without denying functionality.

Nevertheless, the impressive progress in manufacturing and applying the graphene-based nanomaterials proves their capability of changing the technological environment in many fields mostly because of the excellent thermal characteristics. The augmented performances of the materials in thermal management improvement in the electronic devices, reduction of performance in construction Limitations, and improvement of the environmental remediation processes depict the critical importance of graphene in material science. The future of these nanomaterials depends on their further investigations to increase the understanding of their thermal characteristics and further deployment in new areas associated with nanotechnology and smart materials. In a conclusion, graphene odorized nanomaterial velocities of graphene-based nanomaterials is one of the fundamentals of the modern material science. Their record concerning the combination of mechanical sufficiency, orbital conductance and fineness leads to wide applicability in most industries. Newer research advances need to be made in finding out more on the structural complexity and the possible changes that can be made to increase the thermal performance of such revolutionary materials that could find their use in the various forms of applications.

Optical Properties

Due to strong versatile optical properties, graphene-based nanomaterials have attracted considerable attention to perform functions in the photonics, optoelectronics, sensors and energy devices. Such optical properties are inherent due to narrow band gap, two-dimensional component, and excellent response to a wide spectrum range of the electromagnetic radiation by graphene. The importance of deterministic theory of the optical behavior of graphene and its derivatives is in making advanced nanoscale materials and devices that apply light matter interactions at the nanoscale.

Intrinsic Optical Absorption and Transparency

Graphene is made of a monolayer of carbon connected in a hexagonal lattice structure which makes it extremely clear under optical applications with high light absorptions in the broadband area. Graphene also has a % absorption of incident white light over the visible spectrum of about 2.3 percent, which is connected with the fine structure constant (alpha 2 or 1/137). It is such a constant absorption coefficient in a broad spectral extent, including ultraviolet (UV) to infrared (IR) wavelengths, that makes graphene the outstanding material in transparent conductive electrodes and photodetectors. The electrical conductivity and high transparency make graphene more superior to other conventional materials like indium tin oxide (ITO) when applied in flexile and stretchable optoelectronics.

Optical Conductivity and Band Structure Effects

The unusual linear band structure of Dirac points, gapless in graphene, determines its optical conductivity. Broadband absorption of the interband electronic transitions across visible to near-infrared (NIR) regimes are made possible. Unlike the classic semiconductors with narrow band gaps, the zero-band gap of graphene allows the transmission of photons of arbitrarily large energy without threshold. It is universal because of the charge carrier symmetry and the linear dispersion of the charged carriers as massless Dirac fermion.

Tunability via Doping and External Fields

The electrical gating, chemical doping or strain engineering allows the dynamic control of optical properties of graphene. The occupation of electronic states is affected by altering the fermi level that in turn affects interband and intraband transitions:

- Pauli Blocking: raising the Fermi energy by doping or gating yields a reduction of optical absorption inferred that the absorption is telegraphed upon photons of energy lower than twice the Fermi energy.
- Plasmonics: Plasmonic devices in the mid-infrared and terahertz regime allow the doping of graphene to support surface plasmon-polaritons (SPPs) with tight confinement of electromagnetic fields, which permits tunable plasmonic behavior.
- Strain Engineering: Strain tuning changes the band structure, and optical absorption profile, giving an additional degree of control Mechanical strain changes the band structure and optical absorption spectrum, giving a further degree of control Strain tuning changes the band structure and optical absorption profile

Optical Properties of Graphene Derivatives

- Graphene Oxide (GO): GO has a broken sp avoidance network as a result of oxygen functionalities thus changing the optical properties of this material compared with pristine graphene. The most significant optical absorption in GO is observed in the UV region and refers to the 226 nm; n(C)-226 nm; n(O) 220 nm) and that caused by transitions at 155 nm (n(O)-155 nm) associated with oxygen-groups and by the 175 nm (226 nm; n(C)-220 nm; n(O)), and 155 nm transitions connected to aromatic domains. It is also photoluminescent (PL), thus exhibiting the potential of GO to bioimaging and sensing.

• Reduced Graphene Oxide (rGO): Decreased conjugation by reduction occurs between Graphene Oxide and pristine graphene; Therefore, the optical properties are found intermediate between them. rGO will maintain some of the absorption features and has a reduced PL intensity, which is variable according to the reduction degree.
• Carbon Nanotubes (CNTs): The one-dimensional analogues of the graphene, the CNTs can demonstrate the unique optical absorption and emission features due to their chirality and diameter, such as the strongest excitonic effects and near-infrared photoluminescence energy suitable in optoelectronic applications.

Nonlinear Optical Properties

The quite high nonlinear optical behavior of graphene such as saturable absorption, third-harmonic generation and Kerr nonlinearity can be owed to its nonlinear band structure and ultrashort carrier relaxation. Such properties are used in the ultrafast lasers, optical modulators, and frequency converters.

• Saturable Absorption: Graphene has a wide range of operation wavelengths and can functions as a saturable absorber so that some lasers can be mode-locked.
• Third-Harmonic Generation (THG): Harmonics of incident light frequencies can be produced through nonlinear optical processes which can then be used in frequency conversion and photonic integration.

2.5.6 Applications Leveraging Optical Properties

Unique optical properties of nanomaterials based on graphene gave rise to a broad range of applications:

• **Transparent Conductive Films:** High transparency and conductivity makes them to be used in touch screens computer, solar cells and flexible displays.

Photodetectors and Sensors: The broadband absorption and tunable photoconductivity makes graphene a very sensitive photodetector in all UV-to- terahertz ranges.

• **Optical Modulators:** Graphene has fast response as well as electro-optical tunability, both of which are desirable optical modulators in optical communication systems.

Photoluminescent Graphene oxide derivatives have been discussed as useful in fluorescent bio-labeling and in photothermal therapy.

- **Nonlinear Optics and Ultrafast Photonics:** The ultrafast carrier dynamics enabled by graphene stimulates the usage in ultrafast lasers technology.

Compared to non-silicon based in optoelectronic materials, nanomaterials are unique with the combination of high optical transparency throughout the near-infrared spectral region, wide band absorption, optical conductivity with adjustable optical characteristics, and intense nonlinear optical effects. Ability to adjust these properties accurately through chemical doping and external interactions makes graphene a very useful material in upcoming systems of photonic and optoelectronic devices. In addition to that, the graphene derivatives possess a broadened functional optical range, and by further modifying the derivatives, new applications are feasible in sensing, communications, energy conversion, and biomedicine.

METHODS OF GRAPHENE SYNTHESIS

Graphene synthesis has been of the primary interest in the studies of nanomaterials because of the extraordinary properties of graphene and the extensive applications of graphene. Synthesis of high-quality graphene at an efficient, scalable and cost-effective manner is important to the fundamental research, and industrial applications. Several methods of synthesis have now been developed but each with its own strengths and weaknesses, and applicability to varying types of graphene (e.g. single-layer, few-layer, graphene oxide). This part presents a comprehensive description of commonest synthesis methods of graphene where they are classified in terms of fundamental concept. Production of graphene has continued to act as the core focus of research in the wider sector of nanomaterials because of excellent mechanical properties, high electrical and thermal conductivity as well as optical transparency that are distinct and exceptional features that graphene possesses. Such properties have led graphene to the frontiers of materials science, on which its prospective uses in electronics, energy storage, sensors, composites, etc., have been developed. It is, therefore, necessary to come across a dependable, scalable, and economical mode of synthesizing high-quality graphene so that its potential in academic research and commercial production can be unleashed. The production of graphene is a complex issue since the structure, interlayer number, purity, level of defects and size have a direct role to play in the performance of graphene when applied on target. Consequently, there have been many methods of fabrication, improved over time, to make and refine different products and types of graphene, whether in one-layer (SLG, FLG), and graphene oxide (GO), among others, each being optimized to maximize a desired result.

Mechanical Exfoliation

The first isolated technique to remove graphene (called the Scotch tape method) was mechanical exfoliation, which was a key method in the discovery of graphene in 2004 by Novoselov and Geim. This technique is done by the means of peeling thin sheets of graphene through bulk graphite through adhesive tapes.

- Process: Graphite is successively compressed and exfoliated with adhesive tape to transfer to thin flakes of graphene on any substrate including silicon dioxide (SiO_2).
- Pros: Capable of high-quality monolayer or a few layers graphene free of defects with great crystallinity and intrinsic characteristics.
- Shortcomings: Low yield, tiny sized flakes (is typically at a micrometer scale) and it is not scalable into mass production.

Mechanical exfoliation is of use in basic science that demands perfect graphene, but is unfeasible in large-scale production.

Chemical Vapor Deposition (CVD)

CVD Chemical Vapor Deposition is perhaps the most promising method of large-scale graphene manufacture and one of the most popular.

Process: Hydrocarbon gases (e.g. methane, ethylene) are decomposed at high temperatures (usually 900-100^{0}C) in a controlled atmosphere on catalytic metal-based substrates such as copper, nickel or cobalt at elevated temperatures. The diffusion of carbon atoms on or into the surface of the metal is followed by nucleation with formation of layers of graphene. Graphene is transferred onto various substrates of choice by wet or dry transfer technique following growth.

Advantages:

- Continuous films of graphene on large areas with the possibility to control the layer number.
- Semiconductor compatible.
- High crystallinity with a not so high defect density.

Limitations:

- Transfer process can either cause contamination and mechanical damage.
- The price and intricacy of substrate.

- Uniform monolayer coverage by control needs the careful optimization of the process.

CVD graphene has a wide range of usage in transparent conductive electrodes, sensors and flexible electronics.

Reduction of Graphene Oxide (GO)

Graphene oxide is a chemically functionalized precursor to graphene produced by the oxidation of graphite, and contains many oxygens containing functional groups. It can be dissolved in water, and other solvents, allowing a solution processing.

GO synthesis: typically, through oxidation of graphite with powerful oxidants ($KMnO_4$, e.g., H_2SO_4) either in the Hummers or its variations.

- **Reduction Process:** GO is softened chemically either thermally or utilizing electrochemical technique to rebuild the graphene lattice consequently making reduced graphite oxide (rGO). The reductive techniques consist of hydrazine-treatment, thermal annealing or UV-promoted reduction.
- **Advantages:** The future possibility of scale production and affordable grafevering of like product. Coating, printing and fabrication of composites has been done through solution processability.
- **Limitations:** Briefly reduced graphene oxide has remaining defects and oxygen groups. The electrical and mechanical characteristics are lesser compared with pristine graphene. Heterogeneity of structure and little oversight over the extent of reduction.

rGO finds much application in energy storage, sensors, and composite materials in which a high degree of crystallinity is not obligatory.

Epitaxial Growth

Graphene on the SiC substrates grown epitaxially is also another significant approach of generating high-quality of graphene in the electronic industry:

The Process: In the process, SiC wafers are brought to extremely high temperatures (~1200-1600^{0}C) under inert gas, or vacuum, conditions. The silicon atoms evaporate at the surface leaving carbon atoms that reassemble to make graphene layers.

Advantages:

- Generates waver scale graphene on insulating substrate ready to be made into electronics.
- Structural quality is high and the number of defects is low.

Limitations:

- Expensive and high temperature of processing SiC substrates.
- It can be difficult with controlling the number of layers and uniformity.
- The technique is less scalable to the CVD.

The epitaxial graphene has particular promise in high frequency and quantum electronic devices.

Liquid Phase Exfoliation

The liquid phase exfoliation entails the immersion of graphite in appropriate liquids and exfoliated using ultrasonic or shear force to yield graphene flakes in liquid form.

Process: Graphite powders are either sheared under mechanical force or sonicated in solvents like: N-methyl-2-pyrrolidone (NMP), dimethylformamide (DMF) or aqueous solutions of surfactants, which can stabilize the graphene sheets that have been exfoliated.

Advantages:

- Scale is fairly easy and comparatively cheap strategy.
- Manufactures dispersions used in inks, coating and composite.
- Doesn t use harsh chemicals.

Limitations:

- Thickness and size of the flakes are different.
- There is low efficiency of exfoliation and large-scale centrifugation and purification are needed.
- The obtained graphene can have defects and edges with property impacts.

Synthesis of graphene filled inks and a composite filler, used in conductive and mechanical enhancement, is very common using LPE. The production of graphene-based nanomaterials represents a multi-faceted approach designed to fulfill application-specific demand in terms of quality, scaling, cost and structural control. Mechanical exfoliation has set the standard against high quality graphene but large area films can be produced by CVD. Liquid-phase exfoliation Liquid-phase exfoliation of graphite offers a convenient, low cost approach to solution-processable derivatives of graphene. Electronic grade graphene is widened by epitaxial growth on SiC and new techniques. Further development of the technologies of synthesis is central to the introduction of graphene in future devices and materials. Graphene-based nanomaterial synthesis methods range, depending on application-specific needs, in term of quality, scalability, cost and structural control. Mechanical exfoliation as well as CVD is used to produce graphene of high quality, with the latter technique being scalable and with large area films being produced. Graphene oxide reduction as well as liquid phase exfoliation represents cost-effective pathways to solution processed graphene derivatives. Electronic-grade graphene opportunities are increased by epitaxial growth on SiC and nascent methods. Devices and next-generation materials It is the development of synthesis technologies that have enabled the gradual integration of graphene into devices and materials of the next generation.

FUNCTIONALIZATION AND MODIFICATION OF GRAPHENE

The amazing mechanical, electrical, thermal, and chemical physicochemical properties of graphene have enabled one to classify it as an extremely promising material in a wide variety of technological applications. However, the purest form of this material is graphene, which has only two serious shortcomings: it is chemically inert, and dispersible in significantly low concentrations or solvents solubility. These limitations impede application of graphene in practice especially in composite materials, sensors, catalysis and biomedical devices where consistent dispersion, surface reactive, interfaces and interfaces properties are required. As a solution to these issues, however, functionalization and chemical modification of graphene are essential in the creation of reactive sites and the surface chemistry of the material. These adjustments increase processability and interfaces with other substance and also allow the attainment of other functionalities with the intention of maintaining the desired properties of graphene. We can distinguish between functionalization methods (covalent and non-covalent) whose mechanisms are controlled by different processes and whose outcomes are specific.

The modification and functionalization of the graphene-based nanomaterials is a wide-ranging and very dynamically developing study topic, all aimed at improving the characteristics and usage of the nanomaterials. Graphene oxide (GO) and reduced graphene oxide (rGO) play a crucial role with regard to this fact, because the structure and reactivity allows many different modification strategies depending on the intended application. The changes would have been broadly divided into covalent and non-covalent interactions that are used with different purposes and benefit depending on the requirements of a technology. One of the best transformations is covalent functionalization because the bond formed is stable. The irreversibility of this method guarantees long-term serviceability of the altered nanomaterials in the practical field of drugs, sensing, and the manufacture of composite materials. As an example, on the substrate hydro series of functional groups can be added to the graphene surface and alter the mechanical properties and resistance to the chemical environments considerably (Farnsworth et al., 2025). Sun et al. explain that the addition of poly(dopamine) (PDA) onto the graphene nanosheets changes the surface morphology to a rough-structured surface that is applicable in areas like electromagnetic interference (EMI) shielding whereby there is an increase in interaction with electromagnetic waves (Sun et al., 2025). Alternatively, non-covalent functionalization does not affect the electronics properties of graphene. The strategy also involves the use of weaker interactions including pi-pi stacking and Van der Waals forces, and thus it is implementable in avenues where there is need of electrical conductivity and flexibility (Liang, 2025). As an example, layer-by-layer deposition of materials onto the rGO surfaces can result in the humidity sensors that are sensitive because of the robustness of the graphene π-system (Liang, 2025). The resulting nanocomposites are effective and robust and yet can be highly functional as several nanomaterials can be added to the same and these materials do not change the inherent properties of graphene much. GO is a versatile GMM precursor because of its diversity of functional groups which contain oxygen, and make them more reactive towards organic ad inorganic functionality. As an example, chemical reduction of GO creates rGO which may be further modified in relation to the introduction of metal nanoparticles and, this way, may attain a set of synergistic properties of electronic conductivity and thermal conductivity enhancement in a composite (Ikhsan et al., 2025). Such strengthened attributes have the prospective to advance the working of electrochemical sensors and catalytic gadgets, since the tended-to graphene will be an efficient platform to load the nanomaterials, then enlivening the responsiveness, and good quality of recognition (Ran et al., 2025). In addition, increased interest has been shown on how to use carbon materials derived by biomass to generate graphene-like materials via sustainable procedures. Gao et al. underscore that, in this environmental-friendly technology approach, there is a potential to obtain the formation of 2D carbon nanomaterials in the form of graphene, whose potential application areas are battery energy storage systems, i.e., supercapacitors (Gao et al., 2025). This is in line with the general theme of the modified graphene use in newer energy processes, such as thermal energy storage

(TES), where a graphene combination with phase change materials (PCMs) greatly boosted thermal conductivity and stability (Ahmed et al., 2025).

Covalent Functionalization

Covalent functionalization; the strong chemical bond is created between carbon atoms of graphene and the functional groups. This strategy normally alters the sp2 hybridized carbon framework and some of the sp2 carbons are transformed into sp3 centres of hybridization. This may partly interfere with the conjugated system of π-electrons in graphene reducing its electrical conductivity, but it also gives it strong durability against chemicals and good binding of functional groups (Table 2).

Common Covalent Functional Groups and Methods

This representation shows the most important functional groups and chemical modifications that are often propped on graphene surfaces. The epoxy and hydroxyl groups serve as hydrophilic surfaces as well as the reactive sites which are predominantly produced during graphene oxide production. The ability of further chemical conjugation involving biomolecules and polymers is provided by functional groups such as carboxyl and amine groups to increase the versatility of graphene. azide/alkyne reactivity allows specifically targeted molecular grafting by so-called click chemistry. Also, graphene is surface-chemically and electronic-property modified via diazonium salts, halogenation, and thiol-decoration to a wide range of applications (Figure 5).

Figure 5. Functional Groups and Chemical Modifications of Graphene

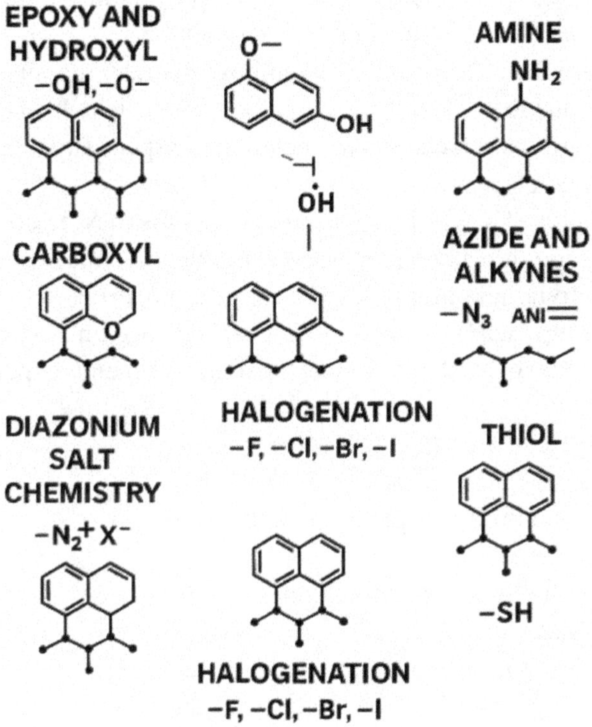

• Epoxy and Hydroxyl Groups (-OH, -O-) Epoxy groups introduced largely with oxidizing processes common in the synthesis of graphene oxide (GO) through Hummers or modified Hummers process. The surface is rich in these groups which contain oxygen; thus, they are hydrophilic and offer locations of reactive reactions that could occur on the surface.

• Carboxyl Groups (COOH): Under normal circumstance presents at the edges of the graphene and the defects resulting after they undergo oxidation. Amidation and esterification of carboxyl groups anchor both conjugation to biomolecules, polymers or other organic species.

• Amine Groups (NH_2): This is added by amidation of carboxyl groups or by direct addition of carboxyl over grafting via the use of diazonium chemistry. Graphene can be made compatible to polymers through amines and also allows covalent attachment to drugs, enzymes or sensors.

• Azide Groups (N_3) and Alkynes: Azide-alkyne cycloaddition takes advantage of the high specificity and high yields of adding a range of functional molecules onto a substrate via a "click" reaction.
• Diazonium Salt Chemistry: Diazonium salts may recombine spontaneously with graphene covalently with carbon atom, which makes aryl or alkyl functional groups attached. The technique is quite flexible to dope surface chemistry and electronics.
• Halogenation (F-, Cl-, Br-, I-): Halogen atoms may be covalently bound to graphene through plasma processes or through chemical processes, resulting in new reactivity and modifying the electric properties.
• Thiol Groups (SH): They can be attached to graphene oxide or reduced graphene oxide so that the additional bonding with metals or polymers can occur.

Effects of Covalent Functionalization:

• Increases solubility in polar, as well as non-polar solvents based on the groups introduced.
• Enhances the ability of compatibility with polymer matrices within composite material.
• It allows biomolecule or catalyst specific binding.
• Reduces slight electrical conductivity because there is an interruption in the π-electron system.
• Affects the mechanical flexibility and the thermal conductivity according to the extent of modification.

4.2.1 Non-Covalent Functionalization

Non-covalent functionalization relies on physical interactions such as π–π stacking, van der Waals forces, hydrogen bonding, and electrostatic interactions to attach molecules onto graphene surfaces without altering the sp^2 carbon network. This has the advantage of maintaining the electrical and mechanical properties of graphene with processes made easier and more functional.

Common Non-Covalent Functional Groups and Agents:

• Aromatic Molecules and Polymers: Graphene attaches aromatic compounds (polycyclic aromatic hydrocarbons, PAHs, pyrene derivatives and conjugated polymers) through π pi-stacking. Such molecules may add solubility, fluorescence or sensorics abilities. Surfactants Ionic and nonionic surfactants have been used to stabilize graphene dispersions in water and organic solvents through the formation of micelle-like structures around sheets.

• Polyelectrolytes and Charged Polymers: These can be combined with graphene oxide through the forces of electrostatic interaction with the graphene oxide itself, or with charged graphite sites on the surface of the graphene substrate. This interaction can be used to achieve controlled dispersion, and layer on layer assembly.
• Biomolecules: Proteins, DNA and enzymes can also bind non-covalently to graphene, with the capability of biological activity being maintained and making biosensing or drug delivery possible.

Effects of Non-Covalent Functionalization:

• Holds the well-behaved electrical and mechanical characteristic of graphene.
• Gives tunable and reversible interactions that are suitable in sensor and catalysis uses.
• Promotes the dispersion devoid of chemical damage.
• Makes it possible to allow multifunctional coating as well as hybrid construction.

Table 2. Functional group in graphene functionalization

Functional Group	Typical Introduction Method	Primary Effects	Applications
Hydroxyl (–OH)	Oxidation (GO synthesis)	Hydrophilicity, reactive site	Composites, sensors
Epoxy (–O–)	Oxidation	Reactivity, cross-linking	Catalysis, membranes
Carboxyl (–COOH)	Oxidation	Binding site for covalent coupling	Bioconjugation, composites
Amine (–NH_2)	Amidation, diazonium chemistry	Surface charge, bonding	Drug delivery, sensors
Azide (–N_3), Alkyne	Click chemistry	Specific conjugation	Targeted functionalization
Halogens (–F, –Cl)	Plasma, chemical halogenation	Electronic tuning, chemical reactivity	Electronics, catalysis
Aromatic molecules	π–π Stacking (non-covalent)	Preserve conductivity, dispersion	Sensors, optoelectronics
Surfactants	Adsorption (non-covalent)	Dispersion stabilization	Composite processing
Biomolecules	Adsorption (non-covalent)	Biocompatibility, sensor function	Biosensing, drug delivery
Heteroatom doping	Chemical vapor deposition, plasma	Electronic structure tuning	Electronics, catalysis

It has also observed that, the surface energy and reactivity of graphene can also be chemically modified, making it adhesive and able to be used with polymer composites. As a testimony to the importance of graphene oxidation control during the creation of super par waterborne polyurethane, Yang et al. showed that it was possible to improve the corrosion resistance of waterborne polyurethane by adding graphene oxide. This result highlights the increased functionality of the graphene of functionalized group in the industrial setting where protection cover is the key factor. Modified graphene application in neuromorphic engineering has also been discussed recently; an example type of use of such changes is the creation of CdS/graphene heterostructures to create integrated sensing-computing devices; her the changes modify the electronic systems with reduced costs to operate (Yang et al., 2025). These discoveries underscore the need to continue exploring ways in which graphene can be functionalized so as to enhance its inherent characteristics, as well as to make it compatible with other technologies. The multifunctionality of graphene is responsible for the spread of modification strategies and the possibility of cross-application. A good example is the hydrogenation of graphene to create a band gap and thus a semiconductor candidate; the electrical properties of the resulting substance widow the possibilities of high-impact electronic applications (Shi et al., 2025). Therefore, further research of the materials science and nanotechnology of graphene functionalization and modification is essential to materials science and nanotechnology. The unsolicited reality in revealing the complete technological capability of graphene is the functionalization and subsequent alteration to address its chemical insensitivity and poor solubility. Covalent methods enable permanent functionalization at high yield, and increased chemical functionalization, at the cost of possibly compromising the electronic behavior of graphene. Non-covalent strategies on the other hand maintain natural properties albeit, they tend to be less chemically stable. Carefully choosing functional groups and pursuing proper synthesis pathways, graphene can be adapted to the needs of composites, sensors, catalysts, biomedical devices, energy storage, etc. The first critical factor defining the impact of graphene commercially and scientifically is the chemical structure of the functionalization, and the resulting material with regard to its physical features. In this respect, it can be indicated that graphene-nanomaterials can be subjected to a wide range of physical and chemical treatments to ensure that they can be optimally used to achieve an enhanced versatility of possible applications. Based on the fact above, the careful combination of the proper functional groups, composite prescriptions, and sound approaches is essential in the development of future research directions on this field considering the evidence provided by modern studies.

APPLICATIONS OF GRAPHENE-BASED NANOMATERIALS

Nanomaterials of graphene have reinvented many scientific and technological fields due to their unique ability to possess the exceptional amount of mechanical strength, high electrical conductivity, excellent thermal conductivity, excellent chemical stability, and a unique optical characteristic. The peculiar combination of these properties makes graphene stand out among traditional materials and allows using it as a building block in pioneering developments in a wide range of fields. Mechanically graphene can also demonstrate excellent tensile strength and elasticity beyond most materials and hence makes a perfect candidate to be used in reinforcing composites and manufacturing of flexible and wearable devices.

Electrically, its ultra-high carrier mobility and conductivity enables faster transport of electrons and may be important to next-generation electronic components, high frequency transistors and ultrasensitive sensors. Graphene has an extremely high known thermal conductivity thermally, and can therefore efficiently dissipate heat, which is important in thermal management, in miniaturized electronics and energy storage devices. Graphene is chemically inert and thus can be used in a long-term setting, but it is also possible to chemically alter the surface of graphene to manipulate its interactions with different molecules, and different substrates, and thereby increase compatibility and further functionality. Optically, graphene displays unique light absorption and transparency properties that allow applications in photodetectors, transparent conductive films (sheet resistance 10^{-18}) and optoelectronic devices. Graphene and its properties are also tunable by controlling the number of layers, doping, functionalization and hybridization to multi-walled carbon nanotubes or to other nanomaterials which further broadens its usage and enables materials to be created with specific performance characteristics catering to the needs of any industry. The flexibility has given rise to some breakthroughs in electronics, energy storage and conversion, composites, sensing technologies, biomedicine, and environmental engineering.

Figure 6. Applications of Graphene-Based Nanomaterials

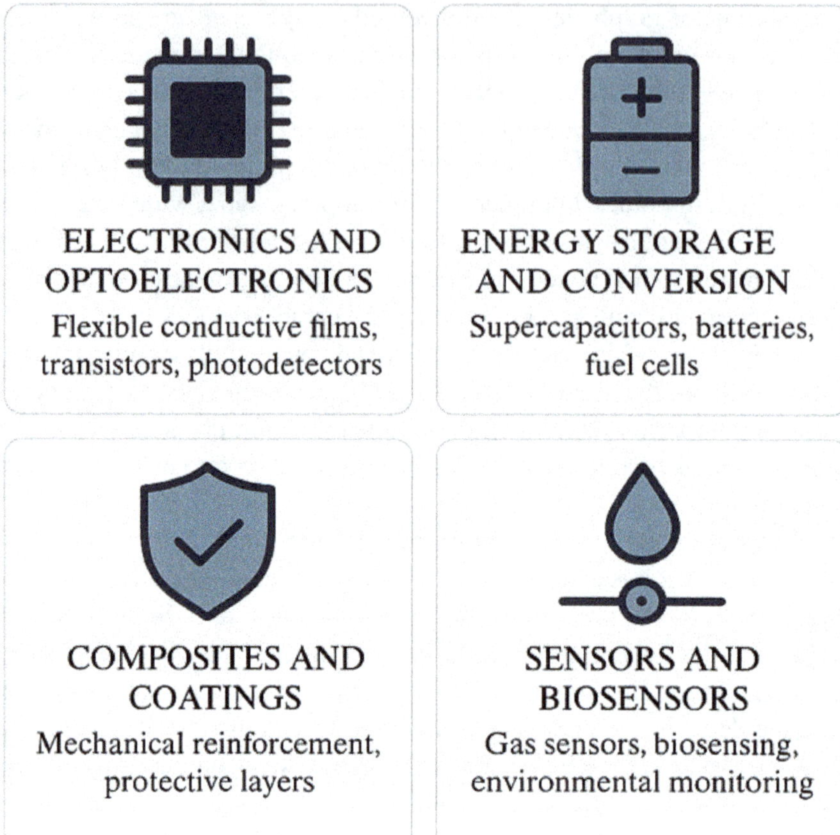

Regardless of such impressive benefits, there are the problems of industrial-level production of high-quality graphene, consistent functionalization without affecting native characteristics, and complex incorporation of graphene-based materials into large-scale manufacturing. The current research aims at addressing these challenges by coming up with new synthesis processes, high-level functionalization mechanism, and new device designs. This section explores the main application fields of graphene and its derivatives with emphasis on main technological achievements, some practical applications and the remaining challenges that have to be overcome to realize the full potential of transformative nanomaterials based on graphene (Figure 6).

Electronics and Optoelectronics

Due to its excellent electrical conductivity, high carrier mobility (up to 200,000 cm 2 / V s in ideal conditions), and optical transparency it is an attractive material in the next-generation electronic and optoelectronic devices.

- Flexible and transparent conductive films: Flexible displays, touchscreens, and organic light-emitting diodes (OLEDs) otherwise use as brittle indium tin oxide (ITO) which is replaced by the graphene film due to its high flexibility and durability.
- field-Effect Transistors (FETs): Graphene FETs are ultra-fast-switching and have low power, making them suitable to high-frequency electronic and sensors.

Photodetectors and Photovoltaics Broadband optical response and light carrier dynamics allow efficient photodetection and photovoltaic energy conversion in solar cells.

- Plasmonics and Terahertz Devices Tunable plasmon resonances are used to develop more advanced plasmon-based systems in areas of communication and sensing.

Energy Storage and Conversion

Nanomaterials based on graphene help to enhance energy devices in terms of facilitating charge storage capacity, conductivity and stability.

- Supercapacitors: High power density, fast charge/discharge coupled with good electrical conductivity and large surface area offered by graphene makes it ideal to be used as supercapacitor electrodes.
- Lithium-Ion Batteries: Graphene increases the conductivity of the electrodes and allows the electrodes to expand and contract during cycling increasing the capacity and increasing the life of the battery.
- Fuel Cells and Catalysis: Graphene as a Functionalized catalyst carrier: Efficient Catalysis Graphene to functionalize catalyst carriers supports catalysts to enhance the oxygen reduction reaction hence new efficiency gain.
- Solar Cells: Addition of the graphene as transparent electrode or a charge transport layer increases the efficiency and flexibility of the solar cells.

Composites and Coatings

Mixing of graphene in polymer, ceramic or metal matrices yields composites that have better mechanical, thermal and barrier performance.

- Mechanical Strengthing: Tensile strength, stiffness, and fracture toughness of composites are augmented through the addition of graphene; aerospace, automotive and construction industries can use composites strengthened using graphene.

Moreover, stability of electronic packaging and high-performance materials: Thermal conductivity of materials based on graphene helps to dissipate heat.

- Corrosion and Wear Resistance: Graphene and coating keeps a surface oxidized and abrasive and increases life of components.

Sensors and Biosensors

High surface area and high electrical conductivity of graphene supplemented by chemical sensitivity enable the creation of highly sensitive and selective sensors.

- Gas Sensors: Trace gas detection: fast response and recovery time towards NO_2, NH_3 and CO.
- Biological Sensors: Graphene with biomolecules- functionalized graphene molecules detect biomolecules such as glucose, DNA, protein and are used to perform diagnostics and monitoring of health.
- Environmental Sensors: Map the pollutants and toxins including high sensitivity and low limit of detection.

Biomedical Applications

Materials based on graphene have potential properties ideally suited to medical application, among them biocompatibility, facile functionalization, and unusual physical behavior.

- Drug Delivery: Graphene can be used as carriers and they allow the specific release of drug.
- Tissue Engineering: Graphene scaffold favors cell growth and differentiation which enhance tissue regeneration.

• Imaging and Diagnostics: Photo luminescence and magnetic qualities of Graphene oxide aid in contrast imaging and bioimaging.
• Antibacterial and Antiviral Agents: Although not a particular narrow field, the efficiency of the Graphene derivative in the sphere of antimicrobials is clear.

Environmental Applications

The adsorption capacity, and catalytic activity of graphene facilizes environmental cleaning.

• Water Purification: Graphene membranes trap contaminants such as impurities like heavy metals, Organic molecules and microbes.
• Air Filtration: The particulate filter and harmful gases are filtered with the help of graphene-based filters.
• Catalytic Degradation: The presence of graphene-backed catalysts fulfills the degradation of pollutants in waste water.

The creation of the graphene nanomaterials, assembling into various nanostructures, had a great impact on various fields of science and technology, mainly due to its refractions with the flows of physicochemical properties. These nanoparticles have been used in biomedical engineering, environmental purification and electronics. The paper will discuss the broad spectrum of graphene nanomaterials usage based on the academic literature that constitutes a rigorous critique of their mechanisms and efficacies. Graphene oxide (GO) and its derivatives have shown significant potential, especially in the improvement of mechanistic features and biocompatibility of polymeric constructions used in bio-medica, specifically with significant importance to dental practice. Recently, it was claimed that flexural strength and fracture toughness of composites reinforced with GO increase significantly thus enhancing the mechanical properties of dental prostheses that have been produced by computer-aided design and manufacturing (CAD/CAM) system (Sachdev et al., 2025).

Besides, the development of laser-induced graphene (LIG) on skeletal muscle tissue engineering has emerged as a breakthrough with LIG showing good biocompatibility. Preliminary tests have indicated that LIG materials have a diminutive negative impact during the development of an embryo in zebrafish models over other types of graphene, but additional experiments are vital to define the safety profile of LIG fully (Ordoño et al., 2025). This proves the ongoing research into the safe uses of different derivatives of graphene in the area of regenerative medicine and tissue engineering. Additionally, the application of nanomaterials based on graphene in the sphere of biosensing is rather promising. Due to the conducting nature of the

Graphene, it is a great candidate to produce very sensitive electrochemical sensors as well as biosensors. The synthesis of a mesoporous CoFe 2 O 4 /reduced graphene oxide (rGO) nanohybrid to detect adrenaline can be primarily seen as one of the best examples of how the material can be effectively used in medical diagnostics to locate neurodegenerative diseases (Verma et al., 2025). Similarly, reduced graphene oxide (rGO) has a large surface area, which has the property of a turntable structure, thus making it applicable in a wide range of sensor technologies, and potentially making it significantly more sensitive to numerous biomolecules (Chemello et al., 2025). Applications like these symbolize the convergence of graphene-based technology/health monitoring-based applications, thus defining new directions in the developments in diagnostic and treatment procedures. Since an environmental viewpoint will be considered, graphene-based nanomaterials cannot be excluded in the wastewater treatment solutions. Their unique physicochemical properties, in particular, their two-dimensional structure and multifaceted porosity make them ideal adsorbents in aqueous. A recent research study based on investigations used a microwave-assisted hydrothermal production of two-adsorptive rGO, which has the potential to be harmful to lot of applications in environmental remediation, and shows the suitability of the material in adsorption and isolation of pollutants. Additionally, another polymer property of graphene oxide is the broad range of contaminants that can be adsorbed.

As an example, the increased capacities of the material make it possible to remove heavy metals and organic wastewater contaminants, thus supporting a positive ecological environment and human health (EL-Shendidy et al., 2025). Another field of application of graphene involves nanostructured energy storage devices (Jamal et al. 64). Graphene is particularly attractive due to its electrical conductivity and large surface area. Combination of graphene with supercapacitors and batteries can experience a substantial increase in the performance of energy and its efficiency. According to the studies, it is possible to combine the capacitance and energy density in the system of energy storage by stacking layers of graphene, which will lead to a more efficient rating (Mistry, 2025). Such endowments also make graphene derivatives great potential antimicrobial coating and treatment of surfaces in healthcare facilities, highlighting the relevance of graphene in the development of medical technology.

The aspect of plasticity of graphene has been viewed in terms of graphene nanocomposites. It is the very high strength of the product that makes it nearly impossible to leave out the introduction of graphene platelets into most compound designs, such as those used in aerospace and in civil engineering. Thus, the mechanical, thermal, and electrical characteristics of such composites are improved over those exhibited by conventional graphene composites, with these characteristics only becoming increased under conditions of large stress (Trai & Sobhy, 2025). These findings affirm the ability of graphene in improving both functionality and safety of engineering materials. In terms of synthesis and characterization, various policies have been put in place so as to generate the most desirable graphene-based polymer matrices (GPMs). Among them, chemical vapor deposition (CVD) and green synthesis are the most common, which are used to produce high-quality graphene with low density of defects, which is the central determinant of the success of the consequent technology (Madhukar, 2025). By nature of the synthesis, the resulting fluxes determine if the obtained material will match the desired physicochemical characteristics that will identify the material as usable in other fields such as nanoelectronics, drug-delivery network and any other advanced technology of the day. Health hazards and risks that medium- and long-term consumers may face due to using the materials of graphene are a comparatively young sphere, which a thorough investigation should be pursued. Although about the future of graphene people continue to make optimistic predictions, further understanding of its relationship with biological systems is urgently needed. The ongoing studies in this area are seeking to clarify the toxicological characteristics of the diverse types of graphene, especially as the substances have been used in extremely sensitive areas of human endeavors like health care and environmental management (Ordoño et al., 2025). To handle these issues, a lot of scientific research is required to reduce the risks and make the technological characteristics of graphene safe to use. To conclude, the range of applications of graphene nanomaterials is very vast and enormous since it is a comparably new branch of technology that can be employed in biomedical informatics, in environmental restoration, in energy storage, and in manufacturing of engineered materials. This is due to the characteristics of graphene, attributed to its inherent versatility, and recent studies aim to make it as functional as possible and closely follow potential risks. It is hardly imaginable that, at this level, new discoveries will not continue on how to apply graphene and thus stop the movement toward the next stage of technological progress with the highlights on graphene science.

CONCLUSION

The grade of material under research has highly superior potential and an improved work with a broad range of scientific and technological divisions specifically with reference to the graphene nanomaterial. This has been partly because it has inherent oxidative, mechanical, thermal, and optical amenities, which in turn are synonymously correlated with the geometry structure of the groups of carbon atoms which constitute patterns as well as the spatial structure of the covalent links. A particularly facilitating approach is to add sp2 -hybridized electrons to the carbon structure, typically achieved by assembling a two-dimensional honeycomb structure whose advantages are traditionally less favorable than bulk -based compounds. As a result, new electronic products, energy-saving systems, sensors, composite materials, and environmental purification products have been considered using these materials. The new invented devices are produced through various production techniques, mechanical exfoliation, chemical vapor deposition (CVD) and liquid-phase exfoliation. All these methods present a trade-off between the quality of the material (and its scalability and cost), and the structural integrity, thus informing the choice of the most appropriate approach to particular applications. Existing production paths should thus be utilized and simplified to meet high-volume production requirements, scalability, and cheapness. There are critical methodologies that can be applied; functionalization and substitution; both strategies serve to maximize surface chemistry of graphene, by providing a platform to regulate the doping and defect density of the crystals, and to perform methods of integration of graphene into composite materials, flexible electronics and approaches to design, which ensures higher level of processability and reduces aggregation. A product of modern technological advancement is graphene nanomaterials. The use of other materials science areas, including improving energy-storage devices (like super-capacitors, batteries), energy-sensitive biosensors, amongst others, has also been certainly recognized as contributing to the so-called third industrial revolution. Other innovations are water purification and elimination of water pollution through non-resistant methods. However, there are multiple issues that should be examined in greater detail and they include: small scale opportunity in manufacturing, variations in the quality of materials and possible health and environmental risks. These issues require a multidisciplinary team effort of chemists, physicists, engineers, and materials scientists to develop a set of well-defined characterization methods. Such combined activities are expected to hasten the adoption of graphene technologies commercially, as well as expand their development further in accordance with industry standards. Applications of the nanomaterials that use graphene have not yet lived up to their potential but there are more inventions going on and technologies discovering an ever-growing range of applications. Particularly, the modern tendencies related to artificial intelligence and Industry (4.0) have a fundamental impact on the sector. It can thus be anticipated that, in the predictable future, graphene will be incorporated into literally all spheres and electronic appliances, therefore, having an overwhelming impact on productivity and well-being. Such advances are likely to stimulate the emergence of a new genealogy of nanotechnologies, to create and increase the knowledge in the spheres of energy, health, environmental stewardship, and expanded management disciplines.

REFERENCES

Abaszade, R. G. (2022). Synthesis and analysis of flakes graphene oxide. *Journal of Optoelectronic and Biomedical Materials, 14*(3), 107–114. DOI: 10.15251/JOBM.2022.143.107

Abaszade, R. G., Babanli, M. B., Kotsyubynsky, V. O., Mammadov, A. G., Gür, E., Kapush, O. A., Stetsenko, M. O., & Zapukhlyak, R. I. (2023). Influence of gadolinium doping on structural properties of carbon nanotube. *Physics and Chemistry of Solid State, 24*(1), 53–158. DOI: 10.15330/pcss.24.1.153-158

Abaszade, R. G., Kapush, O. A., Mamedova, S. A., Nabiyev, A. M., Melikova, S. Z., & Budzulyak, S. I. (2020). Gadolinium doping influence on the properties of carbon nanotubes. *Physics and Chemistry of Solid State, 21*(3), 404–408. DOI: 10.15330/pcss.21.3.404-408

Abaszade, R. G., Kapush, O. A., & Nabiyev, A. M. (2020). Properties of carbon nanotubes doped with gadolinium. *Journal of Optoelectronic and Biomedical Materials, 12*(3), 61–65. https://www.chalcogen.ro/61_AbaszadeRG.pdf. DOI: 10.15251/JOBM.2020.123.61

Abaszade, R. G., Mammadov, A. G., Bayramov, I. Y., Khanmamadova, E. A., Kotsyubynsky, V. O., Gur, E. Y., & Kapush, O. A. (2022). Structural and electrical properties of sulfur-doped graphene oxide/graphite oxide composite. *Physics and Chemistry of Solid State, 25*(2), 256–260. DOI: 10.15330/pcss.23.2.256-260

Abaszade, R. G., Mammadov, A. G., Bayramov, I. Y., Khanmamadova, E. A., Kotsyubynsky, V. O., Gur, E. Y., & Kapush, O. A. (2022). Modeling of voltage-ampere characteristic structures on the basis of graphene oxide/sulfur compounds. *Technical and Physical Problems of Engineering, 14*(2), 302–306.

Abaszade, R. G., Mammadov, A. G., Khanmamedova, E. A., Aliyev, F. G., Kotsyubynsky, V. O., Gür, E., Soltabayev, B. D., Margitych, T. O., & Stetsenko, M. O. (2024). Photoconductivity of functionalized carbon nanotubes. *Digest Journal of Nanomaterials and Biostructures, 19*(2), 837–843. DOI: 10.15251/DJNB.2024.192.837

Abaszade, R. G., Mammadova, A. G., Khanmamedova, E. A., Bayramov, İ. Y., Namazov, R. A., Popal, Kh. M., Melikova, S. Z., Qasımov, R. C., Bayramov, M. A., & Babayeva, N. İ. (2023). Electron paramagnetic resonance study of gadolinium doped graphene oxide. *Journal of Ovonic Research, 19*(2), 259–263. DOI: 10.15251/JOR.2023.193.259

Ahmed, S., Hazarika, S., Tashkandi, M., Armaghani, T., & Kolsi, L. (2025). Enhanced heat diffusion in triangular wavy enclosures using mos_2-water nanofluidics: A comsol simulation study. Zamm - Journal of Applied Mathematics and Mechanics. *Zeitschrift für Angewandte Mathematik und Mechanik, 105*(5). Advance online publication. DOI: 10.1002/zamm.70053

Bawa'neh, H., Lababneh, B., Malkawi, A., & Bozeya, A. (2025). Investigation of the mechanical properties of an aluminum (al 6061-t6)/graphene/bentonite hybrid nanocomposite experimentally and through finite element analysis study. *Materials Advances*, *6*(3), 1031–1041. DOI: 10.1039/D4MA00890A

Chemello, G., Despotelis, K., Paton, K., Clifford, C., Pollard, A., Radnik, J., & Hodoroaba, V.-D. (2025). Approaches for the measurement of lateral dimensions of graphene oxide flakes using scanning electron microscopy. *Nano Express*, *6*(1), 015006. DOI: 10.1088/2632-959X/adae28

Christudoss, A., Sah, K., Vikram, R., Giri, S., Viswanathan, D., Dimkpa, C., & Mukherjee, A. (2025). Tailoring the synthesis route of reduced graphene oxide and its toxicological effects on allium cepa. *ACS Omega*, *10*(20), 20771–20783. DOI: 10.1021/acsomega.5c01919

Ding, Y., Qin, W., Zhu, H., Dai, Y., Hong, X., Han, S., & Xie, Y. (2025). Construction of ZnO/r-GO composite photocatalyst for improved photodegradation of organic pollutants. *Molecules (Basel, Switzerland)*, *30*(5), 1008. DOI: 10.3390/molecules30051008

EL-Shendidy. M., EL-Mekkawi, D., & Hafez, H. (2025). Sustainable and optimized microwave-assisted hydrothermal method for production of dual-adsorptive reduced graphene nanomaterials for wastewater remediation. https://doi.org/DOI: 10.21203/rs.3.rs-6627454/v1

Esmaeili, S., & Hosseini, M. (2025). Buckling analysis of tapered graphene-reinforced composite plates: An analytical approach. *Plastics, Rubber and Composites*, *54*(5), 197–212. DOI: 10.1177/14658011251340776

Farnsworth, A., Hora, Y., Wang, R., Anyaegbu, C., Dharmasiri, B., Henderson, L. C., & Hooper, J. F. (2025). Covalent functionalization of graphene and carbon nanotubes by radical decarboxylation. *ACS Applied Materials & Interfaces*, *17*(26), 38348–38356. Advance online publication. DOI: 10.1021/acsami.5c04671

Figarova, S. R., Aliyev, E. M., Abaszade, R. G., Alekberov, R. I., & Figarov, V. R. (2021). Negative differential resistance of graphene oxide|sulphur compound. *Journal of Nano Research*, *67*, 25–31. . DOI: 10.4028/www.scientific.net/JNanoR.67.25

Gamal, R., Soliman, T., & Khalid, A. (2025). Thermal stability and electrical conductivity of polyvinyl alcohol composites reinforced with Mn_2O_3–rgo. *ChemistrySelect, 10*(1), e202403782. Advance online publication. DOI: 10.1002/slct.202403782

Gao, K., Qi, M., Che, Z., Niu, Q., Guo, W., & Sun, X.. (2025). Bamboo-based carbon fiber/carbon nanosheet composite flexible supercapacitor. *ChemNanoMat : Chemistry of Nanomaterials for Energy, Biology and More, 11*(6), e202500037. Advance online publication. DOI: 10.1002/cnma.202500037

Hkiri, K. (2025). Synthesis and photonic properties of WO3-reduced graphene oxide nanocomposites. *Journal of Physics: Conference Series, 2970*(1), 012002. DOI: 10.1088/1742-6596/2970/1/012002

Hu, R., Wen, S., Chen, Q., Sun, X., Liu, H., Gao, W., & Bai, Y. (2025). Universal construction of electrical insulation and high-thermal-conductivity composites based on the in-situ exfoliation of boron nitride-graphene hybrid filler. *ACS Applied Materials & Interfaces, 17*(4), 6783–6792. DOI: 10.1021/acsami.4c18250

Ikhsan, N., Sutiman, N., & Samsudin, N. (2025). A facile <i>in-situ</i> synthesis and characterizations of silver-graphene oxide (aggo) nanocomposite using chives as a reducing and stabilizing agent. *Journal of Metastable and Nanocrystalline Materials, 41*, 35–42. DOI: 10.4028/p-CcECT2

Jashrapuria, K., & Singh, S. (2025). Biofilm inhibition by laser-induced graphene: Impact of surface texture on rod-shaped e. coli and coccus-shaped staphylococcus. *ACS Applied Materials & Interfaces, 17*(14), 21819–21829. DOI: 10.1021/acsami.5c01748

Kazi, M., Varghese, S., Sarker, N., Aich, N., & Gadhamshetty, V. (2025). Advancing pfas remediation through physics-based modeling of 2d materials: Recent progress, challenges, and opportunities. *Industrial & Engineering Chemistry Research, 64*(4), 1894–1906. DOI: 10.1021/acs.iecr.4c03715

Khalajiolyaie, A., & Jian, C. (2025). Advances in graphene-based materials for metal ion sensing and wastewater treatment: A review. *Environments (Basel, Switzerland), 12*(2), 43. DOI: 10.3390/environments12020043

Khujamberdiev, R., & Cho, H. (2025). Synthesis and characterization of nanoparticles in transforming biodiesel into a sustainable fuel. *Molecules (Basel, Switzerland), 30*(6), 1352. DOI: 10.3390/molecules30061352

Lee, I., Kim, S., Shim, J., Cho, J., & Cho, C. (2025). High flame retardancy of carbon nanotubes reinforced polyelectrolyte multilayered nanocomposites. *Journal of Applied Polymer Science*, *142*(22), e56967. Advance online publication. DOI: 10.1002/app.56967

Li, F., Guo, R., Qiu, Y., Liu, Z., Tao, L., Yang, S., Chen, J., & Yang, H. (2025). Enhanced anticancer activity of graphene oxide quantum dot@cu nanocomposites via cation–π interactions. *Physical Chemistry Chemical Physics*, *27*(14), 7076–7083. DOI: 10.1039/D5CP00352K

Li, J., Wang, M., & Zhao, Y. (2025). Self-assembly of low-dimensional organic graphene-like molecules at interfaces. *Journal of Physics: Conference Series*, *2954*(1), 012007. DOI: 10.1088/1742-6596/2954/1/012007

Liang, H., Guo, L., Niu, Y., Tang, Z., Zhao, Z., Mei, H., Fang, R., Liu, C., & Gong, W. (2025). High-sensitivity, low-hysteresis, flexible humidity sensors based on carboxyl-functionalized reduced-graphene oxide/ag nanoclusters. *Nanomaterials (Basel, Switzerland)*, *15*(11), 800. DOI: 10.3390/nano15110800

Lutomia, D., Poria, R., Kala, D., Singh, A., Gupta, M., Kumar, D., Kaushal, A., & Gupta, S. (2025). Unlocking the potential of 2d nanomaterial-based biosensors in biomarker-based detection of helicobacter pylori. *Materials Advances*, *6*(1), 117–142. DOI: 10.1039/D4MA00546E

Madhukar, P., Kesarwani, R., Pandey, S., Singh, V. K., Rao Gedda, M., Singh, O. P., Shaz, M., Kumar, R., & Sundar, S. (2025). Amine-functionalized graphene quantum dots conjugated with amphotericin b: Synthesis, characterization, and in vitro evaluation for visceral leishmaniasis treatment. *ACS Omega*, *10*(25), 26627–26638. Advance online publication. DOI: 10.1021/acsomega.5c00879

Mammadov, A. G., Abaszade, R. G., Kotsyubynsky, V. O., Gur, E. Y., Bayramov, I. Y., & Khanmamadova, E. A. (2022). Photoconductivity of carbon nanotubes. *Technical and Physical Problems of Engineering*, *14*(3), 155–160.

Ordoño, J., Islam, M., Lantada, A., Echeverry-Rendón, M., & Wang, D. (2025). Evaluation of laser-induced graphene for skeletal muscle tissue engineering applications. *Biomaterials Science*, *13*(12), 3242–3251. DOI: 10.1039/D5BM00174A

Qiu, W., Gao, M., Meng, X., Dong, Y., Liu, Q., Qiu, C., & Song, Z. (2025). Stress impacts of different types of micro- and nanomaterials on vegetable crops. *Environmental Science. Nano*, *12*(6), 3342–3356. DOI: 10.1039/D5EN00237K

Ran, T., Zhou, J., Xu, Y., Li, Z., Chen, C., Wu, T., & Liu, J. (2025). An electrochemical sensor for Cd2+ in water based on RGO-comncnf screen-printed electrodes. *Journal of the Electrochemical Society*, *172*(3), 037521. DOI: 10.1149/1945-7111/adc2a3

Rano, N., & Martsinovich, N. (2025). Stability of sp3 carbons in hydrogenated graphene quantum dots and their electronic and optical properties studied using density functional theory. *The Journal of Physical Chemistry A*, *129*(17), 3790–3806. DOI: 10.1021/acs.jpca.4c07825

Sachdev, H., Soni, M., Gupta, A., Chansoria, H., Patil, M., Sharma, B., Kar, K., & Sharma, R. (2025). Comparative evaluation of flexural strength of pmma reinforced with graphene (g-cam) and monolithic zirconia - an in vitro study. *Journal of Pharmacy & Bioallied Sciences*, *17*(Suppl 1), S238–S241. DOI: 10.4103/jpbs.jpbs_1715_24

Shi, L., Schröder, M., Meyer, H., Peláez, D., Wodtke, A., Golibrzuch, K., Schönemann, A.-M., Kandratsenka, A., & Gatti, F. (2025). Full quantum dynamics study for h atom scattering from graphene. *The Journal of Physical Chemistry A*, *129*(7), 1896–1907. DOI: 10.1021/acs.jpca.4c06712

Song, Y., Xu, J., Sun, Y., Xu, X., Tan, X., Du, M., Hu, J., Shi, Z., Wang, H., & Zhang, X. (2025). Structure and properties of au/n-g composites prepared based on different nitrogen sources. *Journal of Physics: Conference Series*, *2951*(1), 012108. DOI: 10.1088/1742-6596/2951/1/012108

Sun, W., Geng, W., Qian, P., Wang, T., Jing, L., Feng, J., Wang, Y.-H., Wang, Y.-X., Zhou, H.-L., & Geng, H.-Z. (2025). Composite flexible films based on mwcnt and modified graphene for electric heating and emi shielding. *ACS Applied Polymer Materials*, *7*(12), 7798–7810. Advance online publication. DOI: 10.1021/acsapm.5c00373

Tang, Y., Friec, A., Chen, M., & Sun, D. (2025). Designed biomaterial-enhanced cell transplantation for neural tissue engineering. *Aggregate*, *6*(5), e70022. Advance online publication. DOI: 10.1002/agt2.70022

Trai, V., & Sobhy, M. (2025). Vibration and transient response analysis of bi-directional functionally graded graphene platelets reinforced plates resting on a Kerr foundation using a refined finite-element procedure. *Proceedings - Royal Society. Mathematical, Physical and Engineering Sciences*, *481*(2314), 20250297. Advance online publication. DOI: 10.1098/rspa.2025.0297

Verma, R., Yadav, S., Singh, D., & Singh, J. (2025). Fabrication of a mesoporous $CoFe_2O_4$/RGO nanohybrid and laccase interface biosensor for rapid detection of adrenaline for neurodegenerative disease diagnosis. *Materials Advances*, *6*(6), 1988–2001. DOI: 10.1039/D4MA01216J

Weheabby, S., Liu, Z., Pašti, I., Rajić, V., Vidotti, M., & Kanoun, O. (2025). Enhanced electrochemical sensing of methyl parathion using agnps@il/go nanocomposites in aqueous matrices. *Nanoscale Advances*, *7*(8), 2195–2208. DOI: 10.1039/D4NA00919C

Xu, W., Ren, C., Wang, Z., Lochab, B., Li, J., & Zhang, Y.. (2025). Fabrication of graphene oxide/silk protein core-sheath aerogel fibers for thermal management. *Journal of Materials Chemistry. A, Materials for Energy and Sustainability*, *13*(10), 7081–7090. DOI: 10.1039/D5TA00580A

Yang, Q., Hu, J., Li, H., Du, Q., Feng, S., Yang, D., Zhang, Y., & Shen, J. (2025). All-optical modulation photodetectors based on the cds/graphene/ge sandwich structures for integrated sensing-computing. *Advanced Science (Weinheim, Baden-Wurttemberg, Germany)*, *12*(11), 2413662. Advance online publication. DOI: 10.1002/advs.202413662

Yang, Y., Zhang, S., Huang, Z., Wu, H., Hu, D., Liu, H., & Bai, Y. (2025). Improvement of corrosion resistant of waterborne polyurethane coatings by addition of modified graphene oxide. *Surface Topography : Metrology and Properties*, *13*(2), 025008. DOI: 10.1088/2051-672X/add280

Zhang, H., Gan, X., Lu, Z., Li, L., & Lu, L. (2025). Polycarboxylate superplasticizer-modified graphene oxide: Dispersion and performance enhancement in cement paste. *Nanomaterials (Basel, Switzerland)*, *15*(6), 419. DOI: 10.3390/nano15060419

Chapter 2
Characterization Methods for Graphene-Based Nanomaterials

Volodymyr O. Kotsyubynsky
https://orcid.org/0000-0001-6461-937X
Vasyl Stefanyk Precarpathian National University, Ukraine

Ruslan Zapukhlyak
Vasyl Stefanyk Precarpathian National University, Ukraine

V. M. Boychuk
Vasyl Stefanyk Precarpathian National University, Ukraine

V. T. Hoi
Vasyl Stefanyk Precarpathian National University, Ukraine

M. M. Klymyuk
Vasyl Stefanyk Precarpathian National University, Ukraine

ABSTRACT

The chapter is an elaborate description of analytical methods, which are mainly used in the complete characterization of the structural, spectroscopic, thermal and electronic properties of graphene-based nanomaterials. In the study of the structure and morphology of materials like graphene, graphene oxide, and reduced graphene oxide, scholars often apply such techniques as X-ray diffraction (XRD), small-angle X-ray scattering (SAXS), transmission electron microscopy (TEM), high-resolution TEM (HRTEM), atomic force microscopy (AFM) and other methods of porosity analysis.

DOI: 10.4018/979-8-3373-3862-0.ch002

INTRODUCTION

Graphene-derived nanomaterials– including graphene, graphene oxide (GO), and reduced graphene oxide (rGO) – have a distinctive combination of structural, electronic, thermal, and chemical properties, which are the precondition of their applications. These properties are linked to features such as layer thickness, defect concentration, distribution of functional groups, and crystallite size. A complete understanding of these complex, often heterogeneous materials requires a combination of complementary analytical techniques, as no single method can cover all the important aspects. Working with structural and morphological parameter analysis, such as crystallite size, layer stacking, interlayer distance, porosity, and surface texture, one must apply X-ray diffraction (XRD), small-angle X-ray scattering (SAXS), transmission electron microscopy (TEM), high-resolution TEM (HRTEM), and atomic force microscopy (AFM). However, these methods have to be combined with spectroscopic techniques, which give essential details about bonding patterns, elemental composition, and electronic transitions; Raman spectroscopy, X-ray photoelectron spectroscopy (XPS), Fourier transform infrared spectroscopy (FTIR), and UV-Vis spectroscopy. Thermal analysis tools, including thermogravimetric analysis (TGA) and differential scanning calorimetry (DSC), reveal information about thermal stability, degradation profiles, and oxidation behavior. Meanwhile, electrical and electronic characteristics such as conductivity, charge carrier mobility, and dielectric response are assessed through methods like the Four-Point Probe, Electrochemical Impedance Spectroscopy (EIS), and Hall Effect measurements.

The integrating of data from these techniques allows for to develop of a comprehensive understanding of graphene-based nanomaterials, connecting their structure and composition with functional performance. This chapter outlines these characterization methods, highlighting their complementarity role in the evaluation of graphene-based materials.

STRUCTURAL AND MORPHOLOGICAL CHARACTERIZATION

XRD

The XRD analysis is based on Bragg's Law: $n\lambda = 2d \sin\theta$, where λ is the wavelength of the incident X-rays (in many applications Cu Kα radiation with a wavelength of 1.5406 Å), d is the distance between atomic layers, θ is the diffraction angle (or Bragg angle) between the incident X-ray beam and the atomic plane corresponding to the peak position, and n is the order of diffraction (usually $n = 1$). When these conditions are precisely satisfied, constructive interference of X-rays scattered by atomic planes occurs, producing distinct diffraction peaks. The principle of obtaining an XRD pattern involves directing a finely focused, monochromatic X-ray beam onto the sample. Typically, the sample is in the form of powder to ensure random orientation of crystallites. The scattered X-rays are recorded using a detector that moves precisely around the sample through a controlled angular range, producing a plot of X-ray intensity against diffraction angle, commonly denoted as 2θ. The XRD pattern obtained typically consists of distinct diffraction peaks, each corresponding to reflections from specific atomic planes. Peak position (2θ) provides information about interplanar spacing and lattice parameters; peak integral intensity (area) reflects the crystallinity and phase composition; peak width (Full Width at Half Maximum, FWHM) allows determination of average crystallite size and evaluation of lattice strain; peak shape indicates structural disorder and stacking faults (Lee et al., 2021).

The crystallite size in polycrystalline or nanocrystalline samples can be effectively determined from XRD peak broadening using the Scherrer equation: $D = \frac{K\lambda}{\beta \cos\theta}$, where D is the average size of coherently scattering regions (crystallites); K is the dimensionless Scherrer constant, which is determined by the shape of the crystallites, K=0.9 in the spherical crystallites approximation; λ is the X-ray wavelength; β is the FWHM of the diffraction peak (in radians) observed on the diffractogram at an angle of 2θ. For plate-like crystallites, such as rGO, which are essentially thin layers stacked along the c-axis, the relevant diffraction comes from the (00l) reflections (especially the (002) peak). In this case, the average crystallite size extracted from the (002) peak corresponds to the stacking thickness (i.e., thickness perpendicular to the graphene layers). The experimentally measured peak width (β_{exp}) contains contributions from both the instrumental broadening β_{instr} (due to the optics, slits, X-ray source size, and detector resolution) and sample-induced broadening β_{sample} (due to finite crystallite size, microstrain, and defects). Well-crystallized, strain-free standard material (silicon, quartz, or LaB_6) is used for β_{instr} determination.

The mathematical form of this correction depends on the selection of the diffraction peak shape during fitting. The type of fitting function (typically Gaussian, Lorentzian, Pseudo-Voigt Profile) reflects how broadening mechanisms (e.g., crystallite size, strain) and instrumental resolution interact. Gaussian profile is preferable (good for graphite) when instrumental resolution dominates and the sample is well-crystallized: $\beta_{sample} = \sqrt{\beta_{obs}^2 - \beta_{instr}^2}$. Lorentzian profiles typically arise from crystallite size broadening, stacking faults, or defects in disordered systems (good for turbostratic carbons or rGO): $\beta_{sample} = \beta_{obs} - \beta_{instr}$. Most real-world XRD peak shapes are neither purely Gaussian nor purely Lorentzian. They typically result from Gaussian instrumental broadening and Lorentzian sample-related broadening. The pseudo-Voigt function is a weighted linear combination of Gaussian and Lorentzian components: $\beta_{obs} = \eta\beta_L + (1 - \eta)\beta_G$ Where: η is the mixing parameter (0 ≤ η ≤

1), β_L and β_G are Lorentzian and Gaussian contributions. In this case, instrumental and sample components must be separated individually for each component: $\beta_{G,sample} = \sqrt{\beta^2_{G,obs} - \beta^2_{G,instr}} \beta_{L,sample} = \beta_{L,obs} - \beta_{L,instr}$. Most Rietveld refinement programs (e.g., FullProf, TOPAZ) use the pseudo-Voigt function.

Rietveld refinement involves a least-squares fitting procedure, iteratively adjusting theoretical structural and instrumental parameters (lattice parameters, atomic positions, site occupancies, thermal displacement factors, preferred orientation, and instrumental peak profiles) to minimize differences between the calculated and experimentally measured diffraction patterns. Rietveld refinement includes: establishing an initial structural model based on known or hypothesized crystallographic data; generating a theoretical diffraction pattern from the model, crystal symmetry, lattice parameters, atomic positions, and instrumental conditions; adjusting structural parameters to minimize the residual difference between the experimental and theoretical patterns; employing statistical indicators (e.g., weighted profile factor R_w, expected profile factor R_{exp}, and goodness-of-fit $\chi 2$) to evaluate the quality of refinement; obtaining refined structural parameters with minimized uncertainties and high reliability, providing accurate crystallographic details.

The Williamson–Hall method allows for separating the effects of crystallite size D and microstrain in the lattice ε on peak broadening, assuming that the total observed broadening (β_{obs}) is a combination of size-induced and strain-induced effects as: $\beta_{obs} cos\theta = \frac{K\lambda}{D} + 4\varepsilon sin\theta$ with the next linearization of $\beta \cos\theta = f(4 \sin\theta)$ Plot. As D is obtained, the average number N of graphene sheets stacked within a coherent crystalline domain along the c-axis can be estimated by dividing D by the interlayer spacing d_{002}.

The structure of graphite is a combination of sheets of graphene stacked on to each other and bound by van der Waals forces. It exhibits and distinct X-ray diffraction trace (Fig. 1a). The sharp and intense (002) diffraction peak is observed typically at $2\theta \approx 26.4°$, corresponding to an interlayer spacing (d_{002}) of approximately 0.335 nm that directly reflects the regular stacking and well-ordered layered structure of graphite (Li et al., 2007). As a higher-order harmonic reflection of (002), the (004) peak is observed around $2\theta \approx$ 54-55°; it indicates the high crystallographic alignment in the stacked graphene layers. The narrow width (low FWHM) of graphite diffraction peaks indicates high crystallinity, large crystallite sizes, and minimal lattice strain, reflecting well-defined layer stacking and a highly ordered crystalline structure. Any broadening, asymmetry, or shifting of the (002) peak can be directly correlated to structural imperfections, reduced crystallite size, turbostratic disorder, or intercalation effects.

The XRD pattern of graphene typically consist of a weak (002) peak around $2\theta \approx 25.0$-26.5°, corresponding to an interlayer spacing similar to graphite but with significantly reduced intensity and peak sharpness due to the limited number of stacked layers and increased structural disorder reduced intensity and peak sharpness due to the limited number of stacked layers and increased structural disorder (Siburian et al., 2018). Monolayer or bilayer graphene may not have a clear X-ray peak due to the lack of periodic stacking. However, for low-layered graphene and bulk graphene powders, XRD analysis remains useful to confirm the presence of a layered structure, assess crystallinity, and distinguish it from graphite and graphene oxide. The absence of a low-angle peak (001) distinguishes graphene from GO, while a broad, low-intensity peak (002) distinguishes it from highly crystalline graphite.

Graphene oxide production through introducing oxygen functional groups widens the gap between the stacking of graphene sheets. Such structural alteration is witnessed in the GO XRD profile (Fig. 1b) where the sharp diffraction peak normally occurs at a low angle of approximately $2\theta \approx 9\text{-}12°$, that is, (001) reflection (Stobinski et al., 2014).

Additionally, the peak's intensity and width provide insights into the crystallinity and stacking disorder of the oxidized layers. Compared to graphite, which exhibits a strong (002) peak at around 26.4°, the GO pattern shows a significant shift and broadening due to disrupted π–π stacking and reduced order. In some cases, partially oxidized or inhomogeneous GO samples may show both the (001) and residual (002) peaks. XRD is thus used to monitor structural changes during synthesis and drying, track exfoliation behavior, and evaluate the quality of GO before further processing or reduction.

The destruction of oxygen-containing functional groups in the thermal or chemical reduction of graphene oxide produces a partial restoration of the conjugated graphene lattice, and the interlayer spacing is reduced. This transformation is reflected in the X-ray diffraction pattern (Fig. 1c), where the characteristic GO (001) peak at $2\theta \approx 10\text{-}12°$ disappears or is significantly weakened. A broad and less intense peak replaces it at $2\theta \approx 24\text{-}26°$, corresponding to the reflection (002), but with a larger FWHM compared to graphite (Habte et al., 2019). The width and lower intensity of this peak indicate increased disorder, reduced crystallinity, and turbostatic arrangement of graphene layers in rGO (Kotsyubynsky et al., 2024). The interlayer distance typically decreases to about 0.35-0.39 nm, which is larger than in graphite but smaller than in GO, reflecting incomplete reduction and residual functional groups.

In the XRD pattern of rGO, the (100) peak typically appears as a weak and broad reflection around $2\theta \approx 42\text{–}44°$, depending on the degree of in-plane ordering. In contrast to the (002) peak, which indicates the stacking order along the c-axis, the (100) peak provides information about the lattice structure within the basal plane (a-b plane). Its visibility depends on the degree of reduction and the quality of graphite domains; a more pronounced peak (100) indicates better in-plane crystallinity and partial reduction of the graphene lattice. The lateral size of the rGO sheets (average crystallite size in the a-b plane) can be estimated from the diffraction peak (100) on the X-ray diffraction pattern using the Scherrer equation with a shape factor of K = 1.84 (Kumar et al., 2022).

Small-Angle X-ray Scattering

The Small-Angle X-ray Scattering (SAXS) is founded on the elastic scattering of the X-rays at small angles on the inhomogeneities in the electron density (usually the sample is characterized by an interface, a pore, or a particle). Recording the scattered intensity as a function of the scattering angle allows us to measure structure and distribution, size, and the shape of these inhomogeneities at a scale of 1 to 10 nm (Rennhofer et al., 2021). The experimental setup typically includes a high X-ray source, a monochromator or multilayer optic to produce a nearly monochromatic beam, a collimation system to minimize beam divergence (with pinholes or slits), a sample holder (capillary, film, or flow-through cell) under vacuum or ambient conditions, and a 2D or 1D detector positioned downstream. Background subtraction, careful alignment, and calibration with known standards (e.g., silver behenate) are critical for precise data. Data analysis includes determining the scattering vector.$q = \frac{4\pi}{\lambda}\sin(\theta)$, and fitting the scattered intensity profiles I(q) to extract structural parameters. Accordingly to the Guinier approximation, at low q, the scattered intensity I(q) can be described as: $I(q) = I(0) \cdot \exp\left(-\frac{q^2 R_g^2}{3}\right)$, where $I(0)$ is the forward scattering intensity (at q=0 related to the square of the molecular weight), Rg is the radius of gyration (Del Giudice et al., 2022). The radius of gyration Rg is the root-mean-square distance of the scattering elements (e.g., electrons, atoms) from the center of mass. It characterizes the overall size of the particle regardless of its shape: $R_g^2 = \frac{1}{M}\sum_i m_i r_i^2$, where m_i is the mass of the ith element at distance ri from the center of mass, and M is the total mass. The linearization of ln(I(q)) =f(q2) plot at low q (qRg<1) allows for to calculate of Rg of randomly oriented, monodisperse or narrowly distributed scattering centers. Rg quantifies the spatial distribution of electron density. Larger RG means a more extended structure. Rg gives a simple scalar quantity representing size, compactness, and distribution symmetry.

In aggregated or restacked systems like rGO, Rg increases due to interlayer stacking or wrinkling effects. The Guinier approximation for graphene-based nanomaterials assumes random orientation of sheets and negligible interactions with a validity range limited to qRg<1.3 (Kusano et al., 2025). Rg of dispersed GO nanosheets is typically in the range of 5–50 nm, depending on exfoliation level and lateral size, while multilayer aggregates or compact particles may exhibit Rg values >50 nm. Interpretation of Rg must consider sheet morphology, lateral size distribution, and possible folding or stacking effects.

Porod's analysis is applied for ideal two-phase systems (e.g., particles (pores) in a matrix) with smooth and sharp interfaces when the scattering intensity I(q) at high q (qRg≫1) decays as: $I(q) \sim \frac{1}{q^4}$ (Tran et al., 2023). The linearization of Porod plot $I(q) \cdot q^4 = f(q)$ allows (if the electron density contrast between particle (pores) and matrix is known) to calculate specific surface area (S/V, surface area per unit volume) of the sample as where $K = \lim_{q \to \infty} I(q) q^4$ is the Porod constant (from the plateau on the Porod plot). For graphene materials with sharp and smooth interfaces, the Porod law predicts that the intensity I(q) decays as $I(q) \sim q^{-4}$. Fractal analysis in SAXS allows characterization of irregular, self-similar structures such as aggregates, porous networks, and rough surfaces (Cao et al., 2022). These systems often deviate from classical Porod behavior and show power-law scattering: $I(q) \sim q^{-FD}$, where FD is related to the fractal dimension of the system. Mass fractals describe how mass scales with size — typical for aggregated particles: $I(q) \sim q^{-Df}$, where $1 \leq Df \leq 3$ (Df = 1 for linear chain, Df = 2 for branched structures, Df = 3 for densely packed objects). Surface fractals describe rough or complex interfaces (cracked surfaces, rough particle boundaries, or porous solids) with Scattering Behavior as $I(q) \sim q^{-(6-Ds)}$ where $2 \leq Ds \leq 3$ (Df = 2 for smooth surface, Df = 3 for rough surface). For full quantitative analysis and transition from fractal behavior to Porod regime: I(q)=K·P(q)·S(q) where P(q) is a form factor of primary particle or pore (e.g. sphere), S(q)is a structure factor for fractal aggregate: $S(q) = 1 + \frac{D_f \Gamma(D_f - 1) sin\left[(D_f - 1) tan^{-1}(q\xi)\right]}{(q\xi)^{D_f}} (1 + (q\xi)^2)^{(1-D_f)/2}$, where D_f is mass fractal dimension, ξ is cutoff length (size of fractal aggregate), Γ is gamma function.

Figure 1. XRD pattern of (a) synthetic graphite Aldrich, #282863, (b) GO synthesized by the Tour method, (c) rGO obtained NaOH-assisted hydrothermal method (Kotsyubynsky et al., 2024)

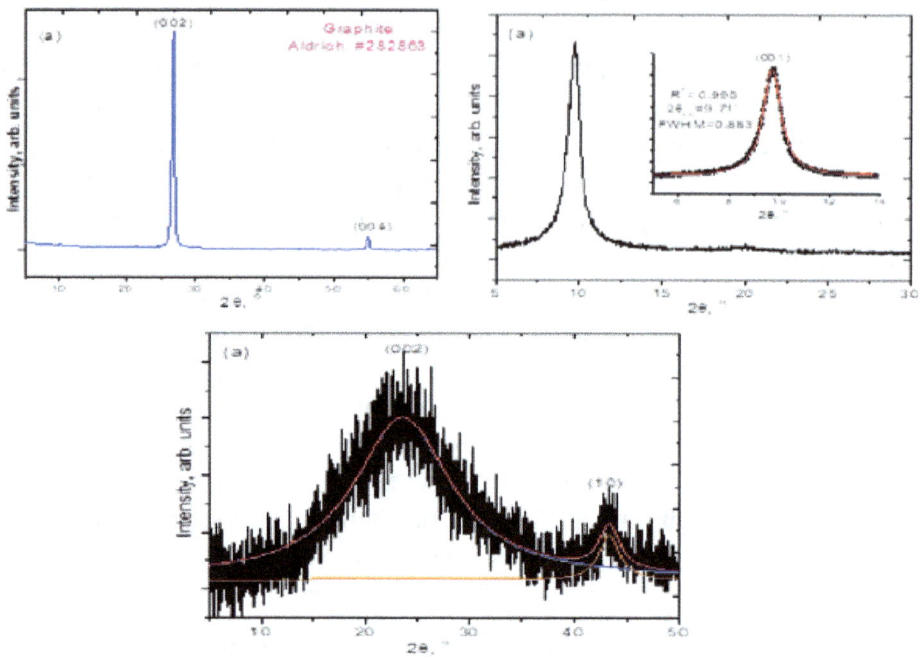

TEM/ HRTEM

Transmission Electron Microscopy (TEM) enables you to observe materials on the atomic level. It operates by launching a highly energetic electron beam (standard 100 to 300 keV) through an extremely thin specimen (less than 100 nm) and recording the electrons that undergo interactions with the specimen to form an image or a diffraction pattern (Egerton, 2005). TEM provides a picture related to the internal structure, crystallography, flaws, and shape of the material. Contrast in TEM is produced by variations in mass-thickness, by diffraction effects, and by phase variation of the beam transmitted. High-resolution transmission electron microscopy (HRTEM) is a technique of TEM, which allows you to observe structures of materials at the atomic level with a point resolution of 0.1-0.2 nm, and even 0.08 nm in modern aberration-corrected instruments (Williams et al., 2009). HRTEM operates by exploiting the phase contrast of the transmitted electron wave, which interacts with the crystal lattice of a thin specimen to produce an interference pattern recorded as an image.

Sample thickness, degree of crystallinity, and the electron beam wavelength influence the visibility of lattice fringes in HRTEM. While HRTEM commonly operates at accelerating voltages between 80 and 300 kV, lower voltages– around 80 kV– are generally chosen for beam-sensitive materials like graphene to reduce knock-on damage.

HRTEM provides not only atomic-level images but also information on interplanar spacings, crystallographic orientation, and local strain fields. However, the technique requires extremely thin, clean, and crystalline samples to avoid multiple scattering and image distortion. Sample drift, charging, and contamination can degrade image quality, and maintaining ultra-high vacuum and stable environmental conditions is essential.

HRTEM can distinguish between monolayer and few-layer graphene by analyzing image contrast and lattice fringe patterns, where increased fringe density and Moiré patterns indicate multilayer stacking. The technique is especially useful for characterizing grain boundaries in polycrystalline graphene, which influence mechanical and electronic properties, and for analyzing the orientation and twist angles between layers in bilayer graphene. Aberration-corrected HRTEM further enhances resolution and contrast, enabling atomic-level imaging with minimal lens distortions and improved signal-to-noise ratio. Operating at lower voltages, such as 80 kV, minimizes knock-on damage in graphene while maintaining sufficient contrast and resolution. However, care must be taken to avoid contamination and beam-induced transformations. Image simulations using multislice methods are often necessary to interpret the contrast in HRTEM images and to correlate experimental data with theoretical models. HRTEM also allows dynamic studies under in situ conditions such as heating or gas exposure, revealing transformations and defect evolution in real time (Robertson et al., 2013; Alam et al., 2017). One of the major advantages of HRTEM is the resolving of individual atomic defects and regions of lattice reconstruction, which are not detectable by bulk spectroscopic methods (Dave et al., 2016). Accordingly, to direct observations while reduction of GO improves graphitization, significant local disorder remains in the form of isolated vacancies, distorted hexagons, and residual oxygen-containing groups. The change in contrast observed in HRTEM allows the calculation of layer numbers and local stacking (Cui et al., 2018).

SAED

The selected area electron diffraction (SAED) technique, commonly used alongside TEM, provides localized crystallographic information by analyzing electron diffraction patterns from specific regions of a sample. SAED relies on the elastic scattering of electrons by periodic atomic planes in crystalline materials, forming diffraction patterns that reveal information about crystal structure, orientation. SAED is applied for phase identification, measurement of *d*-spacings, and distinguishing crystalline/amorphous areas. In graphene-based materials, SAED patterns allow differentiate between single-layer and few-layer graphene (Wang et al., 2008; Reina et al., 2009). A typical SAED pattern of GO displays diffuse rings, reflecting its near-amorphous structure. Upon gradual reduction, the SAED patterns transform, showing sharper and more defined hexagonal diffraction spots, which confirm the restoration of crystalline sp^2 domains. This transition is directly observable and quantifiable through changes in diffraction intensity, ring sharpness, and the appearance of higher-order spots (Ceniceros-Reyes et al., 2022).

While SAED provides valuable localized crystallographic insights from micron-sized areas, it has limitations, such as possible overlap from neighboring crystallites and the requirement for thin, clean, electron-transparent specimens. It is often complemented by high resolution.

AFM

AFM allows for measuring the thickness, lateral dimensions, and surface roughness of carbon nanomaterial flakes or films. For single-layer graphene on substrates (such as SiO_2/Si), AFM can directly measure the step height between the graphene sheet and the substrate (typically ranging from 0.35 nm to about 0.8 nm) (Ishigami et al., 2007). The measured thickness of rGO monolayers typically is in a range of 0.7–1.0 nm compared to 1.0–1.3 nm for GO under ambient conditions (Sedki et al., 2022). AFM allows identification of monolayer, bilayer, and few-layer graphene flakes based on step-height measurements (Kumar et al., 2021). AFM allows direct trace of reduction process by visualizing isolated nanosheets on GO/rGO (typically on substrates like Au, SiO_2, or Si) (Faucett & Mativetsky, 2015). AFM provides information about surface roughness (folds), which are common in exfoliated graphene or rGO nanoflakes (Pacakova et al., 2015). Advanced AFM modes (phase imaging and force modulation) allow for obtaining information about mechanical properties and surface adhesion (Stan & King, 2020). AFM allows for to evaluate of the structural properties of carbon-based materials, especially in the early stages of device fabrication or material transfer.

Porosity and Surface Area

Surface area and porosity study of graphene-based materials is usually done by the application of gas adsorption modes, which yield quantitative increases in terms of specific surface area, pore volume, and pore size distribution. Nitrogen (N_2) adsorption - desorption isotherm at a liquid nitrogen temperature (77 K) of using the Brunauer - Emmett - Teller (BET, BET) theory to determine the surface area and Barrett - Joyner - Halenda (BJH, BJH) method applied to calculate the size distribution of mesopore (2 - 50 nm thermodynamic diameter; IUPAC meaning) is the most used method (Nishi & Inagaki, 2016). Microporosity (pore size ≤2 nm, IUPAC) is evaluated using t-plot or Density Functional Theory (DFT) models applied to the adsorption data (Ravikovitch & Neimark, 2001). In a standard setup, the graphene-based powder is degassed under vacuum at elevated temperatures (typically 120–200 °C) to remove adsorbed species before analysis and adsorption–desorption isotherms recording across a range of relative pressures (P/P_0), from 0.01 to 1.0.

Pristine single-layer graphene theoretically has a BET surface area up to 2630 m^2/g (Stoller et al., 2008), while multilayer graphene and graphene nanoplatelets exhibit lower specific surface area values (typically in the range of 100–600 m^2/g) depending on synthesis method, flake size, thickness, and post-processing treatments (sonication or thermal expansion) (Antunes et al., 2016; Cheng et al., 2016). The reduction in surface area is attributed to van der Waals interactions between graphene sheets, leading to partial restacking during synthesis or drying. Controlling solvent evaporation and introducing structural supports allows for preventing graphene sheet restacking and achieving a BET surface area of over 1300 m^2/g (Shi et al., 2015). At the same time, rGO generally exhibits surface areas ranging from 100 to 800 m^2/g, depending on synthesis method, reduction degree, and thermal treatment, while the specific surface area of composites based on reduced graphene oxide is typically in the range of 100–400 m^2/g (Najmi et al., 2024; Khan et al., 2024). For GO, much lower BET values of surface area are estimated due to stacked layers and blocked pores.

To estimate the size distribution of a mesopore, the BJH method (an estimation made on the analysis of the desorption branch of the isotherm on the Kelvin equation assuming cylinder geometry of the pores) is often adopted. The mesoporosity of rGO arises from sheet stacking disorder and particle aggregation effects. Typical BJH pore size distribution for rGO shows mesopores predominantly in the range of 2–20 nanometers, with specific peaks often observed between 3 and 10 nanometers depending on synthesis method, reduction conditions, and post-treatment processes (Rathnavel et al., 2023; Foschi et al., 2021). The DFT method provides a more accurate and detailed pore size distribution than the BJH method (especially for micropores), as it does not rely on simplified geometric assumptions. In rGO, DFT pore size distribution typically shows micropores in the range of 0.5–2 nm and mesopores, reflecting the heterogeneous structure of the material (Kumar et al., 2021). The presence of micropores is often attributed to intrinsic defects and vacancies in the carbon lattice, while mesopores arise from incomplete restacking and aggregation of graphene sheets (Liu et al., 2024).

SPECTROSCOPIC METHODS

Raman Spectroscopy

Depending on whether the energy was given to or taken by the atomic vibrations, the photons have less (Stokes shift) or have more (anti-Stokes shift) energy than the photon incident on the material. The difference in energy represents one of the vibrational modes of a material. The resulting Raman spectra display intensity as a function of Raman shift, providing a fingerprint of the material. Key features of a Raman spectrum include peak positions, intensity, and peak width, each providing specific structural information about the material. Peak positions correspond to particular vibrational modes of the atoms in the crystal lattice. They are characteristic of certain chemical bonds or lattice vibrations, allowing identification of functional groups or phase composition. The strength of a Raman peak gives the quantity of Raman-active material and can also give the crystallinity; it is usually stronger in well-ordered structures or in higher concentrations of a particular phase. Broader peaks suggest higher levels of disorder or amorphous character, while sharp, narrow peaks indicate well-ordered crystalline regions.

The Raman spectra of graphite (Fig. 2a,b) are characterized by a few well-defined peaks that reflect its highly ordered crystalline structure composed of stacked graphene layers (Yoon & Cheong, 2012). The most intense feature is the G band, located around 1580 cm^{-1}, which corresponds to the E_{2g} in-plane vibrational mode of sp^2-bonded carbon atoms. This peak is a signature of graphitic ordering and is strong and sharp in crystalline graphite.

A second-order peak known as the 2D band (also called G′) appears around 2750 cm^{-1} and is a result of a two-phonon, double-resonance process (Reich & Thomsen, 2004). In order to form a sheet of graphite, the single-layered band formed in graphene is broad and divided into two bands in the presence of more than one graphene layer, as is the case in graphite. The 2D band shape, position, and width tell about the stacking type of graphene layers. The D band, located around 1350 cm^{-1}, has a complex multicomponent structure and is associated with the disorder of the sp^2 carbon lattice. In highly crystalline graphite, the D band is either very weak or absent, indicating low defect density. The ratio of the intensity of the D to G bands (ID/IG) is usually used as an order measure; in crystalline graphite, this ratio is near zero (Pimenta et al., 2007). The displacement of the G band or 2D band positions may give information about mechanical strain and compressive/tensile stress on the lattice. Variations in the G and 2D bands, such as their dispersion with excitation energy or their intensity ratios, can provide insight into the electronic band structure and Fermi level position. Temperature-induced changes in peak positions and widths can be used to study thermal conductivity, phonon behavior, or local heating effects.

In many cases, G and D bands are fitted using two single Lorentz components only (Gutic et al., 2019). However, full analysis is based on the deconvolution of the D band using five Lorentzian components (Kotsyubynsky et al., 2024). The two-dimensional band of the graphene is very sensitive to the layers of the graphene. In monolayer graphene, the 2D band is sharp, symmetric, and stronger than the G band, compared to graphite, where it is broad and split with an intensity that varies across samples. As more and more graphene layers are formed, the shape, width, and intensity ratio of the 2D to G bands change, and this allows distinction between a single layer and a few layers as well as bulk graphite structure (Malard et al., 2009).

Raman spectra of GO (Fig.2c,d) compared to graphene are changed due to the presence of C-O functional groups, broken sp^2 bonding, and the enhanced defect structure. In graphene, the D band is lower or not visible because of a low density of defects, whereas the G-band is high and clear. In contrast, GO shows a significantly stronger D band, reflecting a high level of structural disorder and the disruption of the π-conjugated system by epoxides, hydroxyls, and other functional groups (Tran, 2020).

Figure 2. Raman spectra of (a,b) synthetic graphite Aldrich, #282863, (c,d) GO synthesized by the Tour method, (e,f) rGO obtained NaOH-assisted hydrothermal method (Kotsyubynsky et al., 2024)

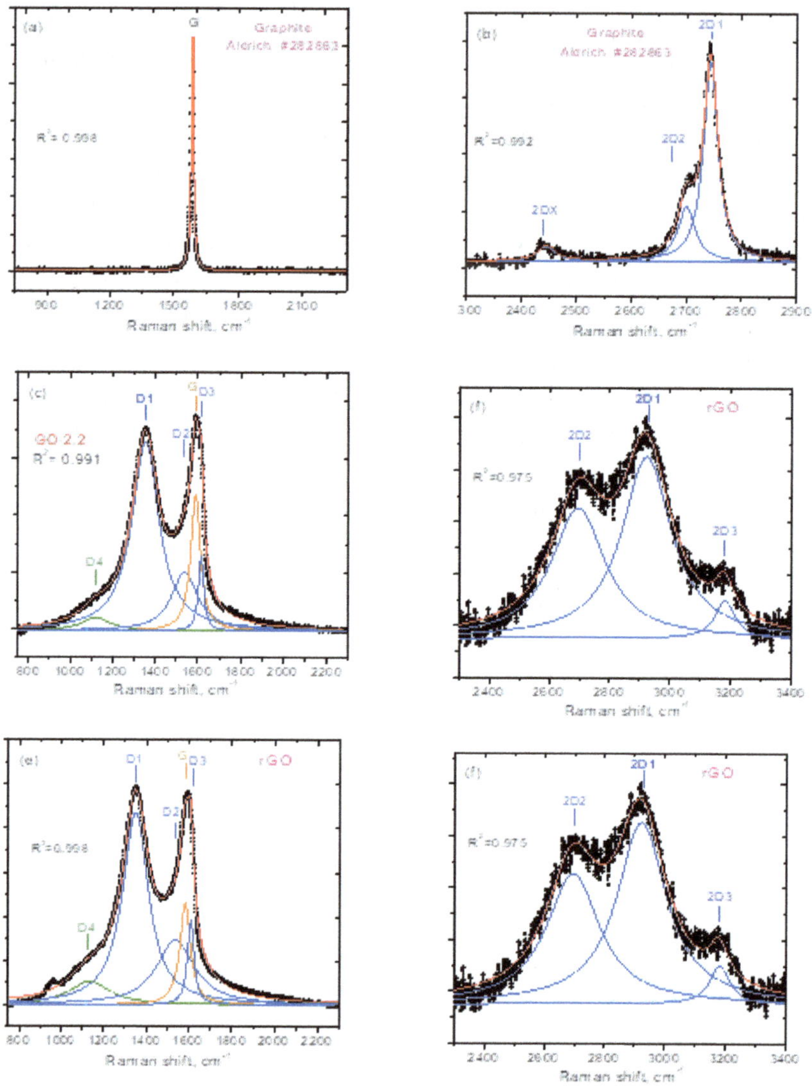

As a result, the I_D/I_G intensity ratio in GO is much higher than in graphene, often exceeding 1.0, indicating extensive sp^3 hybridization and defect formation. Additionally, both the D and G bands in GO are typically broader than in graphene, reflecting reduced crystallinity and heterogeneity in the carbon lattice. The 2D band is either very weak, significantly broadened, or completely absent in GO due to the loss of long-range π-electron coherence.

The Raman spectra of rGO show differences compared to GO, reflecting structural restoration and partial recovery of sp^2 hybridized carbon networks during the reduction process (Fig.2e, f). As a result, the I_D/I_G ratio in rGO may remain high or even increase, not because of more disorder, but due to the formation of smaller, newly created graphitic domains that increase the number of edge-like defects (Ruidíaz-Martínez et al., 2020). Additionally, the G band in rGO may shift slightly toward higher wavenumbers, and both the D and G bands become narrower compared to GO, indicating increased structural ordering. The 2D band, which is weak in GO, may reappear faintly in rGO, although still broadened and of low intensity, reflecting partial re-establishment of long-range π-conjugation. The evolution of the first-order Raman spectra of a monolayer graphene sample, as described in the study by Lucchese et al. (2010) and referenced in Pei and Cheng (2012), follows a clear pattern as defect density increases. Controlled growth of defect concentration (due to ion bombardment) causes the increase of the D band intensity and rising I_D/I_G ratio until the average distance between defects reaches approximately 4–5 nm; further increasing defect density causes a decrease in I_D/I_G due to the breakdown of sp^2 domains and transition toward amorphous carbon (Pei & Cheng, 2012). There is a relationship between the I_D/I_G ratio and the lateral in-plane sizes (L_a (nm)) of the crystallites: $L_a = (2.4 \times 10^{-10})\lambda^4(I_G/I_D)$, where is the wavelength of the excitation radiation (Pimenta et al., 2007). The I_D/I_G ratio for rGO may increase after reduction due to the formation of new small sp^2 domains despite partial restoration of the conjugated structure. This behavior reflects an increase in the number of edges per unit area rather than a simple decrease in defects. Therefore, applying the I_D/I_G ratio to estimate L_a is realible for moderately ordered graphene-like materials only.

FTIR

In FTIR spectroscopy, an interferometer modulates a broad-spectrum IR beam, and the resulting interferogram is mathematically converted into an absorption spectrum using a Fourier transform with high spectral resolution and rapid data acquisition (Khan, 2025). In carbon nanomaterials, FTIR is particularly valuable for detecting oxygen-containing groups such as hydroxyl (–OH), carboxyl (–COOH), carbonyl (C=O), and epoxide (C–O–C), which are prevalent in GO due to oxidative treatment (Scheibe et al., 2010). FTIR analysis allows tracking reduction processes where the gradual disappearance of oxygenated functional group bands indicates the transformation of GO to rGO (Surekha et al., 2020). Noncovalent functionalization of GO- and rGO-based materials, typically with π–π stacking or hydrogen bonding, typically results in shifts or intensity changes of existing bands, such as broadening of the O–H stretching band (~3400 cm^{-1}) or shifts in C=O or C–O bands (Pei & Cheng, 2012). FTIR can detect the interaction between GO or rGO particles and polymer matrices GO as a shift of the O–H and C–O stretching bands (Zhao et al., 2011). The ability to monitor chemical changes and interfacial interactions in GO/rGO composites makes FTIR an important technique for evaluating the surface functionalization and understanding structure–property relationships.

XPS

X-ray Photoelectron Spectroscopy (XPS) is a technique used to analyses the topography of a material in terms of its elemental composition with in-depth information of up to 5-10nm of the material (Bagus et al., 2013). It operates by irradiating a sample with monochromatic X-rays, typically Al Kα (1486.6 eV), causing photoelectrons to be emitted from the core levels of atoms. The kinetic energy of the emitted electrons is measured, and their binding energy is calculated to identify the elements. XPS can precisely estimate the oxidation states of atoms and monitor the presence of surface functional groups (Gengenbach et al., 2021). For pristine graphene, XPS typically shows a dominant C1S peak corresponding to sp^2-hybridized carbon atoms (C=C) centered around 284.4 eV (indicating a high degree of graphitization). The presence of additional peaks in the C 1s spectrum signals the presence of defects or residual functional groups (Kwan et al., 2015). A sharp and symmetric C 1s peak confirms high-quality and defect-free graphene (typically for CVD-grown graphene) (Deokar et al., 2015), while broader or asymmetric peaks suggest disorder or surface contamination (Chen et al., 2020).

The area under the sp^2 C=C peak, relative to the total C 1s peak area, indicates the proportion of conjugated carbon domains present in the sample (Ganguly et al., 2011). In GO, the sp^2 content is significantly reduced due to oxidation, typically showing a dominant presence of C–O and C=O components. Upon reduction to rGO, the relative intensity of the C=C peak increases, reflecting the partial restoration of the sp^2 carbon network. Quantifying the sp^2 fraction allows for assessing the degree of reduction (Xu et al., 2025).

The O 1s peak, typically centered around ~532.0 eV, further confirms the presence of oxygenated species (Liu & Speranza, 2021). The oxygen-to-carbon (O/C) atomic ratio is calculated by integrating the of the O 1s and C 1s peak areas with normalization using respective atomic sensitivity factors (ASF), which account the differences in photoionization cross-sections and detector efficiency: O/C = (O1s peak area / ASF (O)) / (C 1s peak area / ASF(C)) (Wagner et al., 1981). Typical ASF values used for quantification are 0.66 for O 1s and 0.296 for C 1s (based on Scofield's cross sections) (Pinder et al., 2024). carbon framework (Al-Gaashani et al., 2019). Using well-defined peak fitting protocols and consistent sensitivity factors is important to avoid misinterpretation of oxygen functionalities (Kovtun et al., 2019). In doped graphene systems, such as N-, B-, or S-doped graphene, XPS identifies dopant elements, their concentrations, and bonding configurations (Ullah et al., 2020). XPS mapping and depth profiling allow for spatial distribution analysis of chemical species (Hajati & Tougaard, 2010). Advantages of XPS for graphene-based materials include its sensitivity to surface chemistry and ability to resolve multiple oxidation states.

UV–Vis Spectroscopy

UV–Vis spectroscopy measures the absorption of ultraviolet and visible light in the range of 200–1100 nm, which corresponds to electronic transitions. Monolayer graphene shows a nearly featureless absorption spectrum in the UV–Vis range due to its constant optical absorbance (~2.3%) across the visible region, arising from interband π–π^* transitions of sp^2-hybridized carbon atoms (Eigler, 2009). Each additional graphene layer for multilayered graphene reduces transmittance by approximately 2.3% (linear scaling of absorbance with the number is observed for up to 10 layers), consistent with the universal optical absorbance derived from the fine-structure constant ($\pi\alpha \approx 1/137$) (Zhu et al., 2014). This shift is a useful optical marker for tracking the oxidation state of graphene-based materials during synthesis or chemical treatment (Otsuka et al., 2024). GO exhibits a strong absorption peak around 230-235 nm, attributed to π–π^* transitions of aromatic C=C bonds, and a shoulder near 300 nm, corresponding to n–π^* transitions of C=O groups (Sunderrajan et al., 2018). Changes in these peak intensity and position can be used to analyze reduction efficiency and confirm functionalization in dispersions. While UV–Vis spectroscopy does not provide detailed structural or bonding information, it is a rapid method for evaluating the electronic transitions of graphene-based materials.

rGO demonstrates direct allowed transition and bandgap formation as a result of residual oxygen functionalities, structural defects, as well as quantum confinement effects in finite-sized sp^2-domains. The optical bandgap of GO is relatively wide, typically ranging from 2.5 eV to 4.0 eV, depending on synthesis conditions, oxidation level, and dispersion medium (Sharma et al., 2021). The bandgap of rGO correspondingly narrows, often to values in the range of 0.1–2.6 eV, depending on the degree of reduction (Abid et al., 2018). The reducing of the direct bandgap is observed with the increasing thickness of rGO films due to the enhancing of π–π stacking interactions (Aboraia et al., 2024; Eda et al., 2008).

EDS / EDX

Analytical methods Energy-dispersive X-ray spectroscopy (see also: EDX or EDS) is an analysis method typically incorporated with TEM that allows the elemental composition of materials to be determined at high spatial resolutions. When an incident high-energy electron beam in the TEM strikes atoms within the sample, it also excites atoms such that electrons within the inner shell displace electrons in other shells, stimulating characteristic inner-shell emissions of X-rays emitted with a distinct energy characteristic of the atoms in the sample being investigated. A solid-state detector (normally silicon drift detectors (SDDs)) is used to samples these X-rays, and the energy and intensity of the photons emitted provide an energy spectrum of the incident radiation, which is used to determine the elements present. In TEM, EDS can be conducted both under conventional and scanning TEM (STEM) modes, the latter having an improved spatial resolution and more localized analysis, because the probe is very thinly focused. The spatial resolution of EDS in TEM can reach below 1 nm when using probe-corrected instruments, although the actual resolution depends on factors such as beam size, sample thickness, and the interaction volume (Goldstein et al., 2017). EDS is suitable for both qualitative and quantitative analysis, but quantification requires corrections for absorption, fluorescence, and detector efficiency. EDS can detect elements with atomic numbers greater than about 5–6, although lighter elements like boron and carbon can be challenging to quantify accurately due to low X-ray generation efficiency and absorption by the detector window. In graphene-based materials, EDS is used to map the distribution of dopants, oxygen content in graphene oxide (GO), and metal nanoparticles supported on graphene surfaces (Al-Gaashani et al., 2019). It is particularly valuable for analyzing compositional variations across defects, interfaces, or heterogeneous structures. Limitations include overlap of spectral peaks (e.g., Kα of sulfur and Mα of molybdenum), low count rates from thin samples, and beam-induced damage during prolonged acquisition.

EELS

When dealing with carbon nanomaterials in general, EELS can be used to detect and map carbon, oxygen, and nitrogen, high spatial resolution to sample heterogeneous systems, surface functionalities, and edge effects in graphene-based materials. EELS provide information through the analysis of characteristic energy-loss K-edges (Suenaga et al., 2001; Díaz et al., 2001). The oxygen K-edge at approximately 532 eV (as well as the pre-edge and main fine edge structure) is used to reveal details about oxygen bonding of oxygen-containing functional groups (hydroxyl, epoxy, and carboxyl groups) in GO and rGO (D'Angelo et al., 2017). The nitrogen K-edge at approximately 400 eV, (relevant in nitrogen-doped graphene or rGO) allows to identify pyridinic and pyrrolic groups (He et al., 2016). In summary, EELS is good as additional method for characterizing carbon nanomaterials due combination of elemental sensitivity and bonding state information.

THERMAL AND CHEMICAL STABILITY

Thermogravimetric Analysis (TGA)

Thermogravimetric Analysis (TGA) is a technique which rests on the sample heating-up at a constant rate of heating with an inert environment (e.g., nitrogen or argon) or under an oxidative environment (e.g., air or oxygen) with the sample being continuously observed to determine its weight loss (Prime et al., 2009). By using TGA, one is able to analyses the oxygen-containing functional groups content as well as determines the thermal decomposition profiles of the carbon nanomaterials (Bannov et al., 2020). The amount of mass loss at the main stage is directly correlated with the functionalization level of GO surface (Dehghanzad et al., 2016). For rGO with a respectively lower content of oxygen functionalities weight loss is significantly diminished (Liu and Speranza, 2021). The final residue at temperatures above 600–700 °C under oxidative conditions is often attributed to the graphitic carbon structure that resists further degradation (Farivar et al., 2021). The application of different kinetic models (Kissinger (Mageed et al., 2018), Flynn–Wall–Ozawa (Zhang et al., 2018), Coats–Redfern (Ravari et al., 2019), or Vyazovkin (Menezes et al., 2023)) allows to calculate the activation energy of GO/rGO oxidation. The Kissinger method determines the maximum temperature of maximum decomposition rate measured by the various heating rates values and gives an easy extrapolation of activation energy. The FlynnWallOzawa is an is conversional technique and uses TGA data at different heating rates which it uses the conversion degree to calculate the activation energy which gives more elaborate information about the degradation process. The Coats–Redfern method involves fitting the TGA curve to various reaction models. The Vyazovkin method allows the activation energy evaluating as a function of the mass loss fraction at a given temperature. The activation energy of rGO thermal oxidation is in a range of 120–250 kJ/mol, depending on reduction method, residual oxygen content, structural defects, and flake thickness (Feng et al., 2019; Li et al., 2021).

Differential Scanning Calorimetry (DSC)

Differential Scanning Calorimetry (DSC) represents a thermal analytical technique with its basis on measuring the flow of heat that is associated with physical and chemical changes of a material with respect to temperature or time under controlled heating or cooling. DSC in general can reveal the thermal stabilities, phase changes, and energy activation processes through the detection of endothermic and exothermic events, which include but are not limited to melting, crystallization, glass transitions, oxidation, and decomposition. DSC allows for to detect of the exothermic oxidation/reduction events and decomposition processes of GO and rGO. For GO, an exothermic peak attributed to the decomposition of oxygen-containing functionalities is typically observed in the temperature range of approximately 150–300 °C along with the evolution of gaseous CO_2, CO, and (H_2O). The position and intensity of the exothermic peak are directly related to the degree of functionalization: highly oxidized GO shows a more pronounced exothermic peak, while partially reduced or thermally treated GO exhibits a smaller or shifted peak (Bagri et al., 2010). DSC for rGO allows for to examine of its thermal stability and the detection of residual exothermic or endothermic events associated with residual functionalities. Compared to GO, rGO exhibits fewer and weaker exothermic peaks in the DSC profile in a range of 150–400 °C, attributed to the decomposition of remaining functional groups, structural rearrangements, or graphitization processes (Liu & Speranza, 2021). For graphene, DSC is used to assess its thermal stability and oxidation behavior. Unlike GO or rGO, graphene typically does not show exothermic or endothermic peaks below 500–600 °C under an inert atmosphere. An exothermic peak corresponding to thermal oxidation (combustion) for graphene in an oxidative atmosphere (air or oxygen) is observed at temperatures of 600–950 °C, depending on defect density and layers number (Arkhipova et al., 2020). DSC serves as a complementary technique to TGA in the thermal characterization of graphene-based materials.

ELECTRICAL PROPERTIES

The Four-Point Probe Method

Graphene exhibits extremely low sheet resistance due to its continuous sp^2-bonded carbon lattice with highly delocalized π-electrons. This electronic configuration enables in-plane electrical conductivity of pristine monolayer graphene in the range of 10^5–10^5 S/m with extremely high electron mobility (up to 200,000 cm^2·(V·s)) (Bolotin et al., 2008). The absence of structural defects, grain boundaries, and functional group disruptions in pristine graphene leads to minimal scattering of charge carriers, resulting in sheet resistance values 30–100 Ω/sq for mechanically exfoliated graphene (from graphite flakes) in controlled laboratory conditions (Novoselov et al., 2004). Graphene films used as transparent conductive electrodes show sheet resistance compared to indium tin oxide (ITO), with values reported in the range of 150–500 Ω/sq at about 90% transparency (Kim et al., 2009). Multilayer graphene films (deposited on substracts like SiO_2 or Si) exhibit conductivities in a wide range 10^4–10^6 S/m, depending on the preparation details, number of layers, and surface conditions (Tang et al., 2018).

The electrical conductivity with effective electromagnetic interference insulation ability was measured on multilayer graphene paper of up to 20 microns in thickness, flexible and lightweight, which works well as an electromagnetic shielding material to 2×10^5 S/m (Paliotta et al., 2015). CVD or mechanical exfoliation methods yield different defect densities and uniformities affecting the overall conductivity of multilayer graphene (Lv et al., 2024). Conductivity can be reduced by charge traps and roughness in the substrate, especially for multilayer graphene deposited on flexible polymer substrates (Rathanasamy et al., 2021). In-plane conductivity, Multilayer graphene conductivity is high, though not as conductive as in single-layer graphene, because of interlayer and poor stacking resistance. The most commonly used method for assessing rGO electrical conductivity is the Four-Point Probe method for films and sheets, as it minimizes contact resistance and provides reliable in-plane conductivity values (Husnah et al., 2017). rGO electrical conductivity is in a wide range of 10–10^4 S/m depending on reduction method and sample morphology (Iskandar et al., 2017; Rao et al., 2018). Van der Pauw technique, as a kind of Four-Point Probe method, is useful for thin, flat samples with arbitrary shape; it allows accurate measurements of sheet resistance and in-plane electrical conductivity, especially for irregularly shaped or small-area rGO films and flakes (Vázquez-Sánchez et al., 2019). Additionally, two-point probe methods are sometimes applied for smaller samples or films of rGO, where standard four-point probe measurement is impractical due to sample geometry or size limitations, but corrections to separate contact are necessary (Marquez et al., 2016). GO is considered insulating or semi-insulating compared to rGO or pristine graphene. Electrical conductivity values for GO films typically range from 10^{-6} to 10^{-3} S/m, depending on synthesis method, drying conditions, and ambient humidity (Eda et al., 2008). Four-Point Probe Method is the preferred method for GO films or papers, but the Two-Point Probe Method is also used (Sánchez-Trujillo et al., 2023).

GO colloids in water or polar solvents contain negatively charged GO sheets stabilized by oxygen-containing functional groups, which contribute to colloidal stability and ionic conductivity. The electrical conductivity of the GO colloidal solution is measured using a standard two electrode cell arrangement with (platinum or stainless steel electrodes); electrical conductivity values typically range from 10^{-5} to 10^{-2}

S/m, depending on GO concentration, degree of oxidation, pH and ionic strength of the medium and temperature (Hadadian et al., 2014). The measured conductivity of GO colloids is mostly ionic in character and is due to the dissociation of acidic groups (e.g., carboxylic groups) and counter-ion migration in solution, and not due to electronic transport of the GO flakes.

Electrochemical Impedance Spectroscopy (EIS)

Electrochemical Impedance Spectroscopy (EIS) Electrochemical impedance spectroscopy is a procedure of measuring the impedance of a material after applying a small alternating current (AC) voltage across a range of frequencies that is commonly between 10-2 and 107 Hz. These measured impedances are in the form of a real and imaginary component (resistance and capacitance, respectively), which provide insight into the conductive and capacitive response of materials. A Nyquist plot (the dependence of imaginary part (Z'') of impedance on real part (Z') of impedance) is used for graphical representation of EIS data. In a typical EIS setup, graphene or rGO films are deposited on an electrode substrate and immersed in an electrolyte, and the system's impedance response is recorded as a function of frequency. The total impedance is influenced by both the electronic pathways through graphene sheets and ionic transport in the system (Bonanni & Pumera, 2013). Lower impedance at high frequency typically corresponds to higher electronic conductivity. GO samples typically produced large semicircular Nyquist plots with wide diameter arcs, confirming dominant resistive behavior and poor electron transfer, when rGO plots showed smaller semicircles, consistent with enhanced electrical conductivity and reduced interfacial resistance (Casero et al., 2012).

In graphene and rGO, the dielectric response is primarily influenced by the interaction of charge carriers with interfaces and boundaries, interfacial polarization, and defect-related relaxation (Kuang et al., 2018). Graphene typically demonstrates low dielectric loss and minimal capacitive response due to its highly conductive sp^2-carbon lattice, while rGO (as well as reduced graphite oxide) exhibits frequency-dependent dielectric behavior reflecting its partial reduction state and remaining oxygen functionalities (Hong et al., 2017). rGO displays a dielectric constant in the range of $10–10^3$ at low frequencies due to interfacial polarization effects, decreasing at higher frequencies due to relaxation processes (Yuwen et al., 2019). The dielectric constant of GO is typically high at low frequencies, ranging from 10 to 10^4 depending on flake size, degree of oxidation, humidity, and measurement configuration (Prokhorov et al., 2019). The fact that the dielectric properties of GO and rGO materials are frequency dependent has been explained by Maxwell-Wagner-Sillars (MWS) interfacial polarization (occurring due to the difference in conductivity of sp^2 and sp^3) (Likhi et al., 2023). Another mode of dependence is the dielectric constant, which at higher frequencies drops drastically because the interfacial dipoles and bound charges are unable to respond to the quickly alternating electric field, whereas dielectric loss also exhibits a corresponding loss of effects over the frequency range, a sign of relaxation behavior in the trapped charges and the functional groups. GO exhibits significantly higher dielectric loss compared to rGO due to surface functionality and heterogeneous structure comprising both insulating and semiconducting regions. The dielectric loss in GO is dominated by interfacial polarization and charge hopping between localized states; at low frequencies (10^1–10^3 Hz), dielectric loss tangent (tan δ) values are in a range from 3.2 to 4.0, decreasing at higher frequencies (Kumar et al., 2020). Pure rGO exhibits high dielectric loss tangent values at low frequencies, indicating significant interfacial polarization and defect-related dipolar relaxation (Kiranakumar et al., 2024).

Hall Effect

The Silicon on Insulator is developed in charge carrier type (n-type or p-type), carrier concentration, carrier mobility, and Hall coefficient by applying a magnetic field (0.1- 2 Tesla) along the direction of flow of current and the transverse (Hall) voltage, measuring the magnetic field with temperature control with Hall Effect Measurement. For graphene and rGO films, the samples are often deposited on insulating substrates such as SiO_2/Si wafers, with electrodes applied using silver paste, gold contacts, or lithographically patterned metal pads (Rudenko et al., 2022; Eda et al., 2008). In single-layer graphene, Hall measurements typically reveal extremely high charge carrier mobilities (up to 10^6 cm^2/(V·s)) under ambient conditions, with carrier concentration values depending on factors such as doping level and substrate interaction (Ponomarenko et al., 2024). For multilayer graphene, the in-plane charge carrier mobility is lower (approximately from 10^3 to 10^4 cm^2/(V·s) depending on factors such as synthesis method, substrate, and layer number) than that of monolayer graphene due to interlayer interactions and increased defect density (Robinson et al., 2016). Hall Effect measurements on multilayer graphene revealed mobilities above 10^4 cm^2/(V·s) (Novoselov et al., 2004), while reported Hall mobilities for multilayer graphene on SiO_2 substrate are about 3000 cm^2/(V·s) (Nagashio et al., 2009). rGO can exhibit either n-type or p-type behavior influenced by its reduction process, residual functional groups, defect density, adsorbed species, and synthesis environment (Anadkat et al., 2024).

CONCLUSION

To offer an overall view of the graphene-based nanomaterial, the multimethod characterization of its structural, spectroscopic, thermal, and electronic properties is mandatory. The two are divergent and complementary. Further, the most significant data concerning the thermal stability and type of decomposition are obtained through thermogravimetric analysis (TGA) and differential scanning calorimetry (DSC). Examples of the electrical characterizations are, e.g., measurements of conductivity and carrier dynamics, and include the Four-Point Probe, electrochemical impedance spectroscopy (EIS), and measurements of Hall Effects. The combination of the numerous techniques offers a nice and overall indication of characteristically important material properties, including defect density, the stacking sequence of the layers, degree of functionalization, and carrier mobility. This is the type of multifaceted characterization needed in the aerospace engineering element and the additional requirements needed to use materials that have very good strength-to-weight ratios, that are capable of sustained operation in high-temperature environments, and are electronically stable in the conditions and continuation of the operations in which they may be called upon to serve. It provides the basis of the fine-tuning of the graphene-based nanomaterial, and the related development is within the context of refining it to form a lightweight fabric of the structure and enhanced thermal management, and long-term flexibility of the electronics in the next generation anthropomorphic aviation system.

REFERENCES

Abid, , Sehrawat, P., Islam, S. S., Mishra, P., & Ahmad, S. (2018). Reduced graphene oxide (rGO) based wideband optical sensor and the role of Temperature, Defect States and Quantum Efficiency. *Scientific Reports*, *8*(1), 3537. DOI: 10.1038/s41598-018-21686-2

Aboraia, A. M.. (2024). Exploration of the structural rGO thin films and their optical characteristics for optoelectronic device applications. *Journal of Optics*. Advance online publication. DOI: 10.1007/s12596-024-02092-6

Al-Gaashani, R., Najjar, A., Zakaria, Y., Mansour, S., & Atieh, M. A. (2019). XPS and structural studies of high-quality graphene oxide and reduced graphene oxide prepared by different chemical oxidation methods. *Ceramics International*, *45*(11), 14439–14448. DOI: 10.1016/j.ceramint.2019.04.165

Alam, S. N., Sharma, N., & Kumar, L. (2017). Synthesis of graphene oxide (GO) by modified hummers method and its thermal reduction to obtain reduced graphene oxide (rGO). *Graphene*, *6*(01), 1. DOI: 10.4236/graphene.2017.61001

Anadkat, D., Pandya, A., Jaiswal, A., Dungani, S., & Sanchela, A. V. (2024). Experimental comparison between graphene and reduced graphene oxide along with significant conversion of rGO from n-type to p-type. *Journal of Materials Science Materials in Electronics*, *35*(12), 821. DOI: 10.1007/s10854-024-12591-8

Antunes, M., Gedler, G., Abbasi, H., & Velasco, J. I. (2016). Graphene nanoplatelets as a multifunctional filler for polymer foams. *Materials Today: Proceedings*, *3*, S233–S239. DOI: 10.1016/j.matpr.2016.02.039

Arkhipova, E. A., Strokova, N. E., Tambovtseva, Y. A., Ivanov, A. S., Chernyak, S. A., Maslakov, K. I., Egorova, T. B., Savilov, S. V., & Lunin, V. V. (2020). Thermophysical study of graphene nanoflakes by differential scanning calorimetry. *Journal of Thermal Analysis and Calorimetry*, *140*(6), 2641–2648. DOI: 10.1007/s10973-019-09040-8

Bagri, A., Mattevi, C., Acik, M., Chabal, Y. J., Chhowalla, M., & Shenoy, V. B. (2010). Structural evolution during the reduction of chemically derived graphene oxide. *Nature Chemistry*, *2*(7), 581–587. DOI: 10.1038/nchem.686

Bagus, P. S., Ilton, E. S., & Nelin, C. J. (2013). The interpretation of XPS spectra: Insights into materials properties. *Surface Science Reports*, *68*(2), 273–304. DOI: 10.1016/j.surfrep.2013.03.001

Bannov, A. G., Popov, M. V., & Kurmashov, P. B. (2020). Thermal analysis of carbon nanomaterials: Advantages and problems of interpretation. *Journal of Thermal Analysis and Calorimetry*, *142*(1), 349–370. DOI: 10.1007/s10973-020-09647-2

Bolotin, K. I., Sikes, K. J., Jiang, Z., Klima, M., Fudenberg, G., Hone, J., Kim, P., & Stormer, H. L. (2008). Ultrahigh electron mobility in suspended graphene. *Solid State Communications*, *146*(9-10), 351–355. DOI: 10.1016/j.ssc.2008.02.024

Bonanni, A., & Pumera, M. (2013). High-resolution impedance spectroscopy for graphene characterization. *Electrochemistry Communications*, *26*, 52–54. DOI: 10.1016/j.elecom.2012.10.013

Cao, Y., et al. "New structural insights into densely assembled reduced graphene oxide membranes." Advanced Functional Materials 32.42 (2022): 2201535. https://doi.org/.DOI: 10.1002/adfm.202201535

Casero, E., Parra-Alfambra, A. M., Petit-Domínguez, M. D., Pariente, F., Lorenzo, E., & Alonso, C. (2012). Differentiation between graphene oxide and reduced graphene by electrochemical impedance spectroscopy (EIS). *Electrochemistry Communications*, *20*, 63–66. DOI: 10.1016/j.elecom.2012.04.002

Ceniceros-Reyes, M. A., Marín-Hernández, K. S., Sierra, U., Saucedo-Salazar, E. M., Mendoza-Resendez, R., Luna, C., Hernández-Belmares, P. J., Rodríguez-Fernández, O. S., Fernández-Tavizón, S., Hernández-Hernández, E., & Barriga-Castro, E. D. (2022). Reduction of graphene oxide by in-situ heating experiments in the transmission electron microscope. *Surfaces and Interfaces*, *35*, 102448. DOI: 10.1016/j.surfin.2022.102448

Chen, X., Wang, X., & Fang, D. (2020). A review on C1s XPS-spectra for some kinds of carbon materials. *Fullerenes, Nanotubes, and Carbon Nanostructures*, *28*(12), 1048–1058. DOI: 10.1080/1536383X.2020.1794851

Cheng, H., Ye, M., Zhao, F., Hu, C., Zhao, Y., Liang, Y., Chen, N., Chen, S., Jiang, L., & Qu, L. (2016). A general and extremely simple remote approach toward graphene bulks with in situ multifunctionalization. *Advanced Materials*, *28*(17), 3305–3312. DOI: 10.1002/adma.201505431

Cui, T., Mukherjee, S., Cao, C., Sudeep, P. M., Tam, J., Ajayan, P. M., Singh, C. V., Sun, Y., & Filleter, T. (2018). Effect of lattice stacking orientation and local thickness variation on the mechanical behavior of few layer graphene oxide. *Carbon*, *136*, 168–175. DOI: 10.1016/j.carbon.2018.04.074

D'Angelo, D., Bongiorno, C., Amato, M., Deretzis, I., La Magna, A., Fazio, E., & Scalese, S. (2017). Oxygen functionalities evolution in thermally treated graphene oxide featured by EELS and DFT calculations. *The Journal of Physical Chemistry. C, Nanomaterials and Interfaces*, *121*(9), 5408–5414. DOI: 10.1021/acs.jpcc.7b00239

Dave, S. H., Gong, C., Robertson, A. W., Warner, J. H., & Grossman, J. C. (2016). Chemistry and structure of graphene oxide via direct imaging. *ACS Nano*, *10*(8), 7515–7522. DOI: 10.1021/acsnano.6b02391

Dehghanzad, B., Razavi Aghjeh, M. K., Rafeie, O., Tavakoli, A., & Jameie Oskooie, A. (2016). Synthesis and characterization of graphene and functionalized graphene via chemical and thermal treatment methods. *RSC Advances*, *6*(5), 3578–3585. DOI: 10.1039/C5RA19954A

Del Giudice, A.. (2022). Porosity of Nanostructured Carbon Thin Films. In *Nanoporous Carbons for Soft and Flexible Energy Devices. Carbon Materials: Chemistry and Physics*. Springer., DOI: 10.1007/978-3-030-81827-2_8

Deokar, G., Avila, J., Razado-Colambo, I., Codron, J.-L., Boyaval, C., Galopin, E., Asensio, M.-C., & Vignaud, D. (2015). Towards high quality CVD graphene growth and transfer. *Carbon*, *89*, 82–92. DOI: 10.1016/j.carbon.2015.03.017

Díaz, J., Anders, S., Zhou, X., Moler, E. J., Kellar, S. A., & Hussain, Z. (2001). Analysis of the π^* and σ^* bands of the x-ray absorption spectrum of amorphous carbon. *Physical Review. B*, *64*(12), 125204. DOI: 10.1103/PhysRevB.64.125204

Eda, G., Fanchini, G., & Chhowalla, M. (2008). Large-area ultrathin films of reduced graphene oxide as a transparent and flexible electronic material. *Nature Nanotechnology*, *3*(5), 270–274. DOI: 10.1038/nnano.2008.83

Egerton, R. F. (2005). *Physical Principles of Electron Microscopy*. Springer., DOI: 10.1007/b136495

Eigler, S. (2009). A new parameter based on graphene for characterizing transparent, conductive materials. *Carbon*, *47*(12), 2936–2939. DOI: 10.1016/j.carbon.2009.06.047

Farivar, F., Yap, P. L., Karunagaran, R. U., & Losic, D. (2021). Thermogravimetric analysis (TGA) of graphene materials: effect of particle size of graphene, graphene oxide and graphite on thermal parameters. C, 7(2), 41. .DOI: 10.3390/c7020041

Faucett, A. C., & Mativetsky, J. M. (2015). Nanoscale reduction of graphene oxide under ambient conditions. *Carbon*, *95*, 1069–1075. DOI: 10.1016/j.carbon.2015.09.025

Feng, Y., Wang, B., Li, X., Ye, Y., Ma, J., Liu, C., Zhou, X., & Xie, X. (2019). Enhancing thermal oxidation and fire resistance of reduced graphene oxide by phosphorus and nitrogen co-doping: Mechanism and kinetic analysis. *Carbon, 146*, 650–659. DOI: 10.1016/j.carbon.2019.01.099

Foschi, M., Capasso, P., Maggi, M. A., Ruggieri, F., & Fioravanti, G. (2021). Experimental design and response surface methodology applied to graphene oxide reduction for adsorption of triazine herbicides. *ACS Omega*, *6*(26), 16943–16954. DOI: 10.1021/acsomega.1c01877

Ganguly, A., Sharma, S., Papakonstantinou, P., & Hamilton, J. (2011). Probing the thermal deoxygenation of graphene oxide using high-resolution in situ X-ray-based spectroscopies. *The Journal of Physical Chemistry. C, Nanomaterials and Interfaces*, *115*(34), 17009–17019. DOI: 10.1021/jp203741y

Gengenbach, T. R., Major, G. H., Linford, M. R., & Easton, C. D. (2021). Practical guides for x-ray photoelectron spectroscopy (XPS): Interpreting the carbon 1s spectrum. *Journal of Vacuum Science & Technology. A, Vacuum, Surfaces, and Films*, *39*(1), 1. DOI: 10.1116/6.0000682

Goldstein, J. I.. (2017). *Scanning Electron Microscopy and X-ray Microanalysis*. Springer., DOI: 10.1007/978-1-4939-6676-9

Gutic, S. J., Šabanović, M., Metarapi, D., Pašti, I. A., Korać, F., & Mentus, S. V. (2019). Electrochemically Synthesized Ni@ reduced Graphene Oxide Composite Catalysts for Hydrogen Evolution in Alkaline Media-the Effects of Graphene Oxide Support. *International Journal of Electrochemical Science*, *14*(9), 8532–8543. DOI: 10.20964/2019.09.23

Habte, A. T., & Ayele, D. W. (2019). Synthesis and characterization of reduced graphene oxide (rGO) started from graphene oxide (GO) using the tour method with different parameters. *Advances in Materials Science and Engineering*, *2019*(1), 5058163. DOI: 10.1155/2019/5058163

Hadadian, M., Goharshadi, E. K., & Youssefi, A. (2014). Electrical conductivity, thermal conductivity, and rheological properties of graphene oxide-based nanofluids. *Journal of Nanoparticle Research*, *16*(12), 2788. DOI: 10.1007/s11051-014-2788-1

Hajati, S., & Tougaard, S. (2010). XPS for non-destructive depth profiling and 3D imaging of surface nanostructures. *Analytical and Bioanalytical Chemistry*, *396*(8), 2741–2755. DOI: 10.1007/s00216-009-3401-9

He, L., Weniger, F., Neumann, H., & Beller, M. (2016). Synthesis, characterization, and application of metal nanoparticles supported on nitrogen-doped carbon: Catalysis beyond electrochemistry. *Angewandte Chemie International Edition*, *55*(41), 12582–12594. DOI: 10.1002/anie.201603198

Hong, X., Yu, W., & Chung, D. D. L. (2017). Electric permittivity of reduced graphite oxide. *Carbon*, *111*, 182–190. DOI: 10.1016/j.carbon.2016.09.071

Husnah, M., Fakhri, H. A., Rohman, F., Aimon, A. H., & Iskandar, F. (2017). A modified Marcano method for improving electrical properties of reduced graphene oxide (rGO). *Materials Research Express*, *4*(6), 064001. DOI: 10.1088/2053-1591/aa707f

Iskandar, F., Hikmah, U., Stavila, E., & Aimon, A. H. (2017). Microwave-assisted reduction method under nitrogen atmosphere for synthesis and electrical conductivity improvement of reduced graphene oxide (rGO). *RSC Advances*, *7*(83), 52391–52397. DOI: 10.1039/C7RA10013B

Khan, A., Sapakal, S. N., and Kadam, A. "Comparative analysis of graphene oxide (GO) reduction methods: impact on crystallographic, morphological, and optical properties." Graphene and 2D Materials 9 (2024): 101–109. https://doi.org/.DOI: 10.1007/s41127-024-00072-y

Khan, M. M. (2025). *"Fourier transform infrared spectroscopy." Photocatalysts: Synthesis and Characterization Methods*. Elsevier., DOI: 10.1016/B978-0-443-28913-2.00017-4

Kim, K., Zhao, Y., Jang, H., Lee, S. Y., Kim, J. M., Kim, K. S., Ahn, J.-H., Kim, P., Choi, J.-Y., & Hong, B. H. (2009). Large-scale pattern growth of graphene films for stretchable transparent electrodes. *Nature*, *457*(7230), 706–710. DOI: 10.1038/nature07719

Kiranakumar, H. V.. (2024). An impact of RGO on the ZnO nanoparticles: Structural, morphological, electrical, and gas sensing properties. *Sensing Technology*, *2*, 2310479. DOI: 10.1080/28361466.2024.2310479

Kotsyubynsky, V., et al. "NaOH-assisted hydrothermal reduction of graphene oxide." Journal of Physics: Condensed Matter 36.49 (2024): 495701. https://doi.org/.DOI: 10.1088/1361-648X/ad75dc

Kovtun, A., Jones, D., Dell'Elce, S., Treossi, E., Liscio, A., & Palermo, V. (2019). Accurate chemical analysis of oxygenated graphene-based materials using X-ray photoelectron spectroscopy. *Carbon*, *143*, 268–275. DOI: 10.1016/j.carbon.2018.11.012

Kuang, B., Song, W., Ning, M., Li, J., Zhao, Z., Guo, D., Cao, M., & Jin, H. (2018). Chemical reduction dependent dielectric properties and dielectric loss mechanism of reduced graphene oxide. *Carbon, 127*, 209–217. DOI: 10.1016/j.carbon.2017.10.092

Kumar, A., Tan, C.-S., Kumar, N., Singh, P., Sharma, Y., Leu, J., Huang, E.-W., Winie, T., Wei, K.-H., & Tseng, T. Y. (2021). Pentafluoropyridine functionalized novel heteroatom-doped with hierarchical porous 3D cross-linked graphene for supercapacitor applications. *RSC Advances, 11*(43), 26892–26907. DOI: 10.1039/D1RA03911C

Kumar, N., Setshedi, K., Masukume, M., & Ray, S. S. (2022). Facile scalable synthesis of graphene oxide and reduced graphene oxide: Comparative investigation of different reduction methods. *Carbon Letters, 32*(4), 1031–1046. DOI: 10.1007/s42823-022-00335-9

Kumar, P., Penta, S., & Mahapatra, S. P. (2020). Synthesis and Dielectric Relaxation Spectroscopy of Graphene Oxide Nanoparticles: Effects of Frequency, Thickness, and Temperature. *Journal of Electronic Materials, 49*(10), 5801–5807. DOI: 10.1007/s11664-020-08355-9

Kumar, V., et al. "Estimation of number of graphene layers using different methods: A focused review." Materials 14.16 (2021): 4590. https://doi.org/.DOI: 10.3390/ma14164590

Kusano, T., Oyama, N., Yoshida, H., & Tanaka, H. (2025). Varying the rheological behavior of a micellar solution via modified microscopic structures in the presence of graphene oxide. *RSC Advances, 15*(12), 9034–9040. DOI: 10.1039/D5RA00366K

Kwan, Y. C. G., Ng, G. M., & Huan, C. H. A. (2015). Identification of functional groups and determination of carboxyl formation temperature in graphene oxide using the XPS O 1s spectrum. *Thin Solid Films, 590*, 40–48. DOI: 10.1016/j.tsf.2015.07.051

Lee, A. Y., Yang, K., Anh, N. D., Park, C., Lee, S. M., Lee, T. G., & Jeong, M. S. (2021). Raman study of D* band in graphene oxide and its correlation with reduction. *Applied Surface Science, 536*, 147990. DOI: 10.1016/j.apsusc.2020.147990

Li, K., Khanna, R., Zhang, H., Ma, S., Liang, Z., Li, G., Barati, M., & Zhang, J. (2021). Thermal behaviour during initial stages of graphene oxidation: Implications for reaction kinetics and mechanisms. *Chemical Engineering Journal, 421*, 129742. DOI: 10.1016/j.cej.2021.129742

Li, Z. Q., Lu, C. J., Xia, Z. P., Zhou, Y., & Luo, Z. (2007). X-ray diffraction patterns of graphite and turbostratic carbon. *Carbon, 45*(8), 1686–1695. DOI: 10.1016/j.carbon.2007.03.038

Likhi, F. H., Singh, M., Chavan, S. V., Cao, T., Shanbedi, M., & Karim, A. (2023). Effects of film confinement on dielectric and electrical properties of graphene oxide and reduced graphene oxide-based polymer nanocomposites: Implications for energy storage. *ACS Applied Nano Materials*, *6*(13), 11699–11714. DOI: 10.1021/acsanm.3c01674

Liu, M., Wang, L., & Yu, G. (2024). Recent research progress of porous graphene and applications in molecular sieve, sensor, and supercapacitor. *Small*, *20*(42), 2401767. DOI: 10.1002/smll.202401767

Liu, W., & Speranza, G. (2021). Tuning the oxygen content of reduced graphene oxide and effects on its properties. *ACS Omega*, *6*(9), 6195–6205. DOI: 10.1021/acsomega.0c05578

Lv, S., et al. "CVD graphene with high electrical conductivity: empowering applications." 2D Materials 12 (2024): 013003. https://doi.org/.DOI: 10.1088/2053-1583/ad8800

Mageed, A. K., Dayang Radiah, A. B., Salmiaton, A., Izhar, S., & Razak, M. A. (2018). Study the Thermal Stability of Nitrogen Doped Reduced Graphite Oxide Supported Copper Catalyst. *Journal of Cluster Science*, *29*(4), 709–718. DOI: 10.1007/s10876-018-1382-6

Malard, L. M., et al. "Raman spectroscopy in graphene." Physics Reports 473.5-6 (2009): 51-87. https://doi.org/.DOI: 10.1016/j.physrep.2009.02.003

Marquez, C., Rodriguez, N., Ruiz, R., & Gamiz, F. (2016). Electrical characterization and conductivity optimization of laser reduced graphene oxide on insulator using point-contact methods. *RSC Advances*, *6*(52), 46231–46237. DOI: 10.1039/C6RA03630A

Menezes, I. R., Araújo, N. R. S., Araújo, B. C. R., Sakai, T., Lago, R. M., & Sebastião, R. C. O. (2023). Shedding light on the mechanism of graphene oxide thermal decomposition: A kinetic study using isoconversional method and artificial neural network. *Thermochimica Acta*, *721*, 179454. DOI: 10.1016/j.tca.2023.179454

Nagashio, K., Nishimura, T., Kita, K., & Toriumi, A. (2009). Mobility variations in mono-and multi-layer graphene films. *Applied Physics Express*, *2*, 025003. DOI: 10.1143/APEX.2.025003

Najmi, M. I.. (2024). Identification of Crystal Structure, Surface Area, and Magnetic Properties of Mn0.25Fe2.75O4/rGO Nanocomposites. *Journal of Metastable and Nanocrystalline Materials*, *38*, 23–30. DOI: 10.4028/p-GLu976

Nishi, Y., & Inagaki, M. (2016). *Gas adsorption/desorption isotherm for pore structure characterization*. Materials Science and Engineering of Carbon., DOI: 10.1016/B978-0-12-805256-3.00011-8

Novoselov, K. S., Geim, A. K., Morozov, S. V., Jiang, D., Zhang, Y., Dubonos, S. V., Grigorieva, I. V., & Firsov, A. A. (2004). Electric field effect in atomically thin carbon films. *Science*, *306*(5696), 666–669. DOI: 10.1126/science.1102896

Otsuka, H., Urita, K., Honma, N., Kimuro, T., Amako, Y., Kukobat, R., Bandosz, T. J., Ukai, J., Moriguchi, I., & Kaneko, K. (2024). Transient chemical and structural changes in graphene oxide during ripening. *Nature Communications*, *15*(1), 1708. DOI: 10.1038/s41467-024-46083-4

Pacakova, B., Vejpravova, J., Repko, A., Mantlikova, A., & Kalbac, M. (2015). Formation of wrinkles on graphene induced by nanoparticles: Atomic force microscopy study. *Carbon*, *95*, 573–579. DOI: 10.1016/j.carbon.2015.08.043

Paliotta, L., De Bellis, G., Tamburrano, A., Marra, F., Rinaldi, A., Balijepalli, S. K., Kaciulis, S., & Sarto, M. S. (2015). Highly conductive multilayer-graphene paper as a flexible lightweight electromagnetic shield. *Carbon*, *89*, 260–271. DOI: 10.1016/j.carbon.2015.03.043

Pei, S., & Cheng, H.-M. (2012). The reduction of graphene oxide. *Carbon*, *50*(9), 3210–3228. DOI: 10.1016/j.carbon.2011.11.010

Pimenta, M. A., Dresselhaus, G., Dresselhaus, M. S., Cançado, L. G., Jorio, A., & Saito, R. (2007). Studying disorder in graphite-based systems by Raman spectroscopy. *Physical Chemistry Chemical Physics*, *9*(11), 1276–1290. DOI: 10.1039/B613962K

Pinder, J. W., Major, G. H., Baer, D. R., Terry, J., Whitten, J. E., Čechal, J., Crossman, J. D., Lizarbe, A. J., Jafari, S., Easton, C. D., Baltrusaitis, J., van Spronsen, M. A., & Linford, M. R. (2024). Avoiding common errors in X-ray photoelectron spectroscopy data collection and analysis, and properly reporting instrument parameters. *Applied Surface Science Advances*, *19*, 100534. DOI: 10.1016/j.apsadv.2023.100534

Ponomarenko, L. A., et al. "Extreme electron–hole drag and negative mobility in the Dirac plasma of graphene." Nature Communications 15 (2024): 9869. https://doi.org/. Robinson, J., Hollander, M., and Datta, S. Epitaxial Graphene. 2D Materials for Nanoelectronics, 37.DOI: 10.1038/s41467-024-54198-x

Prime, R. B., et al. "Thermogravimetric analysis (TGA)." Thermal Analysis of Polymers: Fundamentals and Applications (2009): 241-317. https://doi.org/.DOI: 10.1002/9780470423837

Prokhorov, E., Barquera-Bibiano, Z., Manzano-Ramírez, A., Luna-Barcenas, G., Kovalenko, Y., Hernández-Landaverde, M. A., Reyes, B. E. C., & Vargas, J. H. (2019). New insights in graphene oxide dielectric constant. *Materials Research Express*, *6*(8), 085622. DOI: 10.1088/2053-1591/ab22f0

Rao, S., Upadhyay, J., Polychronopoulou, K., Umer, R., & Das, R. (2018). Reduced graphene oxide: Effect of reduction on electrical conductivity. *Journal of Composites Science*, *2*(2), 25. DOI: 10.3390/jcs2020025

Rathanasamy, R., Sahoo, S., Lee, J. H., Das, A. K., Somasundaram, M., Palaniappan, S. K., & Sivaraj, S. (2021). Carbon-based multi-layered films for electronic application: A review. *Journal of Electronic Materials*, *50*(4), 1845–1892. DOI: 10.1007/s11664-020-08724-4

Rathnavel, P., Murukesh, C., & Umamaheswari, R. (2023). Fabrication of a novel rGO encapsulated nickel cobalt chalcogenide electrocatalyst as an efficient counter electrode to boost efficiency of dye-sensitized solar cells. *Journal of Materials Science Materials in Electronics*, *34*(11), 1000. DOI: 10.1007/s10854-023-10402-0

Ravari, F., Noori, M., & Ehsani, M. (2019). Thermal stability and degradation kinetic studies of PVA/RGO using the model-fitting and isoconversional (model-free) methods. *Fibers and Polymers*, *20*(3), 472–480. DOI: 10.1007/s12221-019-8606-8

Ravikovitch, P. I., & Neimark, A. V. (2001). Characterization of nanoporous materials from adsorption and desorption isotherms. *Colloids and Surfaces. A, Physicochemical and Engineering Aspects*, *187*, 11–21. DOI: 10.1016/S0927-7757(01)00614-8

Reich, S., and C. Thomsen. "Raman spectroscopy of graphite." Philosophical Transactions of the Royal Society of London. Series A: Mathematical, Physical and Engineering Sciences 362.1824 (2004): 2271-2288. https://doi.org/.DOI: 10.1098/rsta.2004.1454

Reina, A., Jia, X., Ho, J., Nezich, D., Son, H., Bulovic, V., Dresselhaus, M. S., & Kong, J. (2009). Large area, few-layer graphene films on arbitrary substrates by chemical vapor deposition. *Nano Letters*, *9*(1), 30–35. DOI: 10.1021/nl801827v

Rennhofer, H., et al. "Pore development during the carbonization process of lignin microparticles investigated by small angle X-ray scattering." Molecules 26.7 (2021): 2087. https://doi.org/.DOI: 10.3390/molecules26072087

Robertson, A. W., Montanari, B., He, K., Allen, C. S., Wu, Y. A., Harrison, N. M., Kirkland, A. I., & Warner, J. H. (2013). Structural reconstruction of the graphene monovacancy. *ACS Nano*, *7*(5), 4495–4502. DOI: 10.1021/nn401113r

Rudenko, R., Voitsihovska, O., Abakumov, A., Bychko, I., Selyshchev, O., Zahn, D. R. T., & Poroshin, V. (2022). The hopping Hall effect in reduced graphene oxide. *Materials Letters*, *326*, 132932. DOI: 10.1016/j.matlet.2022.132932

Ruidíaz-Martínez, M., Álvarez, M. A., López-Ramón, M. V., Cruz-Quesada, G., Rivera-Utrilla, J., & Sánchez-Polo, M. (2020). Hydrothermal synthesis of rGO-TiO2 composites as high-performance UV photocatalysts for ethylparaben degradation. *Catalysts*, *10*(5), 520. DOI:10.3390/catal10050520

Sánchez-Trujillo, D. J., Osorio-Maldonado, L. V., & Prías-Barragán, J. J. (2023). Temperature dependence of electrical conductivity and variable hopping range mechanism on graphene oxide films. *Scientific Reports*, *13*(1), 4810. DOI: 10.1038/s41598-023-31778-3

Scheibe, B., Borowiak-Palen, E., & Kalenczuk, R. J. (2010). Oxidation and reduction of multiwalled carbon nanotubes—Preparation and characterization. *Materials Characterization*, *61*(2), 185–191. DOI: 10.1016/j.matchar.2009.11.008

Sedki, M., Mirabedini, P. S., Nakama, K., Stephens, G., Groves, M., Lee, I., Neupane, M. R., & Mulchandani, A. (2022). Synthesis of pristine graphene-like behaving rGO thin film: Insights into what really matters. *Carbon*, *186*, 437–451. DOI: 10.1016/j.carbon.2021.10.011

Sharma, N., Tomar, S., Shkir, M., Kant Choubey, R., & Singh, A. (2021). Study of optical and electrical properties of graphene oxide. *Materials Today: Proceedings*, *36*, 730–735. DOI: 10.1016/j.matpr.2020.04.861

Shi, J. L., Tian, G.-L., Zhang, Q., Zhao, M.-Q., & Wei, F. (2015). Customized casting of unstacked graphene with high surface area (>1300 m^2/g) and its application in oxygen reduction reaction. *Carbon*, *93*, 702–712. DOI: 10.1016/j.carbon.2015.06.007

Siburian, R., Sihotang, H., Raja, S. L., Supeno, M., & Simanjuntak, C. (2018). New route to synthesize of graphene nano sheets. *Oriental Journal of Chemistry*, *34*(1), 182. DOI:10.13005/ojc/340120

Stan, G., & King, S. W. (2020). Atomic force microscopy for nanoscale mechanical property characterization. *Journal of Vacuum Science and Technology. B, Nanotechnology & Microelectronics : Materials, Processing, Measurement, & Phenomena : JVST B*, *38*(6), 6. DOI: 10.1116/6.0000544

Stobinski, L., Lesiak, B., Malolepszy, A., Mazurkiewicz, M., Mierzwa, B., Zemek, J., Jiricek, P., & Bieloshapka, I. (2014). Graphene oxide and reduced graphene oxide studied by the XRD, TEM and electron spectroscopy methods. *Journal of Electron Spectroscopy and Related Phenomena*, *195*, 145–154. DOI: 10.1016/j.elspec.2014.07.003

Stoller, M. D., Park, S., Zhu, Y., An, J., & Ruoff, R. S. (2008). Graphene-based ultracapacitors. *Nano Letters*, *8*(10), 3498–3502. DOI: 10.1021/nl802558y

Suenaga, K., Sandré, E., Colliex, C., Pickard, C. J., Kataura, H., & Iijima, S. (2001). Electron energy-loss spectroscopy of electron states in isolated carbon nanostructures. *Physical Review. B*, *63*(16), 165408. DOI: 10.1103/PhysRevB.63.165408

Sunderrajan, S., Miranda, L. R., & Pennathur, G. (2018). Improved stability and catalytic activity of graphene oxide/chitosan hybrid beads loaded with porcine liver esterase. *Preparative Biochemistry & Biotechnology*, *48*(4), 343–351. DOI: 10.1080/10826068.2018.1446153

Surekha, G., Krishnaiah, K. V., Ravi, N., & Suvarna, R. P. (2020, March). FTIR, Raman and XRD analysis of graphene oxide films prepared by modified Hummers method. [). IOP Publishing.]. *Journal of Physics: Conference Series*, *1495*(1), 012012. DOI: 10.1088/1742-6596/1495/1/012012

Tang, D., Wang, Q., Wang, Z., Liu, Q., Zhang, B., He, D., Wu, Z., & Mu, S. (2018). Highly sensitive wearable sensor based on a flexible multi-layer graphene film antenna. *Science Bulletin*, *63*(9), 574–579. DOI: 10.1016/j.scib.2018.03.014

Tran, D. T. (2020). rGO/persulfate metal-free catalytic system for the degradation of tetracycline: Effect of reaction parameters. *Materials Research Express*, *7*(7), 075501. DOI: 10.1088/2053-1591/ab9e47

Tran, T. S., Balu, R., de Campo, L., Dutta, N. K., & Choudhury, N. R. (2023). Sulfonated polythiophene-interfaced graphene for water-redispersible graphene powder with high conductivity and electrocatalytic activity. *Energy Advances*, *2*(3), 365–374. DOI: 10.1039/D2YA00298A

Ullah, S., et al. "Advances and trends in chemically doped graphene." Advanced Materials Interfaces 7.24 (2020): 2000999. https://doi.org/.DOI: 10.1002/admi.202000999

Vázquez-Sánchez, P., Rodríguez-Escudero, M. A., Burgos, F. J., Llorente, I., Caballero-Calero, O., González, M. M., Fernández, R., & García-Alonso, M. C. (2019). Synthesis of Cu/rGO composites by chemical and thermal reduction of graphene oxide. *Journal of Alloys and Compounds*, *800*, 379–391. DOI: 10.1016/j.jallcom.2019.06.008

Wagner, C. D., Davis, L. E., Zeller, M. V., Taylor, J. A., Raymond, R. H., & Gale, L. H. (1981). Empirical atomic sensitivity factors for quantitative analysis by electron spectroscopy for chemical analysis. *Surface and Interface Analysis*, *3*(5), 211–225. DOI: 10.1002/sia.740030506

Wang, G., Yang, J., Park, J., Gou, X., Wang, B., Liu, H., & Yao, J. (2008). Facile synthesis and characterization of graphene nanosheets. *The Journal of Physical Chemistry. C, Nanomaterials and Interfaces*, *112*(22), 8192–8195. DOI: 10.1021/jp710931h

Williams, D. B.. (2009). *The Transmission Electron Microscope*. Springer US., DOI: 10.1007/978-0-387-76501-3

Xu, X., Huang, J., Miao, G., Yan, B., Chen, Y., Zhou, Y., & Cai, W. (2025). Visualizing Thermal Reduction in Graphene Oxide. *Materials (Basel)*, *18*(10), 2222. DOI:10.3390/ma18102222

Yoon, D., & Cheong, H. (2012). Raman spectroscopy for characterization of graphene. In Raman Spectroscopy for Nanomaterials Characterization (pp. 191–214). Springer Berlin Heidelberg. DOI: 10.1007/978-3-642-20620-7_9

Yuwen, C., Liu, B., Zhang, L., Guo, S., & Peng, J. (2019). Synthesis high-qulity graphene oxide and temperature-dependent dielectric properties of reduced graphene oxide. Materials Research Express, 6(9), 0950b4. DOI: 10.1088/2053-1591/ab149f

Zhang, H., Huang, J., Wang, Y., Liu, R., Huai, X., Jiang, J., & Anfuso, C. (2018). Atomic force microscopy for two-dimensional materials: A tutorial review. *Optics Communications*, *406*, 3–17. DOI: 10.1016/j.optcom.2017.05.015

Zhao, X., Zhang, Q., Chen, D., & Lu, P. (2011). Enhanced mechanical properties of graphene-based poly (vinyl alcohol) composites. *Macromolecules*, *44*(7), 2392–2392. DOI: 10.1021/ma200335d

Zhu, S. E., Yuan, S., & Janssen, G. C. A. M. (2014). Optical transmittance of multilayer graphene. *Europhysics Letters*, *108*(1), 17007. DOI: 10.1209/0295-5075/108/17007

Chapter 3
Advancements in Nanomaterials for Sustainable Aviation:
Mechanical Properties and Structural Applications of Graphene-Based Nanomaterials in Aerospace Engines

Mehmet Çağatay Akbolat
https://orcid.org/0000-0001-7579-6478
Gaziantep University, Turkey

ABSTRACT

Aerostructural applications require lightweight materials with high mechanical properties. Graphene and its derivatives, e.g. graphene oxide (GO), reduced graphene oxide (rGO), and graphene nanoplatelets (GNPs), have recently gained attention as promising reinforcement due to their exceptional mechanical, thermal and electrical properties for advanced materials. This chapter presents a comprehensive review of recent advancements in the incorporation of graphene-based nanomaterials into polymer, metal, and ceramic matrix composites for aerospace applications. It explores toughening mechanisms, interfacial bonding, dispersion techniques, and the influence of manufacturing methods on final composite properties. Beyond mechanical performance, the chapter also addresses the multifunctional capabilities of graphene-enhanced composites—including electrical conductivity, damage sensing, and thermal regulation—enabling innovations such as structural batteries, cryogenic hydrogen storage, stealth systems, and morphing aerospace structures.

DOI: 10.4018/979-8-3373-3862-0.ch003

1. INTRODUCTION

Lightweight design is a fundamental need in aerospace engineering, where structural weight minimization contributes to fuel efficiency, payload carrying capacity, and overall performance. Aerostructural components must be engineered to withstand extreme environmental and mechanical conditions in order to achieve these demands without sacrificing functionality or safety. Materials such as polymer matrix composites (PMCs), metal matrix composites (MMCs), and ceramic matrix composites (CMCs) are widely used due to their high strength-to-weight ratios, thermal stability, and corrosion resistance. Nevertheless, the evolving demands of next-generation aerospace systems require significant improvements in the mechanical and functional capabilities of aerostructural materials.

Specifically, enhancements in strength, stiffness, and toughness with the highest electrical and thermal conductivity, fatigue and environmental durability, and long-term sustainability are highly essential. Achieving this level of multifunctionality presents a significant materials engineering challenge. In recent years, nanomaterials have emerged as a promising solution to further enhance composite performance. Among various nanomaterials, graphene has gained significant attention due to its exceptional mechanical, thermal, and electrical properties.

Graphene, a two-dimensional carbon allotrope consisting of a single plane of sp^2-bonded carbon atoms arranged in a hexagonal lattice, has remarkable physical and chemical properties. These include a tensile strength approximately 200 times greater than steel, outstanding thermal conductivity, and high electrical conductivity. Graphene has the potential to impart these superior properties if well integrated into composite matrices, thereby resulting in light-weight, multifunctional composite-based structures for extreme aerospace applications.

It is essential to evaluate the advantages and limitations associated with the selection and treatment of graphene-based nanofillers, such as graphene oxide (GO), reduced graphene oxide (rGO), and graphene nanoplatelets (GNPs), to achieve further improvement in composite material systems with graphene-based fillers. They possess certain advantages associated with compatibility, dispersion, and interfacial bonding in diverse matrix systems. Proper dispersion of graphene within the composite materials is critical for maximum stress transfer uniformity, minimization of agglomeration, and maximization of reinforcement effectiveness. Thus, advanced processing techniques (*e.g.*, solution blending, in situ polymerization, melt mixing, and layer-by-layer assembly) have been developed to maximize graphene incorporation.

The presented chapter provides a comprehensive review of recent advances in the use of graphene-based nanofillers for the incorporation into composite materials. It presents the underlying mechanisms of graphene-induced toughening and strengthening, and filler type-, content, and processing route-dependent effects on final composite material properties. Furthermore, the chapter highlights emerging opportunities and future directions for graphene-enhanced composites in aerospace engineering. It also discusses the potential applications of graphene-reinforced composites, addressing both the challenges and opportunities associated with their integration.

2. GRAPHENE-BASED NANOMATERIALS: PROPERTIES AND CHALLENGES

Graphene (*i.e.*, 2D crystalline sheet including carbon allotropes with a hexagonal lattice) is being employed in an extensive range of engineering applications as it provides high mechanical properties, intrinsic mobility, thermal conductivity, electrical conductivity, and large surface area. Graphene has been utilized in aerostructural materials extensively as contributing to additional improvement in mechanical properties and implementing multifunctionality in the aerostructural materials (Kanti et al., 2025). Graphene is a material family, and it differs depending on structure, surface chemistry, and physical properties. Graphene oxide (GO), reduced graphene oxide (rGO), and graphene-nanoplatelets (GNPs) are mostly used graphene derivatives. There are distinct advantages and disadvantages of all varieties of graphene for case applications (Kausar et al., 2017; Ray, 2015; Yee & Ghayesh, 2023a).

Graphene oxide (GO), synthesized with an oxidative process, consists of oxygen-containing groups in the basal planes and at the edges of graphene sheets. Oxygen groups in graphene sheets provide chemical reactivity, ensuring better interfacial adhesion and better interfacial dispersion in the host materials. The structural defects, caused due to the oxide groups, substantially reduce the electrical conductivity and mechanical properties of GOs (Lava Kumar et al., 2022). Owing to such characteristics, GOs have been utilized in most instances to provide better mechanical behavior in applications wherein strong interfacial adhesion is required, e.g., in structural adhesives. (Khoramishad et al., 2018; Sadigh & Marami, 2018).

Reduced graphene oxide (rGO) is another derivative of graphene, obtained from the partial reduction of GO by the removal of certain oxygen groups. The reduction moderately restores the sp2 carbon network to achieve better mechanical properties than those of GO. However, inclusion of rGO in host materials can lead to a compromise of the strong adhesion and dispersibility achieved with GOs. (Compton & Nguyen, 2010a, 2010b; Phiri et al., 2018).

Graphene nanoplatelets (GNPs) contain multilayer graphene sheet structures with low surface oxidation. GNPs can offer excellent in-plane electrical and thermal conductivity and good mechanical properties. GNPs can also significantly enhance the mechanical properties of aerostructural materials. Their surfaces, being inert, can weaken interfacial bonding and lead to non-homogeneous dispersion within matrix materials. Thus, surface modification or functionalization can be implemented in order to optimize the possible benefit that can be achieved with GNP content in host matrices. (Anwar et al., 2016; Cataldi et al., 2018; Keramati et al., 2016; Worsley et al., 2022; Yee & Ghayesh, 2023b).

Briefly, graphene is one of the candidates to enhance the mechanical properties of aerostructural materials with multifunctional properties of thermal and electric conductivity. The various types of graphene have various advantages and limitations; GO is preferred due to ease of processing, good adhesion, and high dispersibility; rGO offers the best balance of conductivity and support; and GNPs are preferred where peak structure and multifunctional properties matter most. The integration of these materials into aerospace structures represents a critical step toward the development of next-generation lightweight, high-strength, and multifunctional materials.

3. GRAPHENE INTEGRATION IN POLYMER MATRIX COMPOSITES

Polymer composites have been commonly utilized as structure materials in lightweight, high-performance applications, particularly in the field of aerospace, as they offer high specific strength, high stiffness, good fatigue resistance, chemical durability, and tailorability depending on the different loading cases. They have been gradually replacing metals in primary structural elements, whereas they were previously used as secondary structural materials (Dursun & Soutis, 2014; Warren, 2004). However, PMCs often experience poor fracture toughness properties, and they are prone to intra- (*i.e.,* matrix cracking and fiber breakage) and inter-laminar (*i.e.,* delamination) under loading conditions such as impact, flexural, and fatigue loading (Akbolat et al., 2025; Katnam et al., 2019; Mohan, 2013; Qian et al., 2010; Shivakumar et al., 2013). These limitations pose significant challenges when optimizing their performance for demanding structural applications. Toughening strategies with nanofillers, e.g., graphene-based nanofillers, are one of the possible solutions to enhance toughness, strength, and multifunctional performance of PMCs.

PMCs can be categorized depending on the matrix material as thermosets and thermoplastics, both having different properties and features. Thermoset polymers experience cross-linking chemical reactions, resulting in irreversible reactions that offer superior dimensional stability, excellent thermal resistance, and strength resistance to harsh conditions. Thermoset composite materials provide higher strength/stiffness but have lower toughness properties when compared with their thermoplastic counterparts. On the other hand, thermoplastic polymers are recyclable, which can be re-heated, re-shaped, and re-used multiple times. (Mallick, 2008). Additionally, thermoplastic composites provide better damage tolerance and are more suitable for applications requiring high impact resistance and repairability. (Barroeta Robles et al., 2022; Jogur et al., 2018; Reis et al., 2020). The inclusion of graphene fillers in both thermosetting and thermoplastic matrix composite materials offers a promising toughening route to improving mechanical properties and interfacial adhesion, while providing multifunctionality in electrical and thermal conductivity. Consequently, graphene-reinforced thermoset and thermoplastic composites offer a viable pathway toward next-generation structural materials that combine mechanical strength, multifunctional capabilities, and lightweight design.

3.1. Graphene Use in Thermoset Composites

Epoxy resins are commonly used thermoset resins in aerospace engineering applications due to their high mechanical properties, dimensional stability, thermal and environmental resistance, strong fiber-matrix adhesion, and long-term durability when compared to other types of thermosets, e.g., vinyl-ester, phenolic resin, and polyimide resin. Thermoset composites, on the other hand, are susceptible to delamination failure because of their intrinsic brittleness. Therefore, several toughening strategies are employed to improve the low interlaminar fracture toughness of composites, such as interlaminar toughening. (Akbolat et al., 2022, 2025), 3D reinforcement with woven fabrics (J. K. Kim & Sham, 2000; Vieille et al., 2014), through-the-thickness reinforcement methods, *e.g.,* stitching (Caprino et al., 2007; Shiino et al., 2017; Yudhanto et al., 2015), knitting (Leong et al., 2000; Pamuk & Çeken, 2008) and z-pinning (Mouritz, 2007; Pegorin et al., 2015; Yan et al., 2004) and bulk resin toughening (Chang et al., 2024; Guild et al., 2022; H.-Y. Liu et al., 2011; C. Zhou et al., 2014).

The mechanical performance of PMCs can be significantly enhanced with a bulk resin toughening approach. This approach is straightforward to implement using composite manufacturing processes and can enhance matrix-dominated properties, such as transverse strength. Various toughening agents in micro- and macro-size (*e.g.,* rubber particles (Bagheri et al., 2009; Liang & Pearson, 2010; Saghafi et al., 2024), silica particles (Guild et al., 2022; Hsieh et al., 2010; Poonpipat et al., 2017), carbon nanotubes (Gojny et al., 2004; Lyashenko-Miller & Marom, 2017; Qian et al., 2010; G. Wang et al., 2024), etc.) can be employed to improve the mechanical properties of PMCs. In addition to these approaches, using advanced nanomaterials, such as graphene-based fillers, is a viable approach for improving the mechanical properties of composite-based structural components. Numerous studies suggest that graphene incorporation is a potential technique to improve the mechanical properties of thermoset composite laminates (Kanti et al., 2025; Khalid, Kamal, et al., 2023; Tang et al., 2013; Zeinedini, 2024; B. Zhang et al., 2014; Zotti et al., 2022). However, dispersion of graphene is a difficult process for nanocomposite synthesis due to its strong tendency to agglomerate and weak contact with thermoset matrices. To overcome these restrictions, materials researchers have focused on various graphene derivatives, such as GO, rGO, and GNPs, to achieve further improvement in PMCs. GO is the most common derivative utilized to solve dispersion and interface issues. Among them, GO is the most commonly utilized derivative, especially for enhancing dispersion and interfacial bonding in composite systems. For instance, Figure 1 demonstrates the enhancement of interfacial properties in carbon/epoxy composites achieved through GO modification (Pathak et al., 2016a). Functionalization of graphene fillers may result in a loss of intrinsic mechanical characteristics. However, incorporating GOs into epoxy matrices frequently leads to significant mechanical property enhancements compared to the possible improvement that can be achieved with pure graphene incorporation, owing to better dispersion and stronger interfacial contacts with the matrix (Pathak et al., 2016b). In contrast, oxygen-containing functional groups in GOs disturb the sp^2 hybridization of carbon atoms, which is necessary for excellent conductivity in thermoset composites. In this regard, rGO emerges as a promising alternative, offering a balanced improvement in both mechanical and electrical properties (Aradhana et al., 2018; Pramodkumar & Budhe, 2024). Similarly, graphene nanoplatelets (GNPs) effectively enhance mechanical performance while preserving electrical conductivity. The possible fracture mechanisms can be obtained with graphene fillers are given in Figure 2 (Kostagiannakopoulou et al., 2015). This figure shows crack pinning, crack tip bifurcation, and pull-out mechanisms. For graphene-based fillers with multiple layers, such as graphene nanoplatelets (GNPs), the separation of graphitic layers can occur, as seen in Figure 2d.

Figure 1. Fracture surfaces obtained from (a) carbon/epoxy composites and (b) graphene toughened carbon/epoxy composites. The residual matrix on the carbon fibers with a rough surface on the graphene toughened carbon/epoxy composite indicates higher interfacial interactions obtained with graphene modification. (Pathak et al., 2016a).

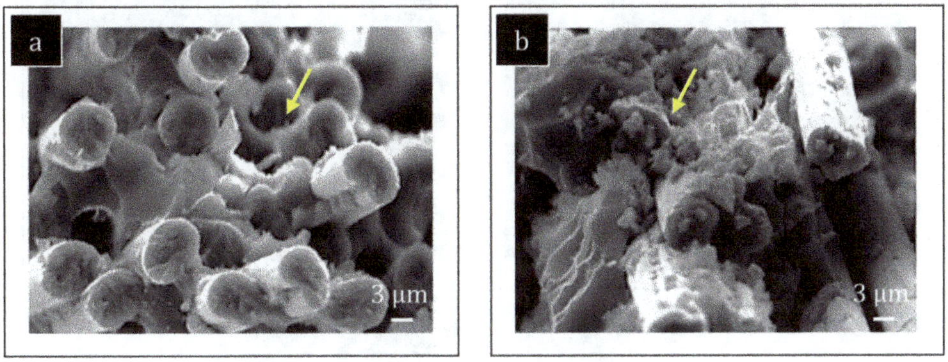

Figure 2. Fracture mechanisms obtained from a PMC with graphene filler modifications: (a) crack bifurcation, (b) crack pinning, (c) graphene pull-out, and (d) separation of graphene layers (Kostagiannakopoulou et al., 2015).

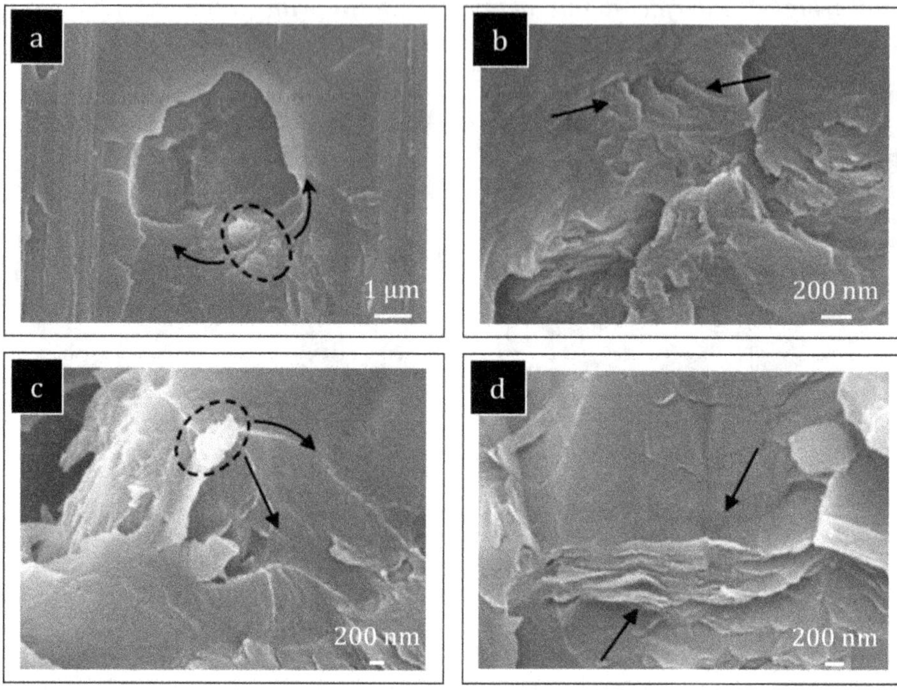

Despite these advantages, graphene-based fillers remain difficult to disperse uniformly, limiting their efficiency when utilized only in bulk resin toughening procedures. Several ways have been used to increase the dispersion of graphene fillers in thermoset resins. The most convenient ways for improving the dispersion of graphene fillers are centrifugal mixing, three-roll milling, high-shear mixing, and ultrasonication (Paredes et al., 2008; M. Y. Shen et al., 2021; Tareq et al., 2021; Zotti et al., 2022). These physical treatment approaches can allow for the cost-effective dispersion of graphene in large-scale production. Solution mixing procedures can also be utilized to distribute graphene fillers in host materials. In this method, the thermoset matrix and graphene are initially dissolved in a solvent, *e.g.*, ethanol (Tang et al., 2013) or acetone (Z. Wang et al., 2021). External stimuli, such as mechanical stirring and ultrasonication, are then applied to achieve a uniform dispersion of graphene within the thermoset matrix. Finally, a hardener is added to the graphene-polymer mixture to facilitate the manufacturing of fiber-reinforced composites. Another important factor affecting the dispersibility and effectiveness of graphene filling is the geometrical characteristics of the graphene sheets, as the geometrical dimensions of the sheets change the aspect ratio (*i.e.*, lateral dimension/thickness) and surface area of graphene sheets. Higher aspect ratio and surface area help polymer composites to achieve further improvement by allowing more uniform dispersion (Kostagiannakopoulou et al., 2021).

Composite manufacturing processes can also influence the effectiveness of graphene-based filling. When fiber-reinforced polymer composite laminates are manufactured with the vacuum-assisted resin infusion method, the mesh ply used for achieving resin flow under a vacuum environment can act as a filter, as shown in Figure 3 (Yao et al., 2021). Additionally, the increase in viscosity caused by the presence of graphene sheets can result in wettability issues in polymer composites. Spray deposition of graphene onto interleaves and interleaves containing graphene-based nanofillers can effectively mitigate the filtering effects occurring with mesh plies in the polymer composite manufacturing processes. (Li, Zhang, et al., 2018; Mulqueen et al., 2022). Spray coating and graphene-based interleaving methods allow composite manufacturing processes with infusion to be used without affecting resin viscosity. On the other hand, the application of the toughening strategies with graphene fillers can be combined with the other types of fillers (*e.g.*, carbon nanotubes. (Y. Liu et al., 2022a; Shin et al., 2012), silica particles (Yu et al., 2016), and rubber particles (Lim et al., 2016), etc.) with the aim of achieving synergistic reinforcement. Also, the combination of interlaminar toughening with graphene fillers is another possible application to enhance, particularly the interlaminar fracture resistance of fiber/epoxy composites by changing the interfacial properties with the help of graphene fillers. (Quan et al., 2020; Wu et al., 2020).

Figure 3. Digital images illustrating the role of the infusion mesh as a filter: (a) graphene-filled epoxy, (b) vacuum-assisted resin infusion process, and (c) infusion mesh filtering graphene fillers (Yao et al., 2021).

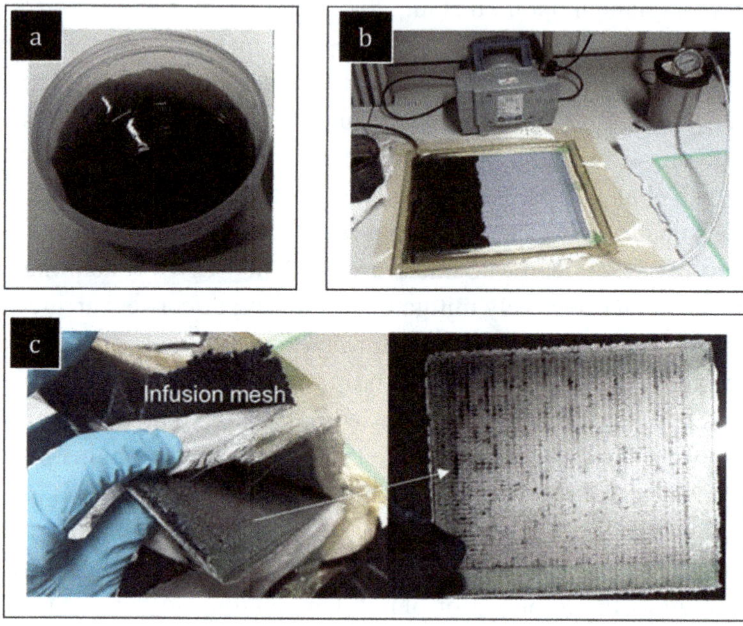

3.2. Graphene Use in Thermoplastic Composites

Thermoplastics are widely employed in polymeric composites due to their ease of processing and adaptability. Compression molding is a popular and effective method of producing thermoplastic composites. Thermoplastic films are laminated with reinforcing fiber textiles and compressed using hot plates. Incorporating graphene-based fillers into thermoplastic composites provides unique production challenges compared to thermoset systems. (Vaidya & Chawla, 2008). Graphene-based fillers are widely employed in the manufacture of thermoplastic composite sheets. This variance is primarily due to changes in film processing processes. Cast film extrusion is a typical method of producing thermoplastic films. This approach involves feeding thermoplastic pellets and graphene fillers into the barrel of a twin-screw extruder. The combination is then processed to generate thin films with integrated graphene content, which enables the production of thermoplastic composites with graphene-based fillers. (Painkal et al., 2025). Vacuum filtration is another method for producing thin nanoparticle-infused membranes. A slurry of graphene nanoparticles is sonicated and centrifuged for better dispersion before being filtered through a medium to form a thin membrane. The membranes can then be inserted between thermoplastic films. (Feres et al., 2025).

The liquid-reactive thermoplastic resin system has recently gained popularity due to its ease of production via vacuum-assisted resin infusion processes. This system is typically built around an acrylic resin, with polymerization triggered by a benzoyl peroxide (BPO) initiator. Graphene fillers can be introduced into liquid thermoplastic matrices using a variety of processes. For example, the surface of reinforcement fibers can be scaled using graphene fillers, and then resin infusion is used to produce the composite (Y. Chen et al., 2024). Another option is to employ liquid resins to co-cure the thermoplastic composite during the production process. In this technique, graphene fillers are mixed into the liquid resin, and polymerization begins with a BPO initiator. The films are then sandwiched between reinforcing fibers, and resin infusion takes place. The process cures at room temperature, followed by heat pressing to improve the final material properties (Khan et al., 2023).

Finally, there are numerous ways to include graphene fillers into thermoplastic composites. These efforts aim to address the problem of a weak interface between the thermoplastic matrix and reinforcing fibers, which is typically caused by a lack of chemical bonding. (Fang et al., 2025). By improving this contact, the mechanical properties of thermoplastic composites can be significantly improved. (Periasamy et al., 2023).

4. GRAPHENE INTEGRATION IN METAL MATRIX COMPOSITES

MMCs are a form of advanced materials with a metallic matrix (*i.e.*, having ductile characteristics) embedded with a secondary phase, such as ceramic particles, fibers, and whiskers, to enhance specific properties of metal materials. MMCs have been adopted in aerospace engineering due to their high mechanical properties, such as strength, hardness, and high elastic modulus. Aluminum (Al), magnesium (Mg), titanium (Ti), and their alloys are the most conveniently used metallic materials in the aerospace industry as matrix materials. MMCs can be processed with powder metallurgy sintering, stir casting, spray atomization, squeeze casting, plasma spraying, and co-deposition techniques. MMCs offer a tailored combination of properties suitable for demanding applications. They can be classified into fiber-reinforced, particulate-reinforced, and whisker-reinforced composites. Particulate-reinforced MMCs differ from the others due to their relatively low cost, ease of processing, and isotropic behavior. (Khalid, Umer, et al., 2023; Meher et al., 2022; Miracle, 2005; Shetty et al., 2022; B. Singh et al., 2023; Srinivasan et al., 2021).

MMCs have gained popularity due to their ability to withstand high-stress and high-temperature environments, such as combustion chambers in jet engines. MMCs are significant material types playing a critical role in the development of lightweight, high-strength components for next-generation engineering systems. One of the most convenient approaches to improve the mechanical properties of MMCs is to use nano-reinforcements. MMCs can be produced using conventional techniques, *e.g.* silicon carbide (SiC), titanium dioxide (TiO_2), alumina (Al_2O_3), titanium carbide (TiC), tungsten carbide (WC), titanium aluminium carbide (Ti_3AlC_2), and barium titanate ($BaTiO_3$), with low weight content prior to the casting process could provide a homogenous distribution of nano-reinforcements, leading to promising mechanical properties such as strength, hardness, creep, high thermal conductivity, fatigue resistance. (Khalid, Umer, et al., 2023; Miracle, 2005).

Beyond conventional nano-reinforcements, graphene is emerging as an attractive filler in MMC applications due to its unique combination of high electrical, thermal, and chemical properties. Its two-dimensional structure and exceptionally high surface area have been utilized to achieve multifunctional performance, expanding the scope of MMC applications. In short, the versatility and compatibility of graphene sheets with various metallic matrices suggest that it could play a significant role in advancing MMC technology toward next-generation materials.

4.1. Strengthening Mechanisms of Graphene Fillers in MMCs

Strengthening mechanisms in metal/graphene composites can be classified into two categories: (a) indirect strengthening mechanisms, which are the result of particle filler incorporation in matrix microstructures and deformation modes, and (b) direct strengthening mechanisms occurring due to the load transfer between particle fillers and metal matrix. The strengthening mechanisms are highly related to the uniform distribution of graphene reinforcements, which leads to an increment in the mechanical performance of MMCs. The strengthening mechanisms that can be attributed to mechanical performance enhancements are refined grains due to the graphene introduction, load transfer, thermal mismatch-induced dislocations, and Orowan looping. (Seyed Pourmand & Asgharzadeh, 2020; Sheinerman, 2022).

Grain refinement occurs when reducing the grain size of the metal matrix (*i.e.,* known as grain refinement), which can significantly enhance mechanical strength. Grain boundaries limit the migration of dislocations and resist plastic deformation. Finer grains have more grain boundaries per unit volume, increasing the resistance to dislocation movement. This improved resistance to dislocation motion results in greater yield strength and hardness. In MMC applications, graphene fillers can promote grain refinement by pinning grain boundaries during solidification or deformation. (Brodova et al., 2022; H. Chen et al., 2024; Fu et al., 2018).

The load transfer mechanism plays an important role in improving mechanical characteristics. As stress is applied, the matrix transfers some of it to the graphene platelets via contact. Due to the high stiffness of graphene fillers, stress transmission plays a crucial role in the overall increase in Young's modulus. The strength of the interfacial bond between the metal matrix and the graphene primarily determines the process's efficiency. A well-bonded interface allows for more efficient load transfer and improved composite performance. (Hidalgo-Manrique et al., 2019; Jiang et al., 2019).

Dislocation hindrance arises when reinforcing particles or microstructural obstacles within the metal matrix obstruct or delay dislocation movement. Strength is improved by any condition that restricts dislocation mobility, which is the basic process of plastic deformation. Dislocations are driven to collect or avoid these obstacles when well-dispersed, stiff graphene reinforcements are present in metal matrix composites, resulting in improved mechanical characteristics. (Qiu et al., 2021; Yang et al., 2021).

Orowan looping occurs when the reinforcing particles have high strength and are massive enough to allow dislocations to pass through, causing them to bend or loop around themselves. As dislocations move through the particles, they leave behind loops, generating a back-stress that limits further dislocation movement. The looping occurs more frequently as the particles grow smaller and closer together, resulting in significant strengthening. This approach works especially well in composites containing small, non-shearable particles like carbides, oxides, and nanoscale graphene fillers. (Sheinerman & Gutkin, 2020; Xiong et al., 2020).

Thermal mismatch strengthening refers to the thermal expansion behavior of the reinforcing phase that differs from that of the metal matrix. When cooling after processing, the mismatch results in internal strains at the matrix-particle interactions. A significant density of dislocations in the matrix could result from these stresses, particularly in the vicinity of the reinforcements. The strength of MMCs is increased by the dislocation networks that form, which serve as extra barriers against plastic deformation. This issue is particularly significant in MMCs that use graphene reinforcements since they frequently expand far less than the metal matrix. (George et al., 2005; J. Wang et al., 2012).

4.2. Effect of Manufacturing Methods on the Mechanical Properties of Graphene-Reinforced MMCs

The manufacturing route of MMCs with graphene filler content influences the mechanical properties of the final products as it dictates graphene dispersion, interface integrity, and structural uniformity. A wide range of studies have shown that different fabrication techniques yield varied improvements in the mechanical properties of MMCs, such as tensile strength, hardness, and ductility, depending on the process parameters.

The mechanical properties of MMCs containing graphene nanofiller have exhibited a promising improvement when utilizing solid-state techniques, such as friction stir alloying and spark plasma sintering. Friction stir alloying, for example, can enable graphite to be immediately exfoliated into multi-layer graphene within a metal matrix due to a strong interface and equal distribution of graphene fillers, resulting in a significant improvement in strength without compromising ductility. (Dixit et al., 2018). Similarly, using spark plasma sintering in conjunction with a hot rolling technique can result in a homogeneous distribution of graphene fillers. By using this process, the tensile strength of final titanium matrix composites containing graphene can increase to approximately 1200 MPa. This increase can be explained by the strong interface that forms between the titanium matrix and graphene filler as a result of Ti carbide nucleation, particle growth, and layer formation. (Cao & Liang, 2020; Mu et al., 2018).

Stir casting and liquid-solid pressure infiltration are two examples of liquid-state processing techniques that are appealing because of their scalability and comparatively easy setup. However, because of the low wettability of carbon-based fillers in molten metals and the strong Van der Waals interactions between graphene layers, these techniques are more likely to result in graphene agglomeration. Performance increases tend to stall or reverse when the graphene concentration surpasses around one weight percent, even though the addition of graphene nanofiller by stir casting usually results in moderate improvements in mechanical characteristics. This is often attributed to agglomeration and the formation of reaction-induced carbide phases (*e.g.*, Al_4C_3), which are brittle and can significantly reduce ductility and properties in the final composite. (Hemalatha et al., 2021; Polayya et al., 2023; Seyed Pourmand & Asgharzadeh, 2020; Venkatesan & Anthony Xavior, 2018). The composition of the carbide and the processing conditions can cause the differential impacts of carbide generation. As was already mentioned, solid-state techniques like spark plasma sintering can improve mechanical properties by forming strong and stable interfacial carbides. In contrast, the uncontrolled formation of brittle phases like Al_4C_3 can act as a crack initiator and degrade the overall mechanical performance of final composites. (Mu et al., 2018; Seyed Pourmand & Asgharzadeh, 2020).

MMCs are fabricated using an innovative additive manufacturing technique called selective laser sintering (SLS) that provides accurate control of phase composition, microstructure, grain size, and crystallographic roughness. In experimental processing of nanocomposite powders, metal powders are frequently blended with graphene or other nanofillers in a dispersion medium, e.g., polyvinyl alcohol (PVA) and deionized water, to ensure uniform distribution. The composite of dry powder is sintered by laser sintering, where the high-energy laser melts selectively the layer of material layer by layer, after the evaporation of the solvent of the dispersion. Because of the very high heating and cooling rates, the process not only enables the generation of complex shapes but also promotes the homogeneous dispersion of the graphene fillers. As a result, significant enhancements in mechanical properties (*i.e.,* including strength, hardness, and fatigue resistance) can be achieved in the final MMCs (Lin et al., 2014; Wen et al., 2019).

In summary, the processing technique has a significant role in determining the mechanical performance of graphene-reinforced MMCs by affecting the microstructural uniformity, interfacial bonding, and graphene dispersion. Because of the uniform dispersion of graphene and the formation of stable interfacial carbides that permit load transmission, solid-state procedures like friction stir alloying and spark plasma sintering can potentially provide an increase in tensile strength and ductility. On the other hand, liquid-state methods, such as stir casting, are vulnerable to filler agglomeration and the development of brittle reaction-induced phases (such as Al_4C_3), which can lead to fracture initiation sites and a decline in mechanical characteristics. Similar to SLS, additive manufacturing provides uniform inclusion of graphene fillers, increases strength, hardness, and fatigue strength, and allows control over fine microstructure and phase composition. These process–property relationships are governed by the load transfer through graphene nanofillers, as shown in Figure 4.

Figure 4. SEM images indicating (a) graphene layers embedded in a metal matrix and (b) pulled-out mechanisms promoting load transfer (X. Zhang et al., 2017).

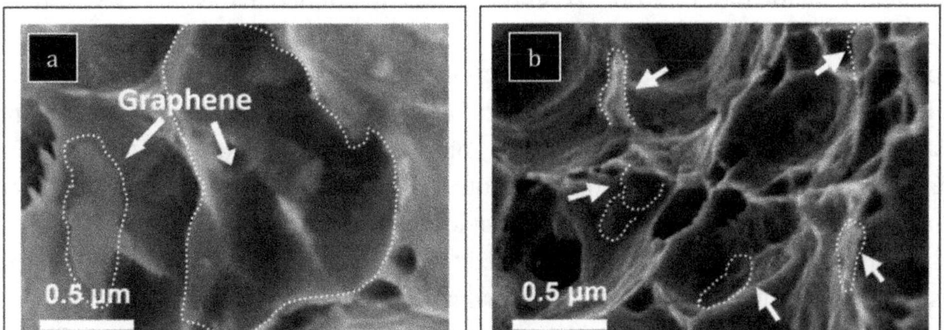

5. GRAPHENE INTEGRATION IN CERAMIC MATRIX COMPOSITES

CMCs have seen growing use in aerospace applications due to their stable mechanical performance under extreme conditions, including high temperatures and corrosive environments. (Rathod et al., 2017). In structural CMCs, alumina (AlO_3) and silicon carbide (SiC) are the most often utilized matrix materials. Continuous, short fibers and nanofillers, which have better mechanical qualities than traditional metallic alloys, are commonly used to strengthen these materials. Because of these characteristics, CMCs can be used in high-performance applications such as structural parts for hypersonic vehicles, thermal protection systems, and hot-section components. (Behera et al., 2020). However, CMCs are inherently brittle, exhibiting limited fracture toughness and poor multifunctional properties—including low thermal and electrical conductivity—which restrict their broader adoption in dynamic aerospace environments that require enhanced fatigue resistance and impact toughness. (K. Cui et al., 2022; Morales-Flórez & Domínguez-Rodríguez, 2022).

Graphene reinforcement is a promising approach for enhancing the mechanical properties of CMCs. Several manufacturing routes have been employed to manufacture CMCs, and the manufacturing route influences the mechanical properties of the final CMC materials. Hot pressing is a widely used fabrication technique for CMCs. Manufacturing CMCs with the hot pressing method can provide dense, high-strength, and uniform microstructures, which are critical for high-performance ceramic-based composite components. Dispersion of graphene fillers is a key parameter for nano-filler reinforced composites, as mentioned in the previous sections regarding PMCs and MCMs. To address this problem, the ball milling procedure is a common application for achieving homogeneous distribution of graphene nano-fillers. In the ball milling procedure, ceramic powders, graphene nano-fillers, and planetary balls are mixed together in a solution, and the mixture is subjected to high-energy mechanical forces that promote uniform dispersion of the fillers, reduce particle size, and enhance interfacial contact between the components (L. Zhou et al., 2020). The dispersion media, *i.e.,* used to promote homogenous dispersion, can also change the interfacial bonding occurring between ceramics and graphene fillers, thus affecting the mechanical properties of the final CMCs (E. Cui et al., 2019). Another parameter affecting the mechanical properties of CMCs is the sintering process. Sintering can be conducted either under pressure or in a pressureless condition. The procedure applied during sintering also influences the morphology of graphene sheets. For example, uniaxial pressure applied during the sintering process avoids the scrolling of graphene sheets, whereas graphene sheets can form tubular structures when pressure is not applied during sintering, as shown in Figure 5. The pressureless manufacturing route reduces mechanical properties, despite resulting in lower density. However, this difference can influence the effectiveness of toughening mechanisms obtained with graphene fillers, and graphene sheets can provide further improvement with a pressureless manufacturing route (Yang et al., 2015). On the other hand, the powder-based manufacturing route with hot pressing allows tailoring the interface between ceramic matrices and graphene fillers. Different types of ceramic powder can be mixed; therefore, strong and weak interfaces with graphene fillers can be developed in the final composite materials. This approach can enhance load transfer and grain refinement, aided by toughening mechanisms with crack bridging and graphene pull-out (X. Wang et al., 2021).

Figure 5. Fracture surfaces obtained from a CMC manufactured with (a) pressured and (b) pressureless sintering routes (Yang et al., 2015).

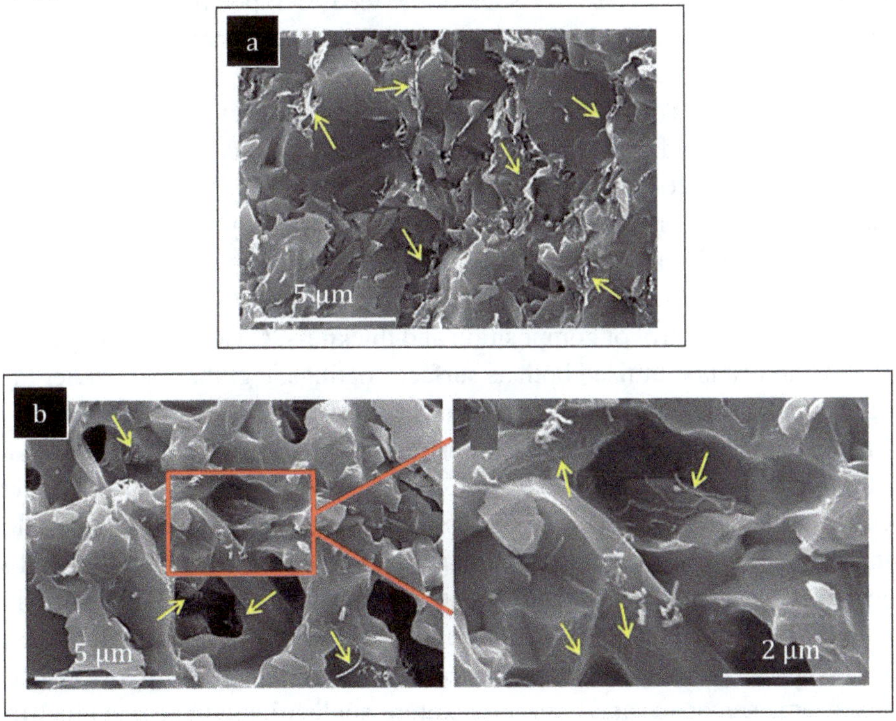

Spark plasma sintering (SPS) is a technique similar to hot pressing but offers the advantage of achieving finer grain sizes and better preservation of graphene structure by minimising degradation mechanisms (*e.g.* oxidation and graphitisation) due to its rapid heating and relatively fast cooling rates. Consequently, CMC manufacturing with spark plasma sintering offers a cleaner interface between ceramic matrices and graphene fillers, which provides enhanced mechanical properties. As spark plasma sintering is a powder-based CMC manufacturing route, the distribution efforts are similar to the hot pressing method, such as ultrasonication, ball milling in an aqueous environment, etc. (E. Cui et al., 2019; X. Wang et al., 2021; Yang et al., 2015). When compared to these methods, CMCs can be reinforced with a higher weight ratio with spark plasma sintering. Although a high loading fraction of graphene-based nano-fillers may result in agglomeration, it can distinctly enhance the multifunctional performance of CMCs, especially with respect to electrical conductivity (X. Wang et al., 2021; Yang et al., 2015).

In addition, advanced manufacturing technologies can be utilised to fabricate CMCs, *e.g.* laser-directed energy deposition (LDED). LDED offers precise control over the deposition process, enabling the fabrication of complex geometries and tailored microstructures. However, the efficiency and quality of this technique are highly dependent on the characteristics of the feedstock powders. Therefore, pre-processing methods such as high-energy ball milling and ultrasonic treatment are commonly employed to improve the homogeneity, dispersion, and surface activation of the powders. These pre-treatment steps are critical to ensuring consistent material properties and optimal performance in the final CMC product fabricated via LDED (Li et al., 2022).

Thin ceramic layers can be effectively manufactured by the tape casting process, a versatile and scalable processing technique widely used in the production of advanced CMCs. The process enables the fabrication of uniform, flat ceramic tapes with precise control of composition and thickness. CMCs created by the tape casting method can be utilised both as surface coatings, *e.g.* thermal barrier layers and as structural elements in their own right, as dictated by the demands of the design and the specific performance criteria of the application (Pereira dos Santos Tonello et al., 2016). In addition, ceramic composites can be laid out in a laminated structure by stacking different tape-cast layers (dos Reis Araújo et al., 2019). Each lamina can be pre-coated with graphene-rich slurries prior to lamination and sintering, which can significantly promote interfacial adhesion and preclude interlayer delamination. The layered approach can allow for the development of functionally graded or anisotropic CMCs with improved structural integrity and performance for use in extreme aerospace and high-temperature applications (Ojalvo et al., 2021).

Graphene-reinforced CMCs represent a significant advancement in the development of high-performance materials for aerospace environments under extreme conditions, with the help of crack deflection, crack bridging, crack branching, and pull-out mechanisms, as seen in Figure 6. While traditional fabrication methods like hot pressing and spark plasma sintering remain effective for producing dense, uniform composites with tailored microstructures, emerging techniques such as laser-directed energy deposition and tape casting provide opportunities for increased geometric complexity and functional integration. Key to the success of these composites is the homogeneous dispersion of graphene fillers and the optimization of interfacial bonding, both of which directly impact the mechanical and multifunctional performance. As research progresses, the integration of graphene into CMCs through precise processing and innovative design holds great promise for enabling the next generation of lightweight, thermally stable, and electrically conductive aerospace materials.

Figure 6. Fracture mechanisms observed in a CMC, including: (a) crack deflection, (b) crack branching, (c) crack bridging, and (d) graphene pull-out (X. Wang et al., 2021).

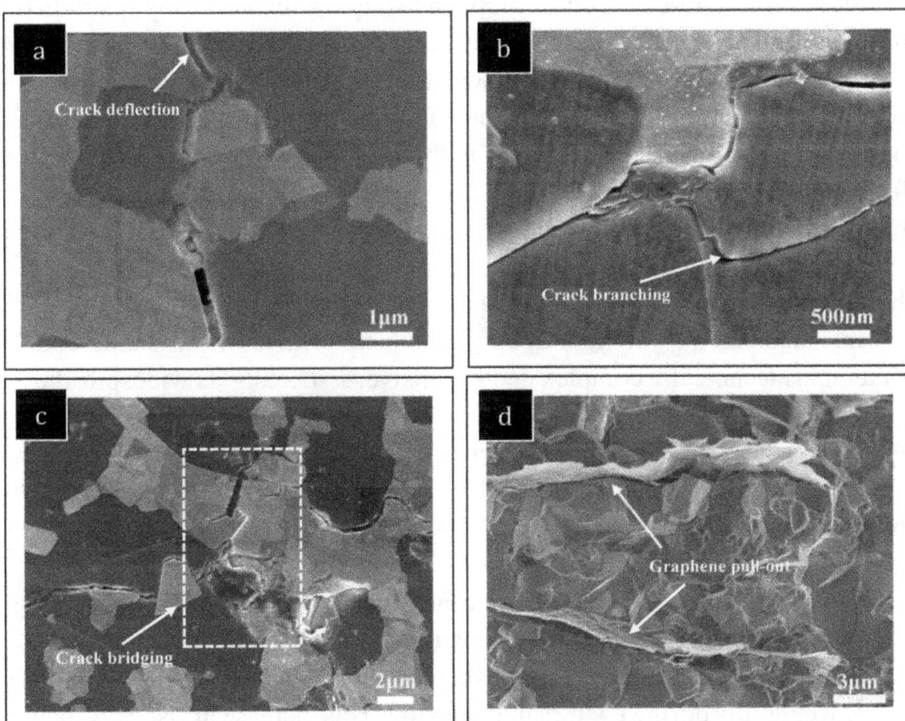

6. GRAPHENE-ENHANCED COMPOSITE MATERIALS FOR AEROSPACE: CHALLENGES AND FUTURE DIRECTIONS

6.1. General Challenges in Graphene-Enhanced Composite Materials

Graphene reinforcing and toughening techniques are a viable solution for improving the mechanical properties of composite materials across all matrix types while also adding multifunctional capabilities. However, certain problems are shared by polymer, metal, and ceramic matrix composites. One of the most extensively reported difficulties is the agglomeration of graphene fillers, which severely restricts their reinforcing ability. Furthermore, inadequate interfacial interaction between graphene and the host matrix can impede efficient load transfer. While functionalization enhances dispersion and interfacial adhesion, it often affects intrinsic mechanical properties of graphene fillers, resulting in trade-offs in the overall multifunctional performance of composites.

Furthermore, traditional manufacturing procedures are not necessarily compatible with graphene integration, necessitating the modification of existing methods or the creation of new dispersion-compatible fabrication techniques. Finally, the large-scale manufacture of high-quality graphene with uniform shape and low flaws remains technically complex and expensive, restricting its widespread use in aircraft applications.

Also, the chapter reviewed all the types of composite material reinforced with graphene fillers. The final marks related to challenges in graphene-reinforced composite material are these:

- *Polymer Matrix Composites:* A crucial factor in obtaining full wetting of fiber textiles is resin viscosity. Because of their large specific surface area, graphene fillers greatly increase resin viscosity, which might hinder resin flow and wettability during PMC manufacturing. This increase in viscosity may also result in an uneven distribution of filler due to a filtering action through flow mesh plies. Although coating methods utilizing graphene fillers might lessen viscosity-related problems, they frequently increase processing complexity and decrease production efficiency. However, there is still a great opportunity to achieve damage-tolerant PMCs with enhanced mechanical and multifunctional performance by integrating graphene reinforcement with interlaminar toughening techniques.
- *Metal Matrix Composites:* The development of reaction-induced brittle phases, such as carbides, is one of the main difficulties in the liquid-state processing of MMCs. When graphene reacts with molten metals like aluminum, unwanted

carbides like Al_4C_3 are created. These carbides drastically reduce ductility and serve as sites for fracture initiation in MMCs. Furthermore, graphene's chemically inert surface limits both interfacial bonding and homogenous filler dispersion by causing poor wettability in molten metal matrices. High processing temperatures make graphene filler integration with metal matrices even more difficult since graphene-based fillers may lose their structural integrity or undergo thermal degradation, especially in uncontrolled casting settings, which would ultimately reduce the reinforcing efficiency in MMCs.

- *Ceramic Matrix Composites:* It is challenging to disperse graphene nanofillers uniformly because agglomeration can produce weak spots that can reduce the mechanical performance of CMCs. Additionally, poor adhesion caused by the interfacial bonding between the graphene fillers and the ceramic matrix can lower interfacial failure under stress and reduce load transfer efficiency. Furthermore, the production of high-performance CMCs using advanced fabrication techniques (*e.g.*, hot pressing, SPS, and LDED) is challenging on an industrial scale due to their complexity, cost, and sensitivity to processing conditions. Despite their thermal stability, CMCs frequently have poor electrical and thermal conductivity, which restricts their range of applications unless they are strengthened by conductive fillers like graphene. However, these fillers may oxidize or graphitize when processed at high temperatures. Furthermore, changes in sintering conditions and pressure can have a substantial impact on the final composite's microstructure and characteristics, making consistent material production even more difficult.

6.2. Future Applications and Directions for Graphene-Enhanced Composites in Aerospace Engineering

For aircraft engineering applications, lightweight designs are essential to increase payload capacity, optimize fuel efficiency, and enhance overall performance. Reducing structural weight without compromising strength, durability, or safety is one of the primary objectives in the construction of aerostructural components. In addition to economic and environmental considerations, striking this balance is essential for fulfilling the demanding performance and operational requirements of high-stress, weight-sensitive situations. In this sense, metal, ceramic, and polymer-based composite materials have proven to be successful in aerostructural applications. Recent research has shown that graphene and its derivatives enhance the mechanical qualities of composites and provide them with additional multifunctional characteristics, expanding their potential use in next-generation aerospace applications.

Because of their exceptional specific strength and stiffness, chemical resistance, fatigue characteristics, and adaptability to particular loading conditions, PMCs have found extensive use in aerospace engineering. As a result, PMCs have progressively taken the place of conventional metals and alloys as the main structural components. However, unlike metallic materials, PMCs are naturally electrically non-conductive, which restricts their multifunctionality. Adding graphene-based nanofillers to PMCs has become a viable way to get around this restriction. These nanofillers offer mechanical reinforcement in addition to improving electrical conductivity (Kandare et al., 2015; Li et al., 2017; Vlassiouk et al., 2015). PMCs can offer lightning protection in aerospace applications owing to their enhanced electrical conductivity, eliminating the need for traditional solutions like copper mesh coverings on the fuselage, while also providing a significant weight reduction. (Das & Yokozeki, 2021; Kumar et al., 2019; B. Wang et al., 2018). Additionally, energy storage capabilities are made possible by the layered structure of PMC laminates in conjunction with graphene-based changes. Because of this, PMCs are attractive options for structural energy storage devices, such as supercapacitors and structural batteries, which combine mechanical load-bearing and energy storage capabilities into a single lightweight material (Asp et al., 2019, 2021; Moyer et al., 2020; Ye et al., 2024). The creation of electroactive polymers, which can be applied to morphing structures, is further aided by the achievement of multifunctionality in polymer-based materials. In this regard, graphene-based modification improves the mechanical and electrical conductivity of aerospace structural elements while also enhancing aerodynamic control through their capacity to change shape and adapt (Lu et al., 2014; Rodriguez et al., 2016). Another important trend in aerospace is the pursuit of alternative and sustainable power sources, with hydrogen emerging as illustrated by Airbus's ZEROe vision, which aims to achieve hydrogen-powered flights by 2035 (*ZEROe: Our Hydrogen-Powered Aircraft | Airbus*). However, cryogenic temperatures and tremendous pressures are needed to store hydrogen in liquid form. Therefore, PMC-based storage tanks need to be able to endure these harsh cryogenic conditions. Matrix cracking in PMCs under operating loads becomes particularly important from a structural standpoint since it can jeopardize safety and durability. By adding graphene-based nanofillers, the resistance of PMCs to matrix cracking can be increased, which can inhibit the hydrogen leakage and provide a lightweight solution for hydrogen storage. This is especially crucial for hydrogen storage tanks, where halting gas infiltration is essential for efficiency and safety (Li et al., 2023; Y. Liu et al., 2022b; X. J. Shen et al., 2015; J. Wang et al., 2024). As PMCs are inherently brittle, PMC-based structural components are also susceptible to dynamic loads like impacts. With the aid of structural health monitoring techniques, PMCs with improved multifunctionality capabilities can enable early detection of fracture mechanisms, such as matrix cracking, fiber breakage, and delamination. This facilitates early damage identification and extends component service life (Du et al., 2017; Li, Liao, et al., 2018; Sam-Daliri et al., 2020; T. Singh & Sehgal, 2021;

Q. Wang et al., 2023). Additionally, graphene nanofillers are essential to the use of stealth technology in aerospace engineering. Two different ways can be used to improve the stealth behavior of structural components: (i) morphological design and (ii) a material-based approach. In the material design approach, the mechanisms of electromagnetic attenuation by the radar absorbing materials are absorption and multiple deflections. Graphene-based nanofillers can improve the microwave absorbing properties if a homogeneous distribution of fillers is provided (Cozzolino et al., 2022; Jayalakshmi et al., 2019; S. H. Kim et al., 2023; Marra et al., 2020). Figure 7 highlights possible advancements achievable with graphene-modified PMCs.

Figure 7. Potential advancements which can be enabled by graphene-modified PMCs: (a) structural battery composites (Moyer et al., 2020), (b) cryogenic hydrogen storage tubes (Composite Materials and Hydrogen Tanks | Innovate Beyond Limits), and (c) stealth technologies (S. H. Kim et al., 2023).

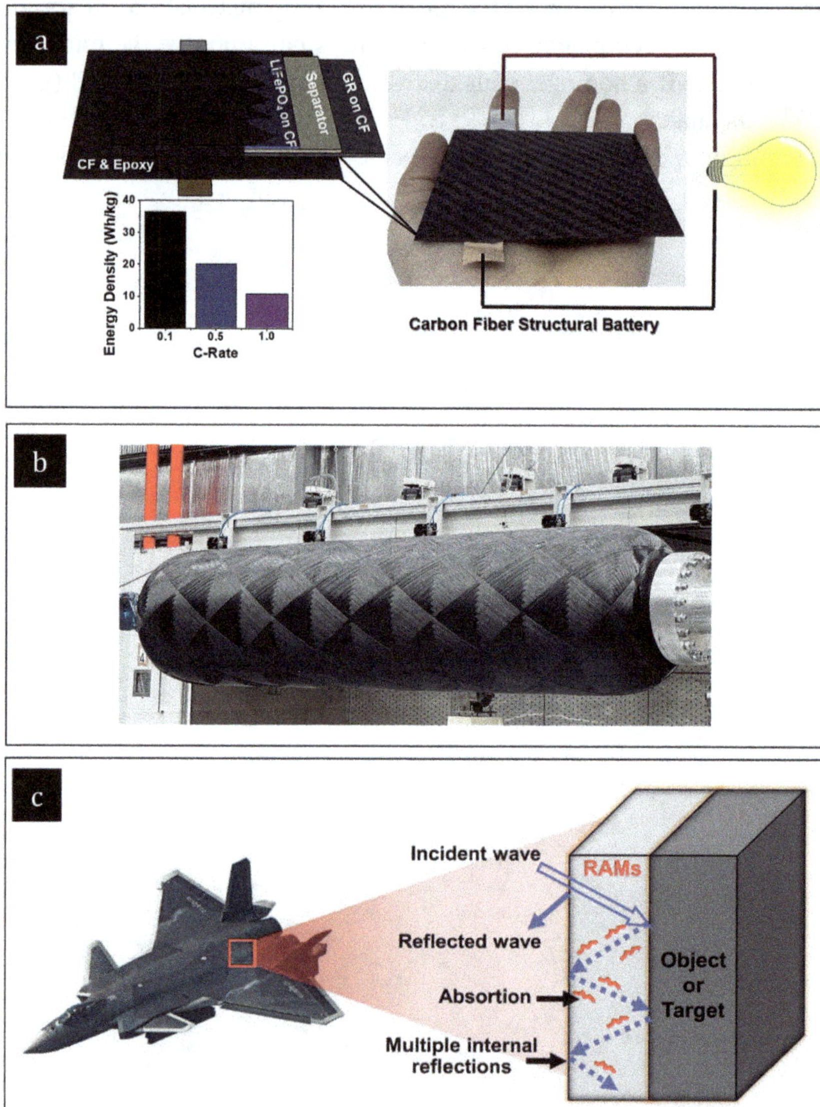

MMCs have traditionally played a significant role in aerostructural applications, owing to their high strength, stiffness, wear resistance, and thermal stability. Nevertheless, MMCs have largely been replaced by PMCs in primary structural applications, dominating areas such as fuselage skins and wing structures. However, their unique properties of MMCs allow the use of high-performance structural components (*e.g.*, landing gear rods and bushings, engine parts, and space panels) subjected to extreme operating conditions, including severe wear and elevated temperatures. Under cryogenic conditions, mechanical parts utilized in space applications, such as liquid hydrogen tanks, must withstand large structural loads. Graphene reinforcement presents a viable approach to improving the mechanical performance of MMCs in this regard (Ashwath et al., 2024; Jayaseelan et al., 2022). At the same time, lightweight design remains a central objective in aerospace engineering. The rise of additive manufacturing with advanced material design has enabled the production of complex cellular structures (see Figure 8), which offer an enhancement in strength and mass reduction. Incorporating graphene into these architectures presents an additional opportunity to further optimize their mechanical properties and weight efficiency for demanding aerospace environments (Oh & Yoo, 2020).

Figure 8. MMC-based cellular structure produced by additive manufacturing (Heinl et al., 2007).

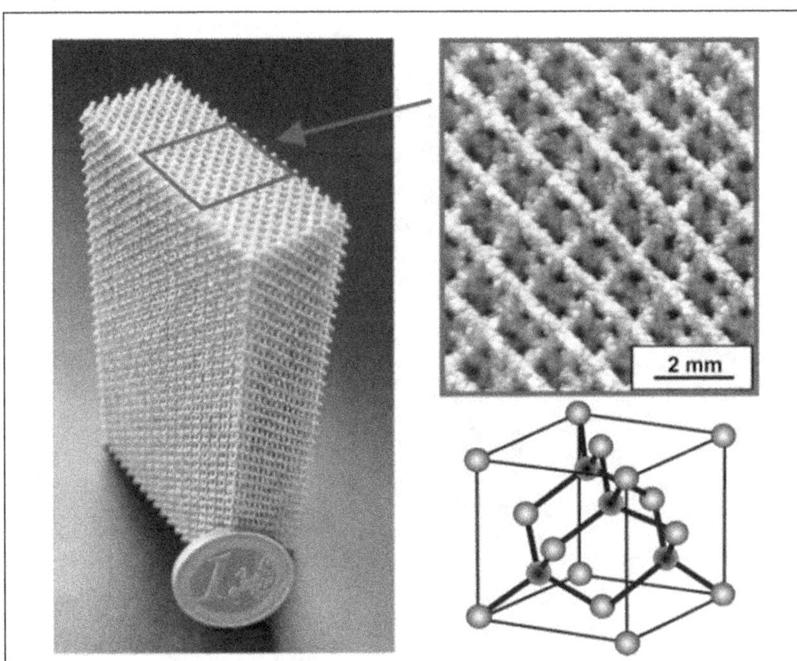

CMCs can endure high temperatures and mechanical strains, such as turbine engine parts, thermal protection systems, and hypersonic vehicle skins. Graphene fillers can improve the performance of CMCs by increasing toughness, thermal stability, and resistance to crack propagation. For instance, Firefly Aerospace replaced the conventional metallic materials used in such applications with a graphene-reinforced CMC coated in ceramic to create a second-stage nozzle extension. (*Firefly Aerospace Awarded AFRL Contract to Develop Ceramic Matrix Composite Rocket Engine Nozzle Extension*). This innovation highlights the potential of graphene-enhanced CMCs to outperform metals in high-temperature aerospace environments. CMCs can also act as protective coatings for metal-based structural elements, such as the hot parts of turbine engines, like HPT blades, combustors, and high-pressure turbine (HPT) vanes, as illustrated in Figure 9 (Zhu, 2018). The longevity, effectiveness, and general performance of aerospace systems operating in harsh environments could be greatly enhanced by the ongoing development of CMCs through graphene filler modification.

Figure 9. Examples of thermal barriers using CMCs in a jet engine, including the combustor, HPT vanes, and HPT blades (Zhu, 2018).

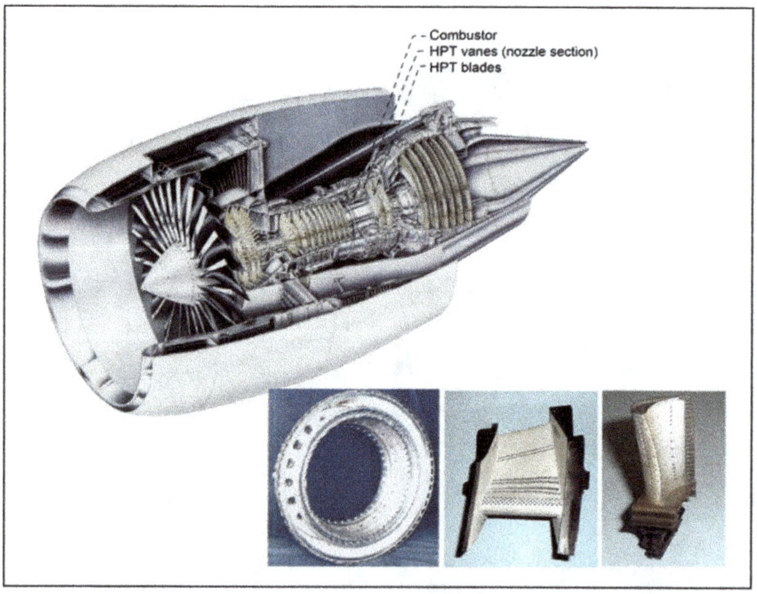

In conclusion, the crucial need for lightweight, high-performance, and multi-purpose materials can be met by the incorporation of graphene-based nanofillers into PMCs, MMCs, and CMCs. In addition to improving mechanical properties like strength, stiffness, and toughness, these cutting-edge composites also provide new features like energy storage, thermal stability, electrical conductivity, and structural health monitoring. Under harsh operating and environmental circumstances, such as high pressures, dynamic loads, and cryogenic temperatures, the unique properties of graphene can help ensure the safety, robustness, and effectiveness of aerostructural components. Furthermore, graphene-enhanced composites can support developments in additive manufacturing and sustainable aviation while also enabling the development of emerging technologies, including morphing structures, hydrogen storage, and stealth systems. As aerospace applications continue to demand higher performance and multifunctionality, graphene-reinforced composites stand out as one of the possible key enablers of next-generation aerospace materials and design strategies.

CONCLUSIONS

Graphene and its derivatives hold exceptional potential to revolutionize composite-based structural components in aerospace applications. Their incorporation markedly improves the mechanical properties of polymer, metal, and ceramic matrix composites while introducing multifunctional capabilities such as enhanced thermal and electrical conductivity, structural health monitoring, and energy storage. These benefits directly support next-generation aerospace demands for lightweight, durable, and multifunctional materials capable of withstanding extreme environments.

However, significant challenges remain before the full potential of graphene-enhanced composites can be realized. The agglomeration of graphene fillers still remains a significant problem and restricts uniform dispersion and undermines reinforcement efficiency. Alongside agglomeration, issues such as poor wettability with host materials, weak interfacial bonding, and the scalability of processing methods limit industrial adoption.

Several solutions have been proposed to overcome these barriers:

- Graphene nanofiller functionalization can enhance interfacial adhesion and dispersion; however, the functionalization may sacrifice intrinsic conductivity and multifunctional qualities.
- Physical dispersion techniques, including ultrasonication, high-shear mixing, and three-roll milling, can promote homogenous distribution.
- Processing innovations, such as spark plasma sintering, friction stir alloying, and additive manufacturing (*e.g.,* selective laser sintering), can offer greater control over microstructure, reduce filler agglomeration, and enhance interfacial bonding.
- Coating and interleaving strategies for PMCs can mitigate wettability and filtering issues during resin infusion.

- Pre-processing of powders (*e.g.* ball milling, surface activation) can help to achieve homogeneity and enhance the mechanical properties of composites.

By systematically addressing agglomeration and related dispersion challenges through these methods, graphene-enhanced composites could reach their full potential as aerospace-grade materials. Future studies should focus on scaling up the production of defect-free graphene, balancing functionalization with multifunctionality, and optimizing the interactions between process, structure, and properties.

Finally, the successful integration of graphene into aerospace composites could enable lightweight structures with superior mechanical resilience, thermal stability, and multifunctional features such as stealth capability and hydrogen storage. As manufacturing methods evolve, these advanced composites will play a central role in shaping the next generation of aerospace systems, including hydrogen-powered aircraft, morphing structures, and thermally adaptive protective skins.

REFERENCES

Akbolat, M. Ç., Katnam, K. B., Soutis, C., Potluri, P., Sprenger, S., & Taylor, J. (2022). On mode-I and mode-II interlaminar crack migration and R-curves in carbon/epoxy laminates with hybrid toughening via core-shell rubber particles and thermoplastic micro-fiber veils. *Composites. Part B, Engineering*, *238*, 109900. DOI: 10.1016/j.compositesb.2022.109900

Akbolat, M. Ç., Wang, S., Katnam, K. B., Potluri, P., & Soutis, C. (2025). Low-velocity impact response of carbon/epoxy laminates with interlaminar hybrid toughening via core–shell-rubber particles and non-woven thermoplastic fiber veils. *Composites. Part A, Applied Science and Manufacturing*, *195*, 108944. DOI: 10.1016/j.compositesa.2025.108944

Anwar, Z., Kausar, A., Rafique, I., & Muhammad, B. (2016). Advances in Epoxy/Graphene Nanoplatelet Composite with Enhanced Physical Properties: A Review. *Polymer-Plastics Technology and Engineering*, *55*(6), 643–662. DOI: 10.1080/03602559.2015.1098695

Aradhana, R., Mohanty, S., & Nayak, S. K. (2018). Comparison of mechanical, electrical, and thermal properties in graphene oxide and reduced graphene oxide filled epoxy nanocomposite adhesives. *Polymer*, *141*, 109–123. DOI: 10.1016/j.polymer.2018.03.005

Ashwath, P., Venkatraman, M., Patel, A., Xavior, M. A., & Batako, A. (2024). Innovation in sustainable composite research: Investigating graphene-reinforced MMCs for liquid hydrogen storage tanks in aerospace and space exploration. *Journal of Materials Research and Technology*, *33*, 4313–4331. DOI: 10.1016/j.jmrt.2024.10.070

Asp, L. E., Bouton, K., Carlstedt, D., Duan, S., Harnden, R., Johannisson, W., Johansen, M., Johansson, M. K. G., Lindbergh, G., Liu, F., Peuvot, K., Schneider, L. M., Xu, J., & Zenkert, D. (2021). A Structural Battery and its Multifunctional Performance. *Advanced Energy and Sustainability Research*, *2*(3), 2000093. DOI: 10.1002/aesr.202000093

Asp, L. E., Johansson, M., Lindbergh, G., Xu, J., & Zenkert, D. (2019). Structural battery composites: A review. *Functional Composites and Structures*, *1*(4), 042001. DOI: 10.1088/2631-6331/ab5571

Bagheri, R., Marouf, B. T., & Pearson, R. A. (2009). Rubber-Toughened Epoxies: A Critical Review. *Polymer Reviews (Philadelphia, Pa.)*, *49*(3), 201–225. DOI: 10.1080/15583720903048227

Barroeta Robles, J., Dubé, M., Hubert, P., & Yousefpour, A. (2022). Repair of thermoplastic composites: An overview. *Advanced Manufacturing: Polymer and Composites Science*, *8*(2), 68–96. DOI: 10.1080/20550340.2022.2057137

Behera, A., Swain, B., & Sahoo, D. K. (2020). Fiber-reinforced ceramic matrix nanocomposites. *Fiber-Reinforced Nanocomposites: Fundamentals and Applications*, 359–368. DOI: 10.1016/B978-0-12-819904-6.00016-5

Brodova, I., Yolshina, L., Razorenov, S., Rasposienko, D., Petrova, A., Shirinkina, I., Shorokhov, E., Muradymov, R., Garkushin, G., & Savinykh, A. (2022). Effect of Grain Size on the Properties of Aluminum Matrix Composites with Graphene. *Metals 2022, Vol. 12, Page 1054, 12*(6), 1054. DOI: 10.3390/met12061054

Cao, H., & Liang, Y. (2020). The microstructures and mechanical properties of graphene-reinforced titanium matrix composites. *Journal of Alloys and Compounds*, *812*, 152057. DOI: 10.1016/j.jallcom.2019.152057

Caprino, G., Lopresto, V., & Santoro, D. (2007). Ballistic impact behaviour of stitched graphite/epoxy laminates. *Composites Science and Technology*, *67*(3–4), 325–335. DOI: 10.1016/j.compscitech.2006.04.015

Cataldi, P., Athanassiou, A., & Bayer, I. S. (2018). Graphene Nanoplatelets-Based Advanced Materials and Recent Progress in Sustainable Applications. *Applied Sciences 2018, Vol. 8, Page 1438, 8*(9), 1438. DOI: 10.3390/app8091438

Chang, W., Rose, L. R. F., Sha, Z., Huang, F., Kinloch, A. J., & Wang, C. H. (2024). Multiscale modeling of nanoparticle toughening in epoxy: Effects of particle-matrix interface, particle size, and volume fraction. *Composites Science and Technology*, *256*, 110788. DOI: 10.1016/j.compscitech.2024.110788

Chen, H., Mi, G., Sun, Y., & Li, P. (2024). Unique grain refinement mechanism of graphene oxide reinforced high-temperature titanium alloy matrix composite. *Materials Today. Communications*, *41*, 110803. DOI: 10.1016/j.mtcomm.2024.110803

Chen, Y., Prasad, V., Yasar, M., Murphy, N., & Ivankovic, A. (2024). Enhancing interfacial performance and fracture toughness of carbon fibre reinforced thermoplastic composites. *Composites. Part A, Applied Science and Manufacturing*, *187*, 108434. DOI: 10.1016/j.compositesa.2024.108434

Composite Materials and Hydrogen Tanks | Innovate Beyond Limits. Retrieved June 27, 2025, from https://www.rayvatek.com/composite-materials-and-hydrogen-tanks/

Compton, O. C., & Nguyen, S. T. (2010a). Graphene oxide, highly reduced graphene oxide, and graphene: Versatile building blocks for carbon-based materials. *Small*, *6*(6), 711–723. DOI: 10.1002/smll.200901934

Compton, O. C., & Nguyen, S. T. (2010b). Graphene oxide, highly reduced graphene oxide, and graphene: Versatile building blocks for carbon-based materials. *Small*, *6*(6), 711–723. DOI: 10.1002/smll.200901934

Cozzolino, F., Marra, F., Fortunato, M., Bellagamba, I., Pesce, N., Tamburrano, A., & Sarto, M. S. (2022). New Sensing and Radar Absorbing Laminate Combining Structural Damage Detection and Electromagnetic Wave Absorption Properties. *Sensors 2022, Vol. 22, Page 8470, 22*(21), 8470. DOI: 10.3390/s22218470

Cui, E., Zhao, J., & Wang, X. (2019). Determination of microstructure and mechanical properties of graphene reinforced Al2O3-Ti(C, N) ceramic composites. *Ceramics International*, *45*(16), 20593–20599. DOI: 10.1016/j.ceramint.2019.07.041

Cui, K., Mao, H., Zhang, Y., Wang, J., Wang, H., Tan, T., & Fu, T. (2022). Microstructure, mechanical properties, and reinforcement mechanism of carbide toughened ZrC-based ultra-high temperature ceramics: A review. *Composite Interfaces*, *29*(7), 729–748. DOI: 10.1080/09276440.2021.2012409

Das, S., & Yokozeki, T. (2021). A brief review of modified conductive carbon/glass fibre reinforced composites for structural applications: Lightning strike protection, electromagnetic shielding, and strain sensing. *Composites Part C: Open Access*, *5*, 100162. DOI: 10.1016/j.jcomc.2021.100162

Dixit, S., Mahata, A., Mahapatra, D. R., Kailas, S. V., & Chattopadhyay, K. (2018). Multi-layer graphene reinforced aluminum – Manufacturing of high strength composite by friction stir alloying. *Composites. Part B, Engineering*, *136*, 63–71. DOI: 10.1016/j.compositesb.2017.10.028

dos Reis Araújo, M., Hotza, D., Janssen, R., & Acchar, W. (2019). Processing and properties of tape-cast alumina/zirconia laminates composites. *Journal of the European Ceramic Society*, *39*(12), 3462–3465. DOI: 10.1016/j.jeurceramsoc.2019.03.034

Du, X., Zhou, H., Sun, W., Liu, H. Y., Zhou, G., Zhou, H., & Mai, Y. W. (2017). Graphene/epoxy interleaves for delamination toughening and monitoring of crack damage in carbon fibre/epoxy composite laminates. *Composites Science and Technology*, *140*, 123–133. DOI: 10.1016/j.compscitech.2016.12.028

Dursun, T., & Soutis, C. (2014). Recent developments in advanced aircraft aluminium alloys. *Materials & Design*, *56*, 862–871. DOI: 10.1016/j.matdes.2013.12.002

Fang, W., Pearson, A., Huang, Y., & Naguib, H. E. (2025). Enhanced carbon fiber interface with thermoplastics via nanostructure surface modification: Failure, morphology and wettability analysis. *Composites. Part B, Engineering*, *291*, 112054. DOI: 10.1016/j.compositesb.2024.112054

Feres, G. D., Ribeiro, B., Costa, M. L., de Paula Santos, L. F., & Botelho, E. C. (2025). Study of the influence of carbon nanotubes, graphene, and hybrid buckypaper on the mechanical behavior of PAEK/carbon fiber composites. *Polymer Composites, 46*(9), 8674–8685. Advance online publication. DOI: 10.1002/pc.29519

Firefly Aerospace Awarded AFRL Contract to Develop Ceramic Matrix Composite Rocket Engine Nozzle Extension. Retrieved June 25, 2025, from https://fireflyspace.com/news/firefly-aerospace-awarded-afrl-contract-to-develop-ceramic-matrix-composite-rocket-engine-nozzle-extension/

Fu, K., Zhang, X., Shi, C., Liu, E., He, F., Li, J., Zhao, N., & He, C. (2018). An approach for fabricating Ni@graphene reinforced nickel matrix composites with enhanced mechanical properties. *Materials Science and Engineering A, 715*, 108–116. DOI: 10.1016/j.msea.2017.12.101

George, R., Kashyap, K. T., Rahul, R., & Yamdagni, S. (2005). Strengthening in carbon nanotube/aluminium (CNT/Al) composites. *Scripta Materialia, 53*(10), 1159–1163. DOI: 10.1016/j.scriptamat.2005.07.022

Gojny, F. H., Wichmann, M. H. G., Köpke, U., Fiedler, B., & Schulte, K. (2004). Carbon nanotube-reinforced epoxy-composites: Enhanced stiffness and fracture toughness at low nanotube content. *Composites Science and Technology, 64*(15 SPEC. ISS.), 2363–2371. DOI: 10.1016/j.compscitech.2004.04.002

Guild, F. J., Tsang, W. L., & Taylor, A. C. (2022). Silica nano-particle filled polymers: Debonding and microstructure. *Composites Science and Technology, 218*, 109202. DOI: 10.1016/j.compscitech.2021.109202

Heinl, P., Rottmair, A., Körner, C., & Singer, R. F. (2007). Cellular titanium by selective electron beam melting. *Advanced Engineering Materials, 9*(5), 360–364. DOI: 10.1002/adem.200700025

Hemalatha, K., Ashwin, R., Santhanam, M., Riyaz, M. M., Sozhamannan, G. G., & Venkatachalapthy, V. S. K. (2021). Experimental study on behaviour of SiC and Graphene on AL7075 hybrid metal matrix composite. *Materials Today: Proceedings, 39*, 26–30. DOI: 10.1016/j.matpr.2020.05.744

Hidalgo-Manrique, P., Lei, X., Xu, R., Zhou, M., Kinloch, I. A., & Young, R. J. (2019). Copper/graphene composites: A review. *Journal of Materials Science, 54*(19), 12236–12289. DOI: 10.1007/s10853-019-03703-5

Hsieh, T. H., Kinloch, A. J., Masania, K., Taylor, A. C., & Sprenger, S. (2010). The mechanisms and mechanics of the toughening of epoxy polymers modified with silica nanoparticles. *Polymer, 51*(26), 6284–6294. DOI: 10.1016/j.polymer.2010.10.048

Jayalakshmi, C. G., Inamdar, A., Anand, A., & Kandasubramanian, B. (2019). Polymer matrix composites as broadband radar absorbing structures for stealth aircrafts. *Journal of Applied Polymer Science, 136*(14), 47241. DOI: 10.1002/app.47241

Jayaseelan, J., Pazhani, A., Michael, A. X., Paulchamy, J., Batako, A., & Hosamane Guruswamy, P. K. (2022). Characterization Studies on Graphene-Aluminium Nano Composites for Aerospace Launch Vehicle External Fuel Tank Structural Application. *Materials 2022, Vol. 15, Page 5907, 15*(17), 5907. DOI: 10.3390/ma15175907

Jiang, Y., Xu, R., Tan, Z., Ji, G., Fan, G., Li, Z., Xiong, D. B., Guo, Q., Li, Z., & Zhang, D. (2019). Interface-induced strain hardening of graphene nanosheet/aluminum composites. *Carbon, 146*, 17–27. DOI: 10.1016/j.carbon.2019.01.094

Jogur, G., Nawaz Khan, A., Das, A., Mahajan, P., & Alagirusamy, R. (2018). Impact properties of thermoplastic composites. *Textile Progress, 50*(3), 109–183. DOI: 10.1080/00405167.2018.1563369

Kandare, E., Khatibi, A. A., Yoo, S., Wang, R., Ma, J., Olivier, P., Gleizes, N., & Wang, C. H. (2015). Improving the through-thickness thermal and electrical conductivity of carbon fibre/epoxy laminates by exploiting synergy between graphene and silver nano-inclusions. *Composites. Part A, Applied Science and Manufacturing, 69*, 72–82. DOI: 10.1016/j.compositesa.2014.10.024

Kanti, P. K., Prashantha Kumar, H. G., Wanatasanappan, V. V., Kumar, A., & Regasa, M. B. (2025). Graphene's Frontier in aerospace: Current applications, challenges, and future directions for space engineering. *Nanoscale Advances, 7*(12), 3603–3618. DOI: 10.1039/D4NA00934G

Katnam, K. B., Dalfi, H., & Potluri, P. (2019). Towards balancing in-plane mechanical properties and impact damage tolerance of composite laminates using quasi-UD woven fabrics with hybrid warp yarns. *Composite Structures, 225*, 111083. DOI: 10.1016/j.compstruct.2019.111083

Kausar, A., Rafique, I., & Muhammad, B. (2017). Aerospace Application of Polymer Nanocomposite with Carbon Nanotube, Graphite, Graphene Oxide, and Nanoclay. *Polymer-Plastics Technology and Engineering, 56*(13), 1438–1456. DOI: 10.1080/03602559.2016.1276594

Keramati, M., Ghasemi, I., Karrabi, M., Azizi, H., & Sabzi, M. (2016). Incorporation of surface modified graphene nanoplatelets for development of shape memory PLA nanocomposite. *Fibers and Polymers, 17*(7), 1062–1068. DOI: 10.1007/s12221-016-6329-7

Khalid, M. Y., Kamal, A., Otabil, A., Mamoun, O., & Liao, K. (2023). Graphene/epoxy nanocomposites for improved fracture toughness: A focused review on toughening mechanism. *Chemical Engineering Journal Advances*, *16*, 100537. DOI: 10.1016/j.ceja.2023.100537

Khalid, M. Y., Umer, R., & Khan, K. A. (2023). Review of recent trends and developments in aluminium 7075 alloy and its metal matrix composites (MMCs) for aircraft applications. *Results in Engineering*, *20*, 101372. DOI: 10.1016/j.rineng.2023.101372

Khan, T., Khalid, M. Y., Andrew, J. J., Ali, M. A., Zheng, L., & Umer, R. (2023). Co-cured GNP films with liquid thermoplastic/glass fiber composites for superior EMI shielding and impact properties for space applications. *Composites Communications*, *44*, 101767. DOI: 10.1016/j.coco.2023.101767

Khoramishad, H., Ashofteh, R. S., Pourang, H., & Berto, F. (2018). Experimental investigation of the influence of temperature on the reinforcing effect of graphene oxide nano-platelet on nanocomposite adhesively bonded joints. *Theoretical and Applied Fracture Mechanics*, *94*, 95–100. DOI: 10.1016/j.tafmec.2018.01.010

Kim, J. K., & Sham, M. L. (2000). Impact and delamination failure of woven-fabric composites. *Composites Science and Technology*, *60*(5), 745–761. DOI: 10.1016/S0266-3538(99)00166-9

Kim, S. H., Lee, S. Y., Zhang, Y., Park, S. J., & Gu, J. (2023). Carbon-Based Radar Absorbing Materials toward Stealth Technologies. *Advanced Science (Weinheim, Baden-Wurttemberg, Germany)*, *10*(32), 2303104. DOI: 10.1002/advs.202303104

Kostagiannakopoulou, C., Loutas, T. H., Sotiriadis, G., & Kostopoulos, V. (2021). Effects of graphene geometrical characteristics to the interlaminar fracture toughness of CFRP laminates. *Engineering Fracture Mechanics*, *245*, 107584. DOI: 10.1016/j.engfracmech.2021.107584

Kostagiannakopoulou, C., Loutas, T. H., Sotiriadis, G., Markou, A., & Kostopoulos, V. (2015). On the interlaminar fracture toughness of carbon fiber composites enhanced with graphene nano-species. *Composites Science and Technology*, *118*, 217–225. DOI: 10.1016/j.compscitech.2015.08.017

Kumar, S. S. A., Uddin, M. N., Rahman, M. M., & Asmatulu, R. (2019). Introducing graphene thin films into carbon fiber composite structures for lightning strike protection. *Polymer Composites*, *40*(S1), E517–E525. DOI: 10.1002/pc.24850

Lava Kumar, P., Lombardi, A., Byczynski, G., Narayana Murty, S. V. S., Murty, B. S., & Bichler, L. (2022). Recent advances in aluminium matrix composites reinforced with graphene-based nanomaterial: A critical review. *Progress in Materials Science, 128*, 100948. DOI: 10.1016/j.pmatsci.2022.100948

Leong, K. H., Ramakrishna, S., Huang, Z. M., & Bibo, G. A. (2000). The potential of knitting for engineering composites—A review. *Composites. Part A, Applied Science and Manufacturing, 31*(3), 197–220. DOI: 10.1016/S1359-835X(99)00067-6

Li, Y., Liao, Y., & Su, Z. (2018). Graphene-functionalized polymer composites for self-sensing of ultrasonic waves: An initiative towards "sensor-free" structural health monitoring. *Composites Science and Technology, 168*, 203–213. DOI: 10.1016/j.compscitech.2018.09.021

Li, Y., Meng, J., Luo, J., Wang, P., Ma, J., Zhao, Z., & Lei, H. (2023). Cryogenic mechanics and damage behaviors of carbon fiber reinforced polymer composites. *Composites. Part A, Applied Science and Manufacturing, 169*, 107484. DOI: 10.1016/j.compositesa.2023.107484

Li, Y., Zhang, D., Ye, Z., Ye, G., He, R., Wang, H., & Cong, W. (2022). The reinforcement mechanisms of graphene oxide in laser-directed energy deposition fabricated metal and ceramic matrix composites: A comparison study. *International Journal of Advanced Manufacturing Technology, 119*(3–4), 1975–1988. DOI: 10.1007/s00170-021-08337-z

Li, Y., Zhang, H., Liu, Y., Wang, H., Huang, Z., Peijs, T., & Bilotti, E. (2018). Synergistic effects of spray-coated hybrid carbon nanoparticles for enhanced electrical and thermal surface conductivity of CFRP laminates. *Composites. Part A, Applied Science and Manufacturing, 105*, 9–18. DOI: 10.1016/j.compositesa.2017.10.032

Li, Y., Zhang, H., Porwal, H., Huang, Z., Bilotti, E., & Peijs, T. (2017). Mechanical, electrical and thermal properties of in-situ exfoliated graphene/epoxy nanocomposites. *Composites. Part A, Applied Science and Manufacturing, 95*, 229–236. DOI: 10.1016/j.compositesa.2017.01.007

Liang, Y. L., & Pearson, R. A. (2010). The toughening mechanism in hybrid epoxy-silica-rubber nanocomposites (HESRNs). *Polymer, 51*(21), 4880–4890. DOI: 10.1016/j.polymer.2010.08.052

Lim, Y. J., Carolan, D., & Taylor, A. C. (2016). Simultaneously tough and conductive rubber–graphene–epoxy nanocomposites. *Journal of Materials Science, 51*(18), 8631–8644. DOI: 10.1007/s10853-016-0122-2

Lin, D., Richard Liu, C., & Cheng, G. J. (2014). Single-layer graphene oxide reinforced metal matrix composites by laser sintering: Microstructure and mechanical property enhancement. *Acta Materialia*, *80*, 183–193. DOI: 10.1016/j.actamat.2014.07.038

Liu, H.-Y., Wang, G.-T., Mai, Y.-W., & Zeng, Y. (2011). On fracture toughness of nano-particle modified epoxy. *Composites. Part B, Engineering*, *42*(8), 2170–2175. DOI: 10.1016/j.compositesb.2011.05.014

Liu, Y., Zou, A., Wang, G., Han, C., & Blackie, E. (2022a). Enhancing interlaminar fracture toughness of CFRP laminates with hybrid carbon nanotube/graphene oxide fillers. *Diamond and Related Materials*, *128*, 109285. DOI: 10.1016/j.diamond.2022.109285

Liu, Y., Zou, A., Wang, G., Han, C., & Blackie, E. (2022b). Enhancing interlaminar fracture toughness of CFRP laminates with hybrid carbon nanotube/graphene oxide fillers. *Diamond and Related Materials*, *128*, 109285. DOI: 10.1016/j.diamond.2022.109285

Lu, H., Yao, Y., Huang, W. M., & Hui, D. (2014). Noncovalently functionalized carbon fiber by grafted self-assembled graphene oxide and the synergistic effect on polymeric shape memory nanocomposites. *Composites. Part B, Engineering*, *67*, 290–295. DOI: 10.1016/j.compositesb.2014.07.022

Lyashenko-Miller, T., & Marom, G. (2017). Delamination fracture toughness of UHMWPE fibers/polyurethane laminates interleaved with carbon nanotube-reinforced polyurethane films. *Polymers for Advanced Technologies*, *28*(5), 606–612. DOI: 10.1002/pat.3848

Mallick, P. K. (2008). *Fiber-reinforced composites : materials, manufacturing, and design*. CRC Press.

Marra, F., Lecini, J., Tamburrano, A., Pisu, L., & Sarto, M. S. (2020). Broadband Electromagnetic Absorbing Structures Made of Graphene/Glass-Fiber/Epoxy Composite. *IEEE Transactions on Microwave Theory and Techniques*, *68*(2), 590–601. DOI: 10.1109/TMTT.2019.2950223

Meher, A., Mahapatra, M. M., Samal, P., & Vundavilli, P. R. (2022). A review on manufacturability of magnesium matrix composites: Processing, tribology, joining, and machining. *CIRP Journal of Manufacturing Science and Technology*, *39*, 134–158. DOI: 10.1016/j.cirpj.2022.07.012

Miracle, D. B. (2005). Metal matrix composites – From science to technological significance. *Composites Science and Technology*, *65*(15–16), 2526–2540. DOI: 10.1016/j.compscitech.2005.05.027

Mohan, P. (2013). A Critical Review: The Modification, Properties, and Applications of Epoxy Resins. *Polymer-Plastics Technology and Engineering*, *52*(2), 107–125. DOI: 10.1080/03602559.2012.727057

Morales-Flórez, V., & Domínguez-Rodríguez, A. (2022). Mechanical properties of ceramics reinforced with allotropic forms of carbon. *Progress in Materials Science*, *128*, 100966. DOI: 10.1016/j.pmatsci.2022.100966

Mouritz, A. P. (2007). Review of z-pinned composite laminates. *Composites. Part A, Applied Science and Manufacturing*, *38*(12), 2383–2397. DOI: 10.1016/j.compositesa.2007.08.016

Moyer, K., Meng, C., Marshall, B., Assal, O., Eaves, J., Perez, D., Karkkainen, R., Roberson, L., & Pint, C. L. (2020). Carbon fiber reinforced structural lithium-ion battery composite: Multifunctional power integration for CubeSats. *Energy Storage Materials*, *24*, 676–681. DOI: 10.1016/j.ensm.2019.08.003

Mu, X. N., Cai, H. N., Zhang, H. M., Fan, Q. B., Zhang, Z. H., Wu, Y., Ge, Y. X., & Wang, D. D. (2018). Interface evolution and superior tensile properties of multi-layer graphene reinforced pure Ti matrix composite. *Materials & Design*, *140*, 431–441. DOI: 10.1016/j.matdes.2017.12.016

Mulqueen, D. W., Sattar, S., Le, T., & Kravchenko, O. G. (2022). Spray deposition of sustainable plant based graphene in thermosetting carbon fiber laminates for mechanical, thermal, and electrical properties. *Polymer Composites*, *43*(12), 8696–8714. DOI: 10.1002/pc.27051

Oh, M. J., & Yoo, P. J. (2020). Graphene-based 3D lightweight cellular structures: Synthesis and applications. *Korean Journal of Chemical Engineering 2020 37:2*, *37*(2), 189–208. DOI: 10.1007/s11814-019-0437-1

Ojalvo, C., Ayllón, M., Ortiz, A. L., & Moreno, R. (2021). Aqueous tape casting of super-hard B4C laminates with rGO-enriched reinforcing interlayers. *Journal of the European Ceramic Society*, *41*(11), 5457–5465. DOI: 10.1016/j.jeurceramsoc.2021.05.001

Painkal, S. K., Balachandran, M., Jayanarayanan, K., Sridhar, N., & Kumar, S. (2025). Synergistic enhancement in mechanical properties of graphene/MWCNT reinforced Polyaryletherketone – carbon fiber multi-scale composites: Experimental studies and finite element analysis. *Advanced Industrial and Engineering Polymer Research*, *8*(1), 20–36. DOI: 10.1016/j.aiepr.2024.02.002

Pamuk, G., & Çeken, F. (2008). Manufacturing of Weft-Knitted Fabric Reinforced Composite Materials: A Review. *Materials and Manufacturing Processes*, *23*(7), 635–640. DOI: 10.1080/10426910802316484

Paredes, J. I., Villar-Rodil, S., Martínez-Alonso, A., & Tascón, J. M. D. (2008). Graphene oxide dispersions in organic solvents. *Langmuir*, *24*(19), 10560–10564. DOI: 10.1021/la801744a

Pathak, A. K., Borah, M., Gupta, A., Yokozeki, T., & Dhakate, S. R. (2016a). Improved mechanical properties of carbon fiber/graphene oxide-epoxy hybrid composites. *Composites Science and Technology*, *135*, 28–38. DOI: 10.1016/j.compscitech.2016.09.007

Pathak, A. K., Borah, M., Gupta, A., Yokozeki, T., & Dhakate, S. R. (2016b). Improved mechanical properties of carbon fiber/graphene oxide-epoxy hybrid composites. *Composites Science and Technology*, *135*, 28–38. DOI: 10.1016/j.compscitech.2016.09.007

Pegorin, F., Pingkarawat, K., Daynes, S., & Mouritz, A. P. (2015). Influence of z-pin length on the delamination fracture toughness and fatigue resistance of pinned composites. *Composites. Part B, Engineering*, *78*, 298–307. DOI: 10.1016/j.compositesb.2015.03.093

Pereira dos Santos Tonello, K., Padovano, E., Badini, C., Biamino, S., Pavese, M., & Fino, P. (2016). Fabrication and characterization of laminated SiC composites reinforced with graphene nanoplatelets. *Materials Science and Engineering A*, *659*, 158–164. DOI: 10.1016/j.msea.2016.02.050

Periasamy, K., Kandare, E., Das, R., Darouie, M., & Khatibi, A. A. (2023). Interfacial Engineering Methods in Thermoplastic Composites: An Overview. *Polymers 2023, Vol. 15, Page 415*, *15*(2), 415. DOI: 10.3390/polym15020415

Phiri, J., Johansson, L. S., Gane, P., & Maloney, T. (2018). A comparative study of mechanical, thermal and electrical properties of graphene-, graphene oxide- and reduced graphene oxide-doped microfibrillated cellulose nanocomposites. *Composites. Part B, Engineering*, *147*, 104–113. DOI: 10.1016/j.compositesb.2018.04.018

Polayya, C., Rao, C. S. P., Veeresh Kumar, G. B., & Surakasi, R. (2023). Synthesis and mechanical properties of graphene nanoparticles reinforced with aluminium alloy matrix composites. *Nanotechnology for Environmental Engineering*, *8*(2), 469–480. DOI: 10.1007/s41204-022-00305-6

Poonpipat, Y., Leelachai, K., Pearson, R. A., & Dittanet, P. (2017). Fracture behavior of silica nanoparticles reinforced rubber/epoxy composite. *Journal of Reinforced Plastics and Composites*, *36*(16), 1156–1167. DOI: 10.1177/0731684417709952

Pramodkumar, B., & Budhe, S. (2024). The effect of graphene oxide on thermal, electrical, and mechanical properties of carbon/epoxy composites: Towards multifunctional composite material. *Polymer Composites*, *45*(7), 6374–6384. DOI: 10.1002/pc.28203

Qian, H., Greenhalgh, E. S., Shaffer, M. S. P., & Bismarck, A. (2010). Carbon nanotube-based hierarchical composites: A review. *Journal of Materials Chemistry*, *20*(23), 4751. DOI: 10.1039/c000041h

Qiu, C., Su, Y., Yang, J., Chen, B., Ouyang, Q., & Zhang, D. (2021). Structural modelling and mechanical behaviors of graphene/carbon nanotubes reinforced metal matrix composites via atomic-scale simulations: A review. *Composites Part C: Open Access*, *4*, 100120. DOI: 10.1016/j.jcomc.2021.100120

Quan, D., Mischo, C., Binsfeld, L., Ivankovic, A., & Murphy, N. (2020). Fracture behaviour of carbon fibre/epoxy composites interleaved by MWCNT- and graphene nanoplatelet-doped thermoplastic veils. *Composite Structures*, *235*, 111767. DOI: 10.1016/j.compstruct.2019.111767

Rathod, V. T., Kumar, J. S., & Jain, A. (2017). Polymer and ceramic nanocomposites for aerospace applications. *Applied Nanoscience 2017 7:8*, *7*(8), 519–548. DOI: 10.1007/s13204-017-0592-9

Ray, S. C. (2015). Application and Uses of Graphene Oxide and Reduced Graphene Oxide. *Applications of Graphene and Graphene-Oxide Based Nanomaterials*, 39–55. DOI: 10.1016/B978-0-323-37521-4.00002-9

Reis, J. P., de Moura, M., & Samborski, S. (2020). Thermoplastic Composites and Their Promising Applications in Joining and Repair Composites Structures: A Review. *Materials 2020, Vol. 13, Page 5832*, *13*(24), 5832. DOI: 10.3390/ma13245832

Rodriguez, J. N., Zhu, C., Duoss, E. B., Wilson, T. S., Spadaccini, C. M., & Lewicki, J. P. (2016). Shape-morphing composites with designed micro-architectures. *Scientific Reports*, *6*(1), 1–10. DOI: 10.1038/srep27933

Sadigh, M. A. S., & Marami, G. (2018). Enhancing fatigue life in adhesively bonded joints using reduced graphene oxide additive: Experimental and numerical evaluation. *International Journal of Adhesion and Adhesives*, *84*, 283–290. DOI: 10.1016/j.ijadhadh.2018.04.001

Saghafi, H., Ahmadi, I., Khamedi, R., Saghafi, H., Saeedifar, M., Brugo, T. M., Maccaferri, E., Ortolani, J., Mongioì, F., Mazzocchetti, L., & Zucchelli, A. (2024). The thickness effect of rubbery nanofibrous mat on modes I–II fracture mechanism of composite laminates. *Thin-walled Structures, 196*, 111556. DOI: 10.1016/j.tws.2024.111556

Sam-Daliri, O., Farahani, M., Faller, L. M., & Zangl, H. (2020). Structural health monitoring of defective single lap adhesive joints using graphene nanoplatelets. *Journal of Manufacturing Processes, 55*, 119–130. DOI: 10.1016/j.jmapro.2020.03.063

Seyed Pourmand, N., & Asgharzadeh, H. (2020). Aluminum Matrix Composites Reinforced with Graphene: A Review on Production, Microstructure, and Properties. *Critical Reviews in Solid State and Material Sciences, 45*(4), 289–337. DOI: 10.1080/10408436.2019.1632792

Sheinerman, A. G. (2022). Plastic deformation and fracture processes in metal/graphene composites: A review. *Critical Reviews in Solid State and Material Sciences, 47*(5), 708–735. DOI: 10.1080/10408436.2021.1965955

Sheinerman, A. G., & Gutkin, M. Y. (2020). Model of Enhanced Strength and Ductility of Metal/Graphene Composites with Bimodal Grain Size Distribution. *Metallurgical and Materials Transactions. A, Physical Metallurgy and Materials Science, 51*(1), 189–199. DOI: 10.1007/s11661-019-05500-w

Shen, M. Y., Liao, W. Y., Wang, T. Q., & Lai, W. M. (2021). Characteristics and Mechanical Properties of Graphene Nanoplatelets-Reinforced Epoxy Nanocomposites: Comparison of Different Dispersal Mechanisms. *Sustainability 2021, Vol. 13, Page 1788, 13*(4), 1788. DOI: 10.3390/su13041788

Shen, X. J., Meng, L. X., Yan, Z. Y., Sun, C. J., Ji, Y. H., Xiao, H. M., & Fu, S. Y. (2015). Improved cryogenic interlaminar shear strength of glass fabric/epoxy composites by graphene oxide. *Composites. Part B, Engineering, 73*, 126–131. DOI: 10.1016/j.compositesb.2014.12.023

Shetty, R., Hegde, A., Shetty SV, U. K., Nayak, R., Naik, N., & Nayak, M. (2022). Processing and Mechanical Characterisation of Titanium Metal Matrix Composites: A Literature Review. *Journal of Composites Science 2022, Vol. 6, Page 388, 6*(12), 388. DOI: 10.3390/jcs6120388

Shiino, M. Y., Pelosi, S., Hilá, M. O., Cioffi, R., & Donadon, M. V. (2017, March). The Role of Stitch Yarn on the Delamination Resistance in Non-crimp Fabric: Chemical and Physical Interpretation. *Journal of Materials Engineering and Performance, 26*(3), 978–986. Advance online publication. DOI: 10.1007/s11665-016-2460-2

Shin, M. K., Lee, B., Kim, S. H., Lee, J. A., Spinks, G. M., Gambhir, S., Wallace, G. G., Kozlov, M. E., Baughman, R. H., & Kim, S. J. (2012). Synergistic toughening of composite fibres by self-alignment of reduced graphene oxide and carbon nanotubes. *Nature Communications*, *3*(1), 1–8. DOI: 10.1038/ncomms1661

Shivakumar, K. N., Panduranga, R., & Sharpe, M. (2013, November). Interleaved Polymer Matrix Composites - A Review. *54th AIAA/ASME/ASCE/AHS/ASC Structures, Structural Dynamics, and Materials Conference*. DOI: 10.2514/6.2013-1903

Singh, B., Kumar, I., Saxena, K. K., Mohammed, K. A., Ijaz Khan, M., Ben Moussa, S., & Shukhratovich Abdullaev, S. (2023). A future prospects and current scenario of aluminium metal matrix composites characteristics. *Alexandria Engineering Journal*, *76*, 1–17. DOI: 10.1016/j.aej.2023.06.028

Singh, T., & Sehgal, S. (2021). Structural Health Monitoring of Composite Materials. *Archives of Computational Methods in Engineering 2021 29:4*, *29*(4), 1997–2017. DOI: 10.1007/s11831-021-09666-8

Srinivasan, V., Kunjiappan, S., & Palanisamy, P. (2021). A brief review of carbon nanotube reinforced metal matrix composites for aerospace and defense applications. *International Nano Letters 2021 11:4*, *11*(4), 321–345. DOI: 10.1007/s40089-021-00328-y

Tang, L. C., Wan, Y. J., Yan, D., Pei, Y. B., Zhao, L., Li, Y. B., Wu, L., Jiang, J.-X., & Lai, G.-Q. (2013). The effect of graphene dispersion on the mechanical properties of graphene/epoxy composites. *Carbon*, *60*, 16–27. DOI: 10.1016/j.carbon.2013.03.050

Tareq, M. S., Jony, B., Zainuddin, S., Al Ahsan, M., & Hosur, M. V. (2021). Fatigue analysis and fracture toughness of graphene reinforced carbon fibre polymer composites. *Fatigue & Fracture of Engineering Materials & Structures*, *44*(2), 461–474. DOI: 10.1111/ffe.13371

Vaidya, U. K., & Chawla, K. K. (2008). Processing of fibre reinforced thermoplastic composites. *International Materials Reviews*, *53*(4), 185–218. DOI: 10.1179/174328008X325223

Venkatesan, S., & Anthony Xavior, M. (2018). Tensile behavior of aluminum alloy (AA7050) metal matrix composite reinforced with graphene fabricated by stir and squeeze cast processes. *Science and Technology of Materials*, *30*(2), 74–85. DOI: 10.1016/j.stmat.2018.02.005

Vieille, B., Casado, V. M., & Bouvet, C. (2014). Influence of matrix toughness and ductility on the compression-after-impact behavior of woven-ply thermoplastic- and thermosetting-composites: A comparative study. *Composite Structures*, *110*(1), 207–218. DOI: 10.1016/j.compstruct.2013.12.008

Vlassiouk, I., Polizos, G., Cooper, R., Ivanov, I., Keum, J. K., Paulauskas, F., Datskos, P., & Smirnov, S. (2015). Strong and Electrically Conductive Graphene-Based Composite Fibers and Laminates. *ACS Applied Materials & Interfaces*, *7*(20), 10702–10709. DOI: 10.1021/acsami.5b01367

Wang, B., Duan, Y., Xin, Z., Yao, X., Abliz, D., & Ziegmann, G. (2018). Fabrication of an enriched graphene surface protection of carbon fiber/epoxy composites for lightning strike via a percolating-assisted resin film infusion method. *Composites Science and Technology*, *158*, 51–60. DOI: 10.1016/j.compscitech.2018.01.047

Wang, G., Liu, X., Wang, M., Yu, M., Zhang, H., & He, L. (2024). Enhancing interlaminar fracture toughness of carbon fiber/epoxy composites toughened by polyethersulfone resin and hybrid carbon nanotubes/graphene oxide. *Polymer Composites*. Advance online publication. DOI: 10.1002/pc.29256

Wang, J., Chang, W., Islam, M. S., Huang, F., Wu, S., Rose, L. R. F., Zhang, J., & Wang, C. H. (2024). Toughening epoxy by nano-structured block copolymer to mitigate matrix microcracking of carbon fibre composites at cryogenic temperatures. *Composites Science and Technology*, *251*, 110548. DOI: 10.1016/j.compscitech.2024.110548

Wang, J., Li, Z., Fan, G., Pan, H., Chen, Z., & Zhang, D. (2012). Reinforcement with graphene nanosheets in aluminum matrix composites. *Scripta Materialia*, *66*(8), 594–597. DOI: 10.1016/j.scriptamat.2012.01.012

Wang, Q., Tian, Y., Duongthipthewa, A., Zhang, J., Liu, M., Su, Z., & Zhou, L. (2023). An embedded non-intrusive graphene/epoxy broadband nanocomposite sensor co-cured with GFRP for in situ structural health monitoring. *Composites Science and Technology*, *236*, 109995. DOI: 10.1016/j.compscitech.2023.109995

Wang, X., Zhao, J., Cui, E., Tian, X., & Sun, Z. (2021). Effect of Interfacial Structure on Mechanical Properties of Graphene Reinforced Al2O3–WC Matrix Ceramic Composite. *Nanomaterials 2021, Vol. 11, Page 1374, 11*(6), 1374. DOI: 10.3390/nano11061374

Wang, Z., Soutis, C., & Gresil, M. (2021). Fracture Toughness of Hybrid Carbon Fibre/Epoxy Enhanced by Graphene and Carbon Nanotubes. *Applied Composite Materials*, *28*(4), 1111–1125. DOI: 10.1007/s10443-021-09906-x

Warren, A. S. (2004). Developments and Challenges for Aluminum - A Boeing Perspective. *MATERIALS FORUM, 28*, 24–31.

Wen, S., Chen, K., Li, W., Zhou, Y., Wei, Q., & Shi, Y. (2019). Selective laser melting of reduced graphene oxide/S136 metal matrix composites with tailored microstructures and mechanical properties. *Materials & Design, 175*, 107811. DOI: 10.1016/j.matdes.2019.107811

Worsley, E. A., Margadonna, S., & Bertoncello, P. (2022). Application of Graphene Nanoplatelets in Supercapacitor Devices: A Review of Recent Developments. *Nanomaterials 2022, Vol. 12, Page 3600, 12*(20), 3600. DOI: 10.3390/nano12203600

Wu, X., Ning, H., Liu, Y., Hu, N., Liu, F., Wang, S., Huang, K., Jiao, Y., Weng, S., Liu, Q., & Wu, L. (2020). Synergistic Delamination Toughening of Glass Fiber-Aluminum Laminates by Surface Treatment and Graphene Oxide Interleaf. *Nanoscale Research Letters, 15*(1), 1–14. DOI: 10.1186/s11671-020-03306-z

Xiong, B., Liu, K., Xiong, W., Wu, X., & Sun, J. (2020). Strengthening effect induced by interfacial reaction in graphene nanoplatelets reinforced aluminum matrix composites. *Journal of Alloys and Compounds, 845*, 156282. DOI: 10.1016/j.jallcom.2020.156282

Yan, W., Liu, H. Y., & Mai, Y. W. (2004). Mode II delamination toughness of z-pinned laminates. *Composites Science and Technology, 64*(13–14), 1937–1945. DOI: 10.1016/j.compscitech.2004.02.008

Yang, Y., Li, B., Zhang, C., Wang, S., Liu, K., & Yang, B. (2015). Fabrication and properties of graphene reinforced silicon nitride composite materials. *Materials Science and Engineering A, 644*, 90–95. DOI: 10.1016/j.msea.2015.07.062

Yang, Y., Liu, M., Zhou, S., Ren, W., Zhou, Q., & Zhang, W. (2021). Strengthening behaviour of continuous graphene network in metal matrix composites. *Carbon, 182*, 825–836. DOI: 10.1016/j.carbon.2021.06.067

Yao, X., Kinloch, I. A., & Bissett, M. A. (2021). Fabrication and Mechanical Performance of Graphene Nanoplatelet/Glass Fiber Reinforced Polymer Hybrid Composites. *Frontiers in Materials, 8*, 773343. DOI: 10.3389/fmats.2021.773343

Ye, J., Ji, X., Liu, Z., Liu, K., Li, J., Wang, R., Wang, J., & Lei, Q. (2024). Carbon fiber reinforced structural battery composites: Progress and challenges toward industrial application. *Composites. Part B, Engineering, 277*, 111411. DOI: 10.1016/j.compositesb.2024.111411

Yee, K., & Ghayesh, M. H. (2023a). A review on the mechanics of graphene nanoplatelets reinforced structures. *International Journal of Engineering Science, 186*, 103831. DOI: 10.1016/j.ijengsci.2023.103831

Yee, K., & Ghayesh, M. H. (2023b). A review on the mechanics of graphene nanoplatelets reinforced structures. *International Journal of Engineering Science, 186*, 103831. DOI: 10.1016/j.ijengsci.2023.103831

Yu, K., Wang, M., Qian, K., Lu, X., & Sun, J. (2016). The synergy effect of Graphene/SiO2 hybrid materials on reinforcing and toughening epoxy resin. *Fibers and Polymers, 17*(3), 453–459. DOI: 10.1007/s12221-016-5862-8

Yudhanto, A., Lubineau, G., Ventura, I. A., Watanabe, N., Iwahori, Y., & Hoshi, H. (2015). Damage characteristics in 3D stitched composites with various stitch parameters under in-plane tension. *Composites. Part A, Applied Science and Manufacturing, 71*, 17–31. DOI: 10.1016/j.compositesa.2014.12.012

Zeinedini, A. (2024). Fracture toughness of graphene/polymer nanocomposites: Well dispersion, agglomeration and toughening mechanisms. *Theoretical and Applied Fracture Mechanics, 131*, 104449. DOI: 10.1016/j.tafmec.2024.104449

ZEROe: our hydrogen-powered aircraft | Airbus. Retrieved June 25, 2025, from https://www.airbus.com/en/innovation/energy-transition/hydrogen/zeroe-our-hydrogen-powered-aircraft

Zhang, B., Asmatulu, R., Soltani, S. A., Le, L. N., & Kumar, S. S. A. (2014). Mechanical and thermal properties of hierarchical composites enhanced by pristine graphene and graphene oxide nanoinclusions. *Journal of Applied Polymer Science, 131*(19), app.40826. Advance online publication. DOI: 10.1002/app.40826

Zhang, X., Shi, C., Liu, E., He, F., Ma, L., Li, Q., Li, J., Bacsa, W., Zhao, N., & He, C. (2017). Achieving high strength and high ductility in metal matrix composites reinforced with a discontinuous three-dimensional graphene-like network. *Nanoscale, 9*(33), 11929–11938. DOI: 10.1039/C6NR07335B

Zhou, C., Wu, S., Yang, B., Gao, Y., Wu, G., & Zhang, H. (2014). Toughening Polystyrene by Core-shell Rubber Particles: Analysis of the Internal Structure and Properties. *Polymers & Polymer Composites, 23*(5), 317–324. DOI: 10.1177/096739111502300505

Zhou, L., Qiu, J., Wang, X., Wang, H., Wang, Z., Fang, D., & Li, Z. (2020). Mechanical and dielectric properties of reduced graphene oxide nanosheets/alumina composite ceramics. *Ceramics International, 46*(12), 19731–19737. DOI: 10.1016/j.ceramint.2020.04.293

Zhu, D. (2018). *Aerospace Ceramic Materials: Thermal, Environmental Barrier Coatings and SiC/SiC Ceramic Matrix Composites for Turbine Engine Applications*. http://www.sti.nasa.gov

Zotti, A., Zuppolini, S., Borriello, A., Vinti, V., Trinchillo, L., Borrelli, D., Caraviello, A., & Zarrelli, M. (2022). Effect of the Mixing Technique of Graphene Nanoplatelets and Graphene Nanofibers on Fracture Toughness of Epoxy Based Nanocomposites and Composites. *Polymers 2022, Vol. 14, Page 5105, 14*(23), 5105. DOI: 10.3390/polym14235105

Chapter 4
Advanced Electrical Behavior of Graphene Nanostructures for Next-Generation Aviation

J. Mangaiyarkkarasi
https://orcid.org/0000-0003-1431-9584
Nadar Mahajana Sangam S. Vellaichamy Nadar College, India

J. Shanthalakshmi Revathy
https://orcid.org/0000-0003-1724-7117
Velammal College of Engineering and Technology, India

ABSTRACT

The emergence of next-generation aviation technologies necessitates materials that combine lightweight properties with exceptional performance and multifunctionality. Graphene-based nanostructures have gained significant attention due to their outstanding electrical, thermal, and mechanical characteristics. This chapter highlights the advanced electrical behavior of graphene and its derivatives such as graphene oxide (GO), reduced graphene oxide (rGO), and graphene quantum dots (GQDs) and their transformative role in aerospace. These materials offer tailored electrical properties suitable for applications including electromagnetic interference (EMI) shielding, lightning strike protection, energy storage, and sensing. Key parameters like layer number, defect density, and functionalization significantly influence their electrical performance. The chapter explores graphene's integration into supercapacitors, batteries, coatings, and structural health monitoring systems, all critical for enhancing safety, efficiency, and sustainability in aviation.

DOI: 10.4018/979-8-3373-3862-0.ch004

1. INTRODUCTION

The aerospace sector is facing the edge of a technological revolution being necessitated by the need to have lighter, faster, as well as energy-efficient aircraft. The very core of this development is the necessity of new high-performance materials that are able to sustain high-performance electronic systems in the most severe operation conditions. One of the most promising materials is graphene, a two-dimensional allotrope of carbon that is famous because of its exceptional electrical, mechanical, and thermal properties. (Wu et al., 2021). Over the past few years, nanostructures of graphene, such as nanoribbons, quantum dots, and heterostructures, have gained novel opportunities in the design of the new-generation aviation technologies. The inherent electron mobility, high conductivity, and programmed electronic characteristics of graphene have made it a perfect material in a variety of aerospace applications, including lightweight avionics, electromagnetic shielding, energy storage, and high-sensitivity sensors. The possibility to control graphene on the nanoscale enables the researcher to customize its electrical characteristics to fit the demanding needs of advanced flight systems. (Sabet, 2024). With the shift to full electric propulsion, autonomous systems, as well as more and more integrated onboard electronics in aviation, the role of graphene nanostructures is even more essential. This chapter explores the electricity of made graphene nanostructures and how it can revolutionize next-generation aviation. Starting with the basic principles of how electrical transport can occur in the graphene-based materials, we then discuss the nanostructure engineering methodology that allows a tailor-made functionality. The aspect of graphene introduction into practical aerospace is discussed, and the focus is on the work of the material under conditions of high-load flight. (Ayyanar & Pramanik, 2023). Lastly, we discuss obstacles and future prospects of implementing the use of graphene-powered electronics in aviation.

2. LITERATURE SURVEY

The seminal work by Geim and Novoselov in the early 2000s initiated a surge of research into graphene, highlighting its extraordinary electrical conductivity and carrier mobility. Recent research has been aimed at the nanoscale modification of the electronic characteristics of graphene. It has been shown that nanoribbons and quantum dots of graphene have size-dependent changes in the size of band gaps, which allow tunable semiconducting phenomena (Yu et al., 2022). Indicatively, a number of studies indicate that the development of edge states and controlled doping can greatly improve the carrier mobility and induce novel transport effects, including quantum confinement effects. The phenomena are essential in the establishment of sensors and electronic parts that are supposed to perform with high reliability even in extreme flight conditions. Furthermore, according to the literature, heterostructures between graphene and other two-dimensional structures (hexagonal boron nitride) increase electronic performance even more. These multi-layered composites are found to enhance electrical maturity and thermal control, a property that has a direct positive effect in aviation systems where weight decrease and heat exchange are vital (Jilani et al., 2020). It has been reported that the performance and power efficiency of field-effect transistors have improved, and the need to include integrated graphene-based electronic systems in these aircraft of the next generation is grounded. Application-wise, much focus has been on the use of graphene nanostructures in avionics. Research has shown that one of the possible applications of graphene-based materials is to create ultra-lightweight, high-performance products, including electromagnetic interference (EMI) shields, sensitive pressure sensors, or even energy storage device components (Panghal et al., 2025). These developments are based on a lot of experimental evidence of the conduct of graphene in the temperature variations and radiation conditions found in aerospace applications. The high resilience of graphene at such a condition is often contrasted with other conventional materials, where it is generally agreed that the superior electrical characteristics of graphene can be applied to create performance improvements in a big way. Computational models have also played a role in determining the electrical behavior of graphene nanostructures under different conditions. Other (but not all) computational methods have shown the impact of defects, functionalization, and strain on the electronic structure of graphene, such as density functional theory (DFT) simulations. Such models have undergone experimental research, which can help better understand the electron transport processes and establish the path to the rational design of graphene-based systems to meet the aviation requirements (Feni et al., 2023). These interconnections of these theoretical, experimental, and applied works point to the radical possibilities of graphene nanostructures in aviation electronics. All the literature indicates that, as techniques of scalability of manufacturing and integration processes continue to evolve, it is possible that in the near future, graphene may transcend limitations of current technology, and therefore, be able to enable next-generation, high-performing aerospace parts. This survey summarizes the development of studies on the electrical properties of graphene and how it can be used in the future to transform avionics with its promising performance, durability, and efficiency.

3. OVERVIEW OF GRAPHENE PROPERTIES AND RELEVANCE TO AVIATION.

The prime properties of graphene and their importance and relevance to aviation are listed in Table 1.

Table 1. Graphene properties and their relevance

Graphene Property	Description	Relevance to Aviation
Exceptional Mechanical Strength	~200 times stronger than steel by weight	Enhances structural strength while reducing weight, improving fuel efficiency
Ultralight and Thin	One-atom-thick, lightweight 2D material	Ideal for light aircraft components and composite materials
High Electrical Conductivity	Better conductor than copper	Supports advanced avionics, de-icing systems, and lightning strike protection
Superior Thermal Conductivity	Excellent heat dissipation	Improves cooling in electronics and engine components
Corrosion Resistance	Chemically stable and impermeable	Acts as a protective coating to extend the lifespan of aircraft parts
Flexibility and Elasticity	Maintains strength even when flexible	Useful for smart skins, flexible sensors, and adaptive structures
Transparency	Nearly transparent in single layers	Applicable in cockpit displays, touchscreens, and transparent antennas
Lightweight Energy Storage	Enables high-capacity, lightweight batteries and supercapacitors	Vital for electric and hybrid aircraft propulsion systems
EMI Shielding	Shields against electromagnetic interference	Protects avionics and communication systems
3D Printing Compatibility	Can be used in printable inks for additive manufacturing	Enables on-demand part production and lightweight custom designs

4. ELECTRICAL TRANSPORT MECHANISMS IN GRAPHENE-BASED MATERIALS

Graphene, a two-dimensional honeycomb structure made of carbon atoms, demonstrates exceptional electrical characteristics, which make it one of the potential next-generation electronic and optoelectronic devices. (Kausar et al., 2023). Its distinct band structure, in which linear dispersion of energy around the Dirac points, gives rise to a number of unusual and highly controllable electrical transport processes. The knowledge of these mechanisms is required to achieve the successful implementation of graphene in devices like sensors, transistors, and conductive composites. There are graphene batteries depicted in Figure 1.

Figure 1. Graphene batteries

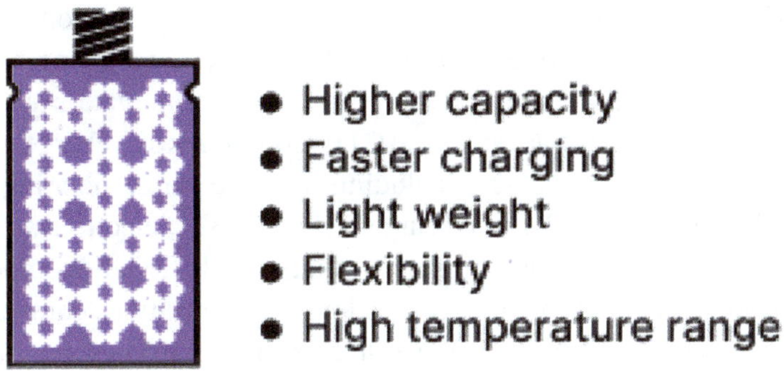

Ballistic Transport: It is one of the most remarkable characteristics of graphene that is capable of supporting ballistic transport at relatively long distances at room temperature. Charge carriers (electrons or holes) do not scatter in the material in ballistic transport when the dimensions of the sample are shorter than the mean free path of the carriers. (Bilisik & Akter, 2022). The effect of this is very low electrical resistance and high carrier mobility - values of more than 200,000 cm 2 / V s have been reported in ideal conditions. This is explained by the fact that graphene is a defect-free lattice and there is no bandgap, which enables electrons to move freely by conventional scattering without any impediment that occurs in traditional semiconductors.

Quantum Tunneling and Klein Tunneling: Graphene charge carriers are massless Dirac fermions, and therefore, they exhibit exotic quantum behavior, including Klein tunneling, a relativistic effect in which electrons are able to tunnel through potential barriers without bouncing back. In contrast to the properties of classical materials, when an electron is reflected by a barrier, in graphene, a sharp and sharp barrier and a normal incidence angle, the electron is perfectly reflected. (Marconcini & Macucci, 2022). This property is important in the design of quantum devices and tunneling transistors based on graphene that can have potential in ultra-fast switching applications.

Diffusive Transport: In more realistic, practical applications, which put graphene onto a substrate or form a part of a composite, imperfections like impurities, grain boundaries, and defects provide centers of scattering. This causes diffusive transport in which the electrons go through a number of scattering events. Nevertheless, even this does not stop graphene from retaining relatively high conductivity because of the delocalized system of π-electrons that enables the efficient transfer of charge. Factors affecting the electrical conductivity under diffusive conditions are the concentration of the carriers, the density of defects, and temperature.

Hopping Transport in Graphene Derivatives: In chemically modified graphene substances, e.g., graphene oxide (GO) as well as reduced graphene oxide (rGO), oxygen-containing functionalities interfere with the π-conjugation and introduce localized states (Eayal Awwad et al., 2021). Consequently, charge transport is through purely hopping processes, including variable range hopping (VRH) or nearest-neighbor hopping (NNH). During such processes, electrons are known to jump between localized states over distances that may differ with temperature and disorder. In spite of the fact that this causes low conductivity as compared to pure graphene, it offers tunability to be applied in flexible electronics and biosensors.

Percolation in Graphene-Based Composites: In a graphene-polymer or graphene-metal composite, transport of electrons is determined by the percolation threshold, the lowest concentration of conductive sheets of graphene required to create a continuous conductive network. The material is insulator-like below the threshold and is conductive like above the threshold. (F. Xu et al., 2022). These composites are frequently used in transport, which combines tunneling between sheets, hopping, and short percolative paths, so that they are useful in antistatic coating, electromagnetic shielding, and flex circuits.

Effects of temperature and Magnetic Field: The transport of graphene is strongly affected by temperature. The quantum effects, or weak localization, or universal conductance fluctuations, might dominate at low temperatures, and thermal activation becomes more important at high temperatures. In the same vein, the fact that a magnetic field applied causes one to observe properties such as the quantum Hall effect, in which the conductivity is quantized, once again testifies to the distinctness of the charge carriers in graphene.

In graphene-based materials, electrical transport is a complicated combination of multiple processes, including ballistic conduction in pure graphene, hopping transport, and percolative transport in composite and derivatives. These varied actions can provide enormous opportunities for the customization of the characteristics of graphene to particular electronic, sensing, and energy-related purposes. (Sharon et al., 2019). Further studies of these mechanisms will advance the development of various industries based on graphene technologies.

5. ENGINEERING GRAPHENE NANOSTRUCTURES FOR TUNABLE ELECTRICAL BEHAVIOR

The electrical properties of graphene that have made it the greatest include a high carrier mobility, ambipolar field effect, ballistic transport due to the unique two-dimensional structure, and Dirac cone dispersion of energy. Nonetheless, pure graphene does not have a bandgap and hence it can only be used in digital electronics to a limited extent. As a solution to this, scientists have come up with a range of ways of creating graphene nanostructures with controllable electrical behavior. Some of the best strategies include doping, edge states manipulation, and heterostructure design.

5.1 Role of Doping

Doping is the deliberate addition of contaminants or other atoms in the lattice of graphene to control its electronic character (Ramezani & Rahmani, 2024). This may be done through chemical doping, electrostatic gating, or substrate interactions. The Fermi level of graphene can be shifted by the introduction of either electron-donating (i.e., alkali metals) or electron-withdrawing (i.e., halogen) species to give n-type or p-type behavior. This tuning plays an important role in the design of graphene-based diodes and transistors. In bandgap engineering, although undoped graphene is gapless, the creation of a small bandgap is possible under some conditions, particularly in the presence of other such modifications as strain or substrate patterning. This makes graphene more appropriate for digital switching. The doped graphene is also more environmentally sensitive, which also works in their favor when it comes to use in chemical and biological sensors, where a slight variation of the electrical conductivity will show the presence of the target molecules. Therefore, doping is a potent means of controlling the conductivity of graphene, its carrier density, and even the bandgap of this substance, which is invaluable in the creation of nanoelectronic devices (Bhattacharya et al., 2019).

5.2 Edge States in Graphene Nanostructures

Nanostructures of graphene, Nanoribbons, and quantum dots. Nanostructures of graphene exhibit edge-dependent quantum confined electrical properties based on the nature of the carbon edges and the quantum confinement. The two major forms of edges in graphene, zigzag and armchair, are characterized by different electronic properties. Zigzag edges contain localized edge states that are magnetic in nature and have a large density of states at the Fermi level, which could render them metallic. On the other hand, the edges of armchairs can be either metallic or semiconducting based on the ribbon width. Figure 2 depicts Zig Zag edges.

Figure 2. Zig-Zag edges.

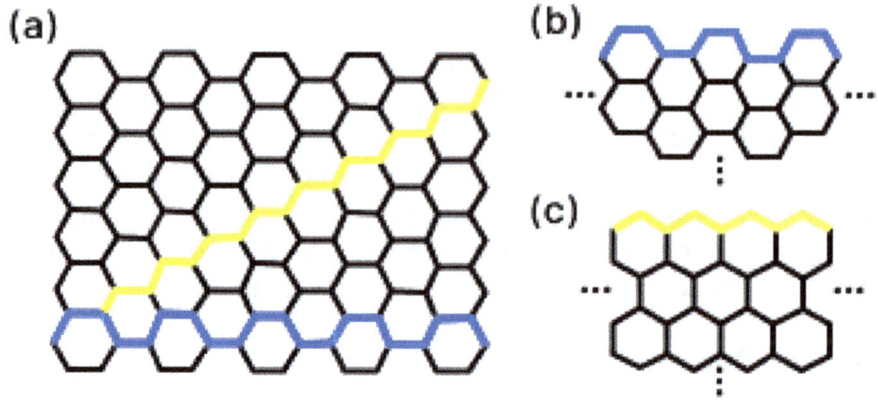

Quantum confinement: As a scale-down to nanoribbons is done, the electrons are confined laterally, resulting in the appearance of a bandgap which becomes wider as the ribbon is narrow. This can be used to fine-tune the electronic band structure to suit certain purposes.

Edge functionalization: The electrical response can also be further customized by chemical modification of the edge atoms. As an example, the local electronic density can be altered by hydrogenation or oxidation and affect the behavior of edge states. The geometry and chemistry of edges are paramount in the fabrication of graphene-based transistors and spintronic devices, and the localized edge states are functional.

Edge-shaped nanostructures are a category of nanomaterials that have sharp edges, corners, or boundaries at the atomic or molecular scale. These edge-dominated nanostructures, unlike flat or smooth nanostructures, have distinct physical and chemical characteristics that render them extremely useful in a broad variety of applications. Such structures are common in nanoplates, nanosheets, nanocubes, nanorods, and nanoprisms. An example of such materials is graphene and transition metal dichalcogenides (TMDs) such as $MoS_{2,}$ which typically display reactive edges that have a large impact on their behavior (Jiang et al., 2025). The growth and shape of edge-shaped nanostructures must be precisely controlled, which may be done by such techniques as chemical vapor deposition (CVD), hydrothermal or solvothermal synthesis, seed-mediated growth, as well as exfoliation. All of these techniques enable the researcher to adjust the size, thickness, and edge structure of the final nanomaterials. As an example, 2D materials with edge terminations can be grown by CVD, and well-faceted nanocrystals such as gold nanoprisms can be growth by seed-mediated growth. The incredibly high surface energy and reactivity of the edge-shaped nanostructures are also among the most astonishing facts due to the occurrence of under-coordinated atoms at the edges. These locations have better catalytic activity than atoms on flat surfaces. All this is in the case of materials such as MoS_2, where the edge sites have been identified to be the catalysts behind the catalytic reaction of the hydrogen evolution reaction (HER), and are useful in energy conversion. Equally, the platinum nanoparticle with sharp edges has shown better results in fuel cell catalysis owing to enhanced concentration of active sites. Besides catalysis, the edge-shaped nanostructures have unique electronic and optical properties. If edge states exist, the electronic band structure can be adjusted, and this can affect the conductivity and mobility of charge carriers. In 2D materials, the edge effects can oftentimes lead to quantum confinement, which changes the absorption and emission characteristics of the material. These are very desirable properties of optoelectronic devices and sensors. Edge sites in sensing applications offer higher sensing binding sites to analytes than other binding sites, thereby increasing sensitivity and selectivity (Polynskaya et al., 2022). The higher surface area and high capacity of the charge transfer of the edges result in enhanced functioning of detecting gases, biomolecules, and environmental contaminants. In addition, the nanostructures in the form of edges have been used in storing energy, like in batteries and supercapacitors. The concentration of active sites in the edges also enhances quicker ion diffusion and redox reaction rate, which enhances charge/discharge rate, as well as the efficiency. Electrode materials, such as those made of edge-rich graphene and TMDs, are highly useful due to their conductivity and structural integrity. Although they have potential, edge-shaped nanostructures have several difficulties. It is technically challenging to synthesize controlled edge orientation and density materials. Also, the stability of the structural of these edges is a matter of concern since in the course of operation, it is likely to be degraded or oxidized. The description of edge structures on the atomic scale is also challenging and needs more complex microscopy methods, e.g., transmission electron microscopy (TEM) or scanning tunneling microscopy (STM). In the future, edge-shaped nanostructures have a bright future. The development of computing techniques, such as machine learning, will be useful in designing and predicting optimum edge configurations in particular applications.

Moreover, approaches to selective functionalization of site of edges are currently being developed, allowing to be more precise fabrication of sensors and catalysis. (Lebedeva et al., 2024). The edge-shaped nanostructures will be transformative in the next-generation technologies, with the incorporation of such materials into flexible and wearable electronics. Further study to enhance synthesis, stability, and scalability will be important to realize their potential.

5.3 Heterostructure Design

Graphene can be employed together with other two-dimensional materials to form van der Waals heterostructures, thus enabling a fine-tuned control of charge flow, interband alignment, and interlayer interactions. (Nikfar et al., 2022). Figure 3 shows van der Waals heterostructures.

Figure 3. Vander Waals heterostructures

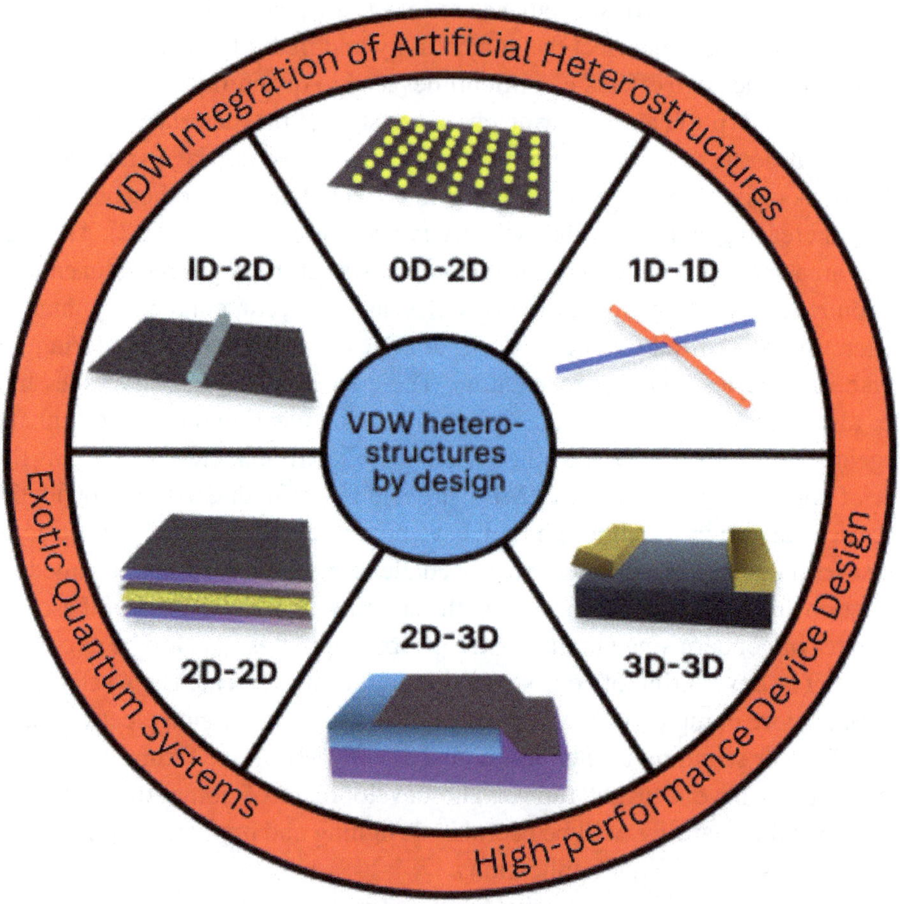

Layered 2D materials: Layered materials, including hexagonal boron nitride (hBN), molybdenum disulfide (MoS_2), or transition metal dichalcogenides (TMDs), become the object of study when overlaid onto graphene; the new 2D material becomes subject to the influence of interlayer tunneling, charge transfer, and moire superlattices. These effects make the designing of systems with customized band gaps and new optoelectronic characteristics possible.

Tunnel junctions and superlattices: Heterostructures can be fabricated as resonant tunneling diodes or quantum well, with conductive channels formed by layers of graphene and carrier control by layers of carrier barriers.

Twistronics: Superconductivity and Mott insulating phases are instances of novel quantum phases in which layers of graphene are stacked at magic angles (e.g., magic-angle twisted bilayer graphene). This introduces possibilities that had never been imagined with regard to the design of quantum devices.

Heterostructures allow scientists to now control the electronic properties of graphene with mathematical precision and come up with new devices that are well beyond the capabilities of monolayer graphene alone. This ability to tune the electrical characteristics of graphene through doping, engineering of edge-states, and heterostructure architecture has expanded its areas of application in electronics, optoelectronics, and quantum technologies. It is these classes of engineering that make the present-day research of graphene important to enable the customization of the functionality of graphene to specific applications in nanoelectronics and other applications.

6. INTEGRATION OF GRAPHENE INTO AVIONICS AND AEROSPACE SYSTEMS

The remarkable physical and chemical characteristics of graphene, which include high electrical conductivity, mechanical strength, low density, and thermal stability, make it a perfect choice in a number of aerospace applications. Graphene is increasingly being incorporated into the avionics and aerospace systems, particularly in the conductive composites, sensorial devices, and future generation energy storage systems. These applications are intended to improve the performance, safety, efficiency, and weight reduction in aircraft and spacecraft systems.

6.1 Conductive Composites

Weight reduction is one of the priorities of aerospace engineering. Graphene-based conductive composite reinforced by graphene is a highly potent solution based on its elevated strength-to-weight ratios and multi-functional characteristics (Zarepour et al., 2023).

Structural and electrical integration: Graphene, once included in polymer or carbon fiber composites, not only enhances mechanical strength and fatigue resistance, but also creates electrical conductivity. This dualism can be applied in the production of materials that can be used as structural components and as the electrical systems of the aircraft. Graphene has a high mechanical strength, electrical conductivity, thermal stability, and lightweight, which makes it a multidisciplinary process to integrate its structural and electrical properties into avionics and aerospace systems. This is done through a meticulous design of materials, engineering of compatibility with aerospace components, and functional adaptation to withstand extreme conditions that are experienced in aerospace environments. Here is an elaboration on the way this integration is implemented:

6.1.1 Material Fabrication and Formulation

Most frequently, graphene is produced through chemical vapor deposition (CVD), mechanical or liquid-phase exfoliation, or reduction of graphene oxide (GO). A quality monolayer of few-layer films of graphene is more desirable when used in aerospace applications because they have high electrical and structural properties on a large scale. Graphene is converted into various forms, which may include films, coatings, composites, or foams, depending on the application, to meet the aerospace component it is to be incorporated into.

6.1.2 Integration into Structural Components

Graphene is also being employed to reinforce structural composites, especially carbon fiber reinforced polymer (CFRPs) very popular in aircraft fuselages, wings, and other repositories. With the inclusion of a graphene nanosheet or graphene oxide in a small proportion of the composite matrix, mechanical strength, stiffness as well and fracture toughness can be boosted tremendously. (Kanti et al., 2025). This is done by a homogeneous interspersion of the graphene with the polymer resin and by a good interfacial bonding with the matrix, typically by functionalizing the surface. In addition to this, the composites reinforced by graphene have better damage tolerance and fatigue strength, which is crucial to the long-term stability of aerospace structures. The materials can also be used to lightweight the overall aircraft or spacecraft without affecting the strength, which increases the fuel efficiency.

6.1.3 Electrical and Thermal Integration

Graphene has one of the most notable advantages in that it has remarkable electrical conductivity. Graphene films and coatings are incorporated in many other electronic systems of the aircraft to act as lightning strike shields, electromagnetic interference (EMI) shields, and thermal conductors. In the lightning strike protection, a one-mole-thin conductive film of graphene is sprayed on the exterior composite skin of the aircraft. This layer assists in the dissipation of electrical charge very fast without causing damage to structures underneath. In the case of EMI shielding, graphene has high electron mobility that enables it to absorb electromagnetic wave frequencies and shield sensitive avionics equipment. Graphene is under consideration in avionics circuitry as a possible material in flexural and miniaturized antennas, sensors, interconnects, and transistors (J. Wang et al., 2019). The flexibility of electronics with graphene is the most beneficial to conformal surfaces in aircraft interiors and UAVs, where space and weight are of primary importance. Graphene can also be used in thermal integration. Graphene coatings and composites are applied to create high thermal conductivity to increase the heat dissipation in electronic modules and satellite components with high power and, consequently, increase the system reliability in high-temperature conditions.

6.1.4 Sensor and Monitoring Applications

The sensitivity of graphene to strain, pressure, gas, and temperature renders it the perfect sensor to be used in embedded sensors on aerospace structures. These sensors are made of graphene and incorporated in the airframe/engine components to monitor structural health in real time (SHM). These systems are able to identify microcracks, stress concentration, temperature change, or gas leakages, and predictive maintenance can be performed, and the operation of flights safer.

6.1.5 Integration Challenges and Solutions

However, there are a number of challenges associated with incorporating graphene in aerospace. These include:

Scalability: There is still a bottleneck in synthesizing large-scale, defect-free films of graphene to industrial standards.

Dispersion and compatibility: It is not easy to obtain uniform dispersion of graphene in composites without aggregation, and this influences mechanical and electrical properties.

Interfacing with traditional aerospace materials: Graphene will have to establish stable and durable interfaces with metals, ceramics, and polymers utilized in aerospace applications.

Regulatory and qualification hurdles: Aerospace materials have to be strongly tested and qualified, and graphene-based systems are in the initial phase of many applications.

To overcome such concerns, scientists are designing functionalized derivatives of graphene, superior techniques of nanocomposite processing, and hybrid material systems incorporating graphene with other nanomaterials. Additive manufacturing, or 3D printing based on graphene-infused inks, is also a new technique to allow complex geometries and parts with multiple functions. The new field of integrating graphene into avionics and aerospace systems joins the disciplines of advanced material science and aerospace engineering. Structurally, it augments the composite materials with regard to their mechanical performance and minimizes weight. Its technical advantages include lightning protection, EMI shielding, thermal control, and flexible electronics, which are only possible with electrical power. With the advancement of research and overcoming the difficulties in manufacturing, graphene is expected to be one of the cornerstone materials in the creation of new aerospace platforms of the future.

Lightning strike protection (LSP): Aircraft are susceptible to lightning strikes, which would need conductive materials to safely discharge the energy. The conventional metal meshes are heavy, whereas the graphene-based composite has lightweight products with similar or better applicability in conducting away electrical energy.

Electromagnetic interference (EMI) shielding: In avionics, a high level of shielding is necessitated for sensitive equipment against EMI. Composites fabricated using graphene also serve as efficient EMI shields that avoid inconvenience in the communication, navigation, and control systems without causing any major weight. The materials are particularly useful in advanced generation aircraft and unmanned aerial vehicles (UAVs), where the selection of material has a direct effect on performance and efficiency.

6.2 Graphene-Based Sensors

Graphene with high surface area, conductivity, and sensitivity to changes in the environment would thus be good for aerospace sensor technology.

Strain and pressure sensors: Flexible strain and pressure sensors made of graphene can be implanted in the aircraft structure and offer real-time strain, pressure, or deformation data. Early detection of micro-cracks or stress concentrations with these sensors can be used to predictive maintainability and minimize the threat of structural failures. (Ariaeenejad et al., 2021).

Gas and chemical sensors: Graphene technology is useful in the detection of low-concentration gases such as CO_2, O_2, NOx, and volatile organic compounds (VOCs) in cabin air quality monitoring and environmental sensing of aerospace missions.

Temperature and humidity sensors: Graphene is sensitive and thermally stable, so it can be used in high-performance thermal sensors, which operate quite effectively between the extreme temperatures found in aerospace use.

Radar and RF sensors: With its excellent radio frequency (RF) transparency and tunability, graphene is also being explored for stealth technologies and advanced radar systems, potentially enabling more compact and lightweight electronic packages. (Asmat et al., 2020).

The implementation of such sensors in airframes and systems can make the planes smart to carry out real-time diagnostics, enhance their safety, and efficiency. Graphene, which is a two-dimensional carbon material and has remarkable electrical, mechanical, and thermal characteristics, is undergoing a remarkable focus in the aerospace industry, especially in the creation of the next generation of sensors. The operating conditions are extreme in the aerospace environment: high altitude and temperature, high mechanical stress, which require not only a high sensitivity and precision of sensor systems but also lightweight, durable, and multifunctional sensor systems. Graphene, due to its peculiarities, has already shown itself as an outstanding material to meet these harsh demands (O'Neill et al., 2021). Among the strongest benefits of the next-generation sensors based on graphene is the high surface-to-volume ratio of the sensors, which significantly increases their sensitivity. The property enables the graphene sensors to detect even the trace levels of gases, chemicals, or even biological molecules and thus makes them suitable for environmental monitoring in aircraft cabins, fuel systems, and life-support systems in spacecraft. Indicatively, gas sensors using graphene have demonstrated excellent detection limits of dangerous substances, including carbon monoxide, volatile organic compounds, and ammonia, at parts-per-billion (ppb) concentration. They have fast response and recovery time, which is also useful in real-time monitoring. Besides sensing chemically, graphene has also been very efficient in mechanical sensing. Incorporated into the structural elements of an aircraft or a spacecraft, graphene-based strain and pressure sensors will be able to observe mechanical deformations, vibrations, and stress accumulations. These sensors offer important information to their structural health monitoring (SHM), where engineers can identify micro-cracks, delamination, or fatigue, prior to catastrophic failure. This predictive maintenance feature is essential for the provision of safety, minimization of downtime, and the longevity of the operational life of aerospace vehicles.

The outstanding electrical conductivity of graphene and its ability to respond to strain or temperature, or pressure, also enable it to be user-friendly in temperature and thermal sensors. Graphene-based thermal sensors are capable of recording rapid variations and giving a warning of heat overload or thermal stress in high-speed aircraft or spacecrafts (where the temperature gradients can be very harsh) (Tang et al., 2023). In other fields outside structural and environmental monitoring, graphene is also being considered in biosensing to aerospace medicine, and astronaut health monitoring. The biocompatibility of graphene and its selective binding with biomolecules enable it to serve such applications as biomarker detection of fatigue, stress, dehydration, or radiation exposure. These implantable or wearable biosensors would prove of paramount importance in keeping the astronauts healthy on long missions in space. The incorporation of graphene-based sensors in aerospace is also consistent with the tendency of the industry towards tiniaturisation and multipurposeness. The lightness and flexibility of graphene enable it to be used in the fabrication of flexible electronic skins or an array of sensors that can be integrated into the wing of an aircraft, fuselage, or spacecraft hull, and do not impose a large amount of mass on the structure. These sensors can withstand extreme environments, such as high vacuum, radiation, and drastic changes in temperature that conventional materials cannot withstand in such situations. Although this has some benefits, there are still issues with the mass application of graphene sensors in aerospace. These are scalable and reproducible production of high-quality graphene, long-term stability and endurance under operating stress, as well as the incorporation of graphene sensors into the currently used aerospace electronics and communication systems. However, these barriers are gradually being defeated by continuous research and technological innovations.

Sensors that are made of graphene are a new breakthrough in aerospace innovation. They have unparalleled sensitivity, flexibility, and resilience, which makes them suitable for a huge variety of aerospace sensing needs, such as environmental and structural monitoring, biomedical, and thermal sensing. (Yang et al., 2023). With the maturity of production methods and the enhancement of the mechanism of integration, the development of graphene sensors will be central in the development of intelligent, adaptive, and safer aerospace systems.

6.3 Energy Storage Devices

Efficient and lightweight energy storage solutions are increasingly becoming very significant in aerospace systems, particularly with the tendency to move towards electrification and hybrid propulsion. (H. Zhang et al., 2021). This is an area where graphene is transforming the world as it is applied in batteries and supercapacitors..

Graphene-enhanced batteries: Graphene can be used as an additive or anode material in lithium-ion batteries to enhance charge capacity, cycle life, and charge/discharge rates. These are essential to avionics backup systems, UAVs, and electric airplanes that need a small energy storage that is powerful.

Supercapacitors: Graphene supercapacitors provide advantages in speed of storing and releasing energy, having high power density and high thermal stability. They are best suited to the systems where rapid supply of energy is needed, i.e., emergency actuators, de-icing systems, or electric propulsion bursts.

Structural energy storage: Multi-functional materials structures are also being developed for aerospace. Structural elements built on graphene would be able to carry a load and be batteries or act as capacitors at the same time, which would decrease aircraft weight and complexity even more.

The use of graphene as a part of the energy storage system will enable aerospace platforms to be more efficient in energy consumption, lightweight, and durable, which is of great significance in space missions and prolonged flights of UAVs. The insertion of graphene into the avionics and aerospace systems will be a revolution towards a more lightweight, intelligent, and efficient aircraft and spacecraft. The use in conductive composites, sensors, and energy storage is already proving to have great advantages, and further research and development are bringing even greater use. With the development of materials science, graphene remains one of the foundations of future aerospace technology.

7. PERFORMANCE UNDER EXTREME AERONAUTICAL CONDITIONS

Graphene is a material that has a high potential in aerospace applications due to its ability to develop into a high electrically conductive material, high mechanical strength, and due to its weightless nature. Its operation under extreme aeronautical conditions, like operating at high altitudes, extreme temperatures, and radiation conditions, however, needs to be looked at closely. Graphene and graphene-based materials can be greatly affected by the conditions of their performance and thus improved to be integrated in the next generation of avionics and aerospace systems.

7.1 High-Altitude Effects on Electrical Performance

The environmental conditions at high altitudes, such as those experienced by aircraft, are quite different at sea level. The differences are capable of influencing the electrical characteristics of materials, including graphene.

Reduced atmospheric pressure: The atmospheric pressure is lower in high altitudes, and this may affect the charging properties of graphene-based supercapacitors and batteries. Although graphene is structurally strong to pressure variations because it is made of atoms, the components of energy storage systems or sensor devices could slightly change their performance. (Varghese et al., 2015).

Oxygen and moisture availability: The reduced amounts of oxygen at high altitudes may influence the oxidation of graphene. This oxidation may change the electronic characteristics of graphene and graphene oxide-based composites, which may result in the alteration of conductivity. As an example, the conductors or composites of graphene can have their resistance enhanced by oxidation at high altitudes unless it is properly shielded.

Vacuum effects in space missions: Space near vacuum conditions may result in outgassing or the release of trapped gases in materials. Although graphene is not very prone to this, the composite materials that contain graphene can be subject to physical degradation with time unless the composite design is directed to inhibit this occurrence.

7.2 Temperature Effects on Electrical Performance

One of the factors that makes graphene be considered in thermal management and electronics is that it has good thermal conductivity. Nevertheless, the maximum temperature limits of the electroless performance of graphene and graphene-based materials may be influenced by extreme temperature variations, such as the high temperature of atmospheric re-entry or the extreme low temperature of high-altitude conditions. (Chen et al., 2017).

Low-temperature behavior: When graphene is subjected to very low temperatures (such as in high-altitude or space environments), the electrical resistance is usually low because the charge carriers are highly mobile. However, with a lower temperature, certain graphene-based sensors or batteries may experience limited charge capacity or slower charge/discharge rates, especially when graphene is incorporated in composites or when it is mixed with other low-temperature-sensitive materials. The rigidity of the material can also be enhanced by the low temperatures, which may affect the functionality of the flexible graphene-based sensors or coating.

High-temperature behavior: When the temperature is high, e.g., in high velocity aerodynamic heating or spacecraft re-entry, this may be a challenge. Although graphene can resist temperatures up to approximately 1000 0 C in non-reactive environments, it can be oxidized and then degraded to lose its electric characteristics at high temperatures in the presence of oxygen or other reactive gases. (Chen et al., 2022). The resistivity of graphene may go up, and this will be the case with the conductivity of the composites or any electronics with graphene that will be exposed to such circumstances. Nevertheless, the high thermal conductivity of graphene can be used to cool down the heat, and with the correct engineering, it is also capable of operating in high temperatures relatively stable.

7.3 Radiation Effects on Electrical Performance

Aerospace systems, particularly those in low Earth orbit (LEO) or beyond, are exposed to cosmic radiation and solar radiation. The effects of these radiations may have a substantial impact on the performance of graphene, particularly in electronic and sensing schemes.

Ionizing radiation: The cosmic rays and solar particles are ionizing radiation that can interact with graphene, creating displacement damage in the lattice structure, in outer space. This may lead to modification of its electrical properties, including high resistance or low carrier mobility. The effects of radiation on the electronic properties of graphene are based on the dose of radiation, the thickness of the material, and the extent of defects or impurities in the graphene lattice. The formation of defects in the graphene structure when it is exposed to heavy ion radiation may enhance the electron scattering, hence impeding electrical performance. (R. Zhang et al., 2024).

Electromagnetic radiation and shielding: Graphene has the potential to block/ shield electromagnetic radiation, which is also a major benefit, particularly in the fabrication of radiation-resistant avionics and electronic gadgets. Nevertheless, prolonged exposure to strong radiation in space may create changes in the capacity of graphene to protect delicate avionics successfully. This can be alleviated by incorporating graphene with other materials in order to produce radiation-resistant composite materials with high shielding performance over time.

Radiation-induced conductivity changes: The effect of radiation on graphene is not homogeneous. Radiation may cause chemical changes to graphene oxide or functionalized graphene composites to change their electrical conductivity. As an illustration, radiation could also oxidize graphene or produce other electronic states, which affect the charge transport processes.

Graphene is a material with high versatility that has a high potential in the aerospace industry, although its functionality in the harsh aeronautical environment, including high altitudes, high temperatures, and radiation, should be critically assessed. Though graphene has proven to be a strong material in most of these conditions, its possible degradation modes, including oxidation in high altitude, changes in resistivity with temperature extremes, and radiation-induced defects, have to be taken into account when designing the material and integrating it into a system (Shamsaei et al., 2018). As more and more improvements are made on graphene engineering, such as the surface coatings, composite material structures, and even radiation hardened graphene structures, the incorporation of graphene into aerospace and avionics systems can be engineered to be performance and long life even in the most adverse environments.

8. ENVIRONMENTAL IMPACT ASSESSMENTS AND REGULATORY STUDIES

Reasons for environmental and regulatory analysis matter for graphene in aviation.

Graphene parts have the potential to save a lot of energy and weight in airframes, thermal, and electrical/avionics subsystems, which have the potential to lower fuel burn and resultant operational emissions. (Ramezani & Rahmani, 2024). Yet, these advantages should be balanced with the environmental costs of the graphene production process, occupational and environmental exposure risks in the course of production/use/maintenance, and the changing regulation of the nanomaterials, which influence commercialization and certification. (L. Wang, 2024).

Life-cycle assessments (LCAs) exhibit a large range of environmental footprint variability based on production route (chemical exfoliation, CVD, laser-induced routes, biomass-derived routes). Most of the methods are dominated by greenhouse gas, water, and toxicity footprints by energy use and precursor chemicals; even some emerging routes (biomass-derived, laser-induced graphene) have reduced impacts in initial studies, but need validation on a massive scale.

Use-phase benefits can offset production impacts (but context matters)

Examples of savings in the use-phase of aviation (fuel, maintenance) include weight reduction, enhanced electrical/thermal operation, and multifunctional integration. LCAs with the complete aircraft system indicate that even those materials with greater impacts of production can be helpful in total when they allow valuable operational efficiency improvements, but outcomes strongly depend on the substituting design and realistic service lifetimes.(Sabet, 2024).

End-of-life, recycling, and persistence

The graphene-containing composites do not have well-defined end-of-life pathways (recycling, incineration, landfill). Such a small form of the particle (i.e., fragments, nanoplatelets) can remain in place in graphene and can alter thermal/combustion behavior or particle release during mechanical recycling or crash/maintenance processes. Data gaps remain large.

Human health and ecotoxicology

It has been reviewed that some forms of graphene and functionalization may cause pulmonary toxicity, oxidative stress, and ecotoxic effects on aquatic organisms, but the risk differs substantially depending on particle size, chemistry (oxide vs pristine graphene), dose, and route of exposure. Strong exposure assessment is needed to convert risk hazard results to actual risk to the workers, maintainers, and the environment.

9. Graphene's role in communication systems and sensors, which are emerging applications in aerospace

The combination of electrical, mechanical, and optical characteristics has made graphene a material gaining recognition as one of the potential transformative materials in next-generation aerospace communication systems and sensor platforms. Graphene has very high carrier mobility, low resistivity and tunable band structure, therefore it is an interest candidate device of high-frequency communication devices that require high speed transmission of signals with minimum losses. Graphene-based antennas and transceivers are under consideration in satellite and aircraft communications as the ability to work with millimeter-wave and terahertz frequencies, which are critical to fast data transmission, secure communications, and advanced radar systems (Hirankittiwong et al., 2025). In contrast to traditional metallic antennas, the conductivity of graphene can be dynamically gated with electrostatic gate or chemically functionalized to allow reconfigurable and miniaturized antenna designs to be developed based on the size-constrained design space of aerospace platforms. Similar to communications, graphene also has great potential in the aerospace sensing technologies. Graphene-based sensors can record gases, biomolecules, and physical stimuli in very low concentration due to its high surface-to-volume ratio, a high level of sensitivity to perturbations of charge, and its functionalization with molecular receptors. This is translated into lightweight and versatile sensors in the aerospace industry that can deliver real-time measurements on the quality of air in the cabin, fuel leakage, or structural health of their airframes and composite material. Moreover, optical transparency and plasmonic response at terahertz and infrared frequencies can also be exploited in graphene to create new photonic sensing and imaging devices as well as to incorporate new adaptive coatings to stealth and thermal management in the avionics. Critically, these graphene-supported devices not only hold better performance, but also respond to the wider aerospace need to become lighter and more reliable since their thin, flexible and mechanically tough form factors are easy to add in the existing systems (S. Xu, 2024). Ongoing research also also point out the integration of graphene sensors and wireless communication nodes that will lead to fully networked aircraft and spacecrafts that will be able to perform autonomous monitoring and diagnostics. These benefits notwithstanding, a number of obstacles are still on the way to the large-scale adoption of graphene, such as scalable large-area graphene production, stability over time in aerospace environments, and meeting of high electromagnetic compatibility and safety requirements. However, the overlapping of graphene communication and sensing devices is one of the most promising frontiers to aerospace innovation, as it can change the way aircraft and spacecraft communicate and interact with the environment and keep their missions safe.

9. FUTURE PROSPECTS AND CHALLENGES IN AEROSPACE GRAPHENE ELECTRONICS

Scalability Challenges

Lab vs. Industry Scale: Graphene of high quality can be produced with a lot of ease in laboratories but it is very hard to reproduce in large scale and in industry.

Uniformity Issues: Aerospace is a process where consistency in the quality of the material is needed and this is not feasible in the current scalable production processes.

Emerging Solutions: Roll-to-roll manufacturing and chemical vapor deposition (CVD) processes are under development to be applied on a large scale but they require enhancement in cost-effectiveness and quality management.

Manufacturing Integration

Compatibility Concerns: Graphene has to be made a part of the conventional aerospace elements, and this demands significant changes in the current manufacturing procedures.

New Materials Development: Innovations like graphene-based inks, films, and composites are in progress to make integration smoother.

Environmental Durability: Aerospace applications require materials with the ability to work under extreme conditions (temperature, pressure, radiation), and graphene-based components should be put through the test regarding these conditions.

Regulatory and Safety Considerations

Lack of Standards: Graphene is a fairly emerging material and is yet to attain complete regulatory and safety standards within the aerospace industry.

Certification Needs: There is a strong demand for standardized testing and certification processes to verify the safety, reliability, and performance of graphene-based electronics.

Risk Aversion: The risk-aversion in the aerospace industry is very great since the industry cares about safety, and, therefore, it slows down before full validation is done.

Future Prospects

Ongoing Research: Academic-industrial cooperation is overcoming scalability and integration challenges.

Growing Investment: More investment in nanotechnology and aerospace innovation is facilitating the emergence of practical use (Cui et al., 2023).

Potential Impact: An effective implementation of graphene may result in less heavy, more efficient in consumption and more durable aerospace systems, such as, advanced satellite electronics, high-performance avionics, new sensors, and communication systems

Although the issues of scalability, manufacturing, and regulation are quite severe, potential application of graphene in aerospace is quite great. These challenges can be crossed with further innovation and partnerships and a technological breakthrough in aerospace electronics can be achieved.

10. CONCLUSION

Graphene nanostructures have founded their place as a game changer in the next generation aviation that presents revolutionary progress in electrical characteristics, devices performance, and material integration. In this chapter, we have discussed the extraordinary electrical conducting qualities, large carrier mobility, tunable bandgap applications, and excellent thermal and mechanical stability of graphene all which make it a candidate material in the next generation aerospace applications. The integration of graphene in the aviation sector is a fresh horizon in the technology of producing high effective, low-weight, and multi-capability electronic parts. Such applications as ultra-sensitive sensors, high-frequency transistors, flexible and low-powered electronics are no longer mere concepts; they are coming into reality with further advancement in nanostructuring, material synthesis and integration methods. The ability of graphene to increase the electromagnetic shielding, signal integrity, and high-temperature electronics is especially useful in the harsh conditions of aerospace deployment. Notwithstanding these promising prospects, they are accompanied by drawbacks such as the problem of mass production, material consistency, environmental friendliness, and that of aerospace-quality products. Nevertheless, these barriers are gradually overcome in the course of current studies and interdisciplinary work. As a result of dedicated innovation in the fabrications methods, materials engineering, and regulatory adaptation, moving an experimental stage to realistic aerospace systems is becoming more feasible. In the future, graphene nanostructures will dominate the re-invention of the electronics in aviation. Not only will their adoption open up new possibilities, but also spur sustainable, efficient, and high-performance designs both in the commercial and defense aviation industries. Graphene is one of the driving forces of the technological change as we approach smarter, safer, and more responsive aircraft systems. To sum up, the high-order electrical performance of graphene nanostructures can be seen as a radical breakthrough in the field of material science, which has immense effects on the future of aviation. This is a continuous process but the trend is evident graphene is transforming the skies of the future.

11. REFERENCES

Ariaeenejad, S., Motamedi, E., & Hosseini Salekdeh, G. (2021). Application of the immobilized enzyme on magnetic graphene oxide nano-carrier as a versatile bi-functional tool for efficient removal of dye from water. *Bioresource Technology, 319*, 124228. DOI: 10.1016/j.biortech.2020.124228 PMID: 33254455

Asmat, S., Husain, Q., Shoeb, M., & Mobin, M. (2020). Tailoring a robust nanozyme formulation based on surfactant stabilized lipase immobilized onto newly fabricated magnetic silica anchored graphene nanocomposite: Aggrandized stability and application. *Materials Science and Engineering C, 112*, 110883. DOI: 10.1016/j.msec.2020.110883 PMID: 32409040

Ayyanar, C., & Pramanik, S. (2023). Effect of graphene on self-healing performance of hydroxyapatite/polydimethysiloxane composites. *Materials and Manufacturing Processes, 38*(9), 1068–1080. DOI: 10.1080/10426914.2022.2146714

Bhattacharya, S., Agarwal, A. K., Prakash, O., & Singh, S. (Eds.). (2019). *Sensors for Automotive and Aerospace Applications*. Springer Singapore., DOI: 10.1007/978-981-13-3290-6

Bilisik, K., & Akter, M. (2022). Graphene nanocomposites: A review on processes, properties, and applications. *Journal of Industrial Textiles, 51*(3, suppl), 3718S–3766S. DOI: 10.1177/15280837211024252

Chen, L., Yang, Q., Yang, X., Liu, Z., & Song, Q. (2022). Numerical investigation of heat transfer performance of graphene-doped anti-/deicing component. *Thermal Science and Engineering Progress, 28*, 101098. DOI: 10.1016/j.tsep.2021.101098

Chen, L., Zhang, Y., & Wu, Q. (2017). Effect of Graphene Coating on the Heat Transfer Performance of a Composite Anti-/Deicing Component. *Coatings, 7*(10), 158. DOI: 10.3390/coatings7100158

Cui, X., Zhang, N., Huang, M., Gao, G., Liu, S., & Liu, C. (2023). Polyurethane-based nanocomposite film with thermal deicing capability and anti-erosion for wind turbine blades protection in extreme environments. *Journal of Materials Chemistry. A, Materials for Energy and Sustainability, 11*(44), 23844–23853. DOI: 10.1039/D3TA05595G

Eayal Awwad, K. Y., Yousif, B. F., Fallahnezhad, K., Saleh, K., & Zeng, X. (2021). Influence of graphene nanoplatelets on mechanical properties and adhesive wear performance of epoxy-based composites. *Friction, 9*(4), 856–875. DOI: 10.1007/s40544-020-0453-5

Feni, F., Jahan, M., Zhao, R., Li, G., Zhao, G.-L., & Mensah, P. F. (2023). Multifunctional composite with hybrid carbon fiber and carbonaceous coconut particle reinforcement. *Frontiers in Materials, 10*, 1278222. Advance online publication. DOI: 10.3389/fmats.2023.1278222

Hirankittiwong, P., Chomchok, T., Chakraborty, S., Prajongtat, P., Singh, D. P., Hongkarnjanakul, N., Channumsin, S., Ghosh, S., & Chattham, N. (2025). Advanced engineered nanostructures for aerospace technology: A review. *Results in Engineering, 26*, 105381. DOI: 10.1016/j.rineng.2025.105381

Jiang, W., Liu, Z.-F., Wu, Q.-P., & Xiao, X. (2025). Role of edge reconfiguration in generating corner states in Zigzag graphene nanoribbons. *Physica E, Low-Dimensional Systems and Nanostructures, 172*, 116260. DOI: 10.1016/j.physe.2025.116260

Jilani, A., Hussain, S. Z., Khan, A. A. P., Khan, A., Othman, M. H. D., & Ansari, M. O. (2020). Graphene-based material for self-healing: mechanism, synthesis, characteristics, and applications. In *Self-Healing Composite Materials* (pp. 163–175). Elsevier., DOI: 10.1016/B978-0-12-817354-1.00009-0

Kanti, P. K., Prashantha Kumar, H. G., Wanatasanappan, V. V., Kumar, A., & Regasa, M. B. (2025). Graphene's Frontier in aerospace: Current applications, challenges, and future directions for space engineering. *Nanoscale Advances, 7*(12), 3603–3618. DOI: 10.1039/D4NA00934G PMID: 40386820

Kausar, A., Ahmad, I., Eisa, M. H., & Maaza, M. (2023). Graphene Nanocomposites in Space Sector—Fundamentals and Advancements. *C, 9*(1), 29. DOI: 10.3390/c9010029

Lebedeva, I. V., Vyrko, S. A., Sinitsa, A. S., Ratkevich, S. V., Popov, A. M., Knizhnik, A. A., Poklonski, N. A., & Lozovik, Y. E. (2024). Magnetic and electronic properties of 1D hybrid nanoobjects composed of alternating polycyclic hydrocarbon regions and double carbon chains. *Computational & Theoretical Chemistry, 1241*, 114913. DOI: 10.1016/j.comptc.2024.114913

Marconcini, P., & Macucci, M. (2022). Transport Simulation of Graphene Devices with a Generic Potential in the Presence of an Orthogonal Magnetic Field. *Nanomaterials (Basel, Switzerland), 12*(7), 1087. DOI: 10.3390/nano12071087 PMID: 35407205

Nikfar, M., Taati, E., & Asghari, M. (2022). On the theoretical and molecular dynamic methods for natural frequencies of multilayer graphene nanosheets incorporating nonlocality and interlayer shear effects. *Mechanics of Advanced Materials and Structures, 29*(20), 2873–2883. DOI: 10.1080/15376494.2021.1880675

O'Neill, C., Johnson, M. B., DeArmond, D., Zhang, L., Alvarez, N., Shanov, V. N., & White, M. A. (2021). Thermal conductivity of 3-dimensional graphene papers. *Carbon Trends*, *4*, 100041. DOI: 10.1016/j.cartre.2021.100041

Panghal, A., Kumar, Y., Mishra, P. K., Mathur, A., & Srivastava, A. K. (2025). Graphene-Based Polymer Composites for Aerospace, Electronic, Energy, and Biomedical Applications. In *Multifunctional Materials* (pp. 81–104). Wiley., DOI: 10.1002/9781394234134.ch4

Polynskaya, Y. G., Lebedeva, I. V., Knizhnik, A. A., & Popov, A. M. (2022). Optimal model of semi-infinite graphene for ab initio calculations of reactions at graphene edges by the example of zigzag edge reconstruction. *Computational & Theoretical Chemistry*, *1214*, 113755. DOI: 10.1016/j.comptc.2022.113755

Ramezani, M. J., & Rahmani, O. (2024). A review of recent progress in the graphene syntheses and their applications. *Mechanics of Advanced Materials and Structures*, •••, 1–33. DOI: 10.1080/15376494.2024.2420911

Sabet, M. (2024). Unveiling advanced self-healing mechanisms in graphene polymer composites for next-generation applications in aerospace, automotive, and electronics. *Polymer-Plastics Technology and Materials*, *63*(15), 2032–2059. DOI: 10.1080/25740881.2024.2365282

Shamsaei, E., de Souza, F. B., Yao, X., Benhelal, E., Akbari, A., & Duan, W. (2018). Graphene-based nanosheets for stronger and more durable concrete: A review. *Construction & Building Materials*, *183*, 642–660. DOI: 10.1016/j.conbuildmat.2018.06.201

Sharon, M., Rodriguez, A. S. L., Sharon, C., & Gallardo, P. S. (Eds.). (2019). *Nanotechnology in the Defense Industry*. Wiley., DOI: 10.1002/9781119460503

Tang, C., Wang, Y., Li, Y., Zeng, S., Kong, L., Li, L., Sun, J., Zhu, M., & Deng, T. (2023). A review of graphene-based temperature sensors. *Microelectronic Engineering*, *278*, 112015. DOI: 10.1016/j.mee.2023.112015

Varghese, S. S., Lonkar, S., Singh, K. K., Swaminathan, S., & Abdala, A. (2015). Recent advances in graphene based gas sensors. *Sensors and Actuators. B, Chemical*, *218*, 160–183. DOI: 10.1016/j.snb.2015.04.062

Wang, J., Mu, X., & Sun, M. (2019). The Thermal, Electrical, and Thermoelectric Properties of Graphene Nanomaterials. *Nanomaterials (Basel, Switzerland)*, *9*(2), 218. DOI: 10.3390/nano9020218 PMID: 30736378

Wang, L. (2024). Achieving Electric Aviation with Advanced Battery Technology and Nanomaterials. *E3S Web of Conferences, 553*, 02015. DOI: 10.1051/e3sconf/202455302015

Wu, K., Niu, Y., Zhang, Y., Yong, Z., & Li, Q. (2021). Continuous growth of carbon nanotube films: From controllable synthesis to real applications. *Composites. Part A, Applied Science and Manufacturing, 144*, 106359. DOI: 10.1016/j.compositesa.2021.106359

Xu, F., Wu, X., Wang, H., Liu, H., & Ye, Z. (2022). Experimental research on moulding of graphene/PEKK composite powder by spark plasma sintering technology. *Applied Physics. A, Materials Science & Processing, 128*(2), 122. DOI: 10.1007/s00339-022-05262-0

Xu, S. (2024). Nanostructured Materials for Energy Storage in Future Electric Aircraft. *E3S Web of Conferences, 553*, 02021. DOI: 10.1051/e3sconf/202455302021

Yang, Y., Yin, Z., & Zhang, W. (2023). Graphene-based physical sensors for humidity, temperature and strain detections. *Microelectronic Engineering, 279*, 112055. DOI: 10.1016/j.mee.2023.112055

Yu, H., Chen, C., Sun, J., Zhang, H., Feng, Y., Qin, M., & Feng, W. (2022). Highly Thermally Conductive Polymer/Graphene Composites with Rapid Room-Temperature Self-Healing Capacity. *Nano-Micro Letters, 14*(1), 135. DOI: 10.1007/s40820-022-00882-w PMID: 35704244

Zarepour, A., Ahmadi, S., Rabiee, N., Zarrabi, A., & Iravani, S. (2023). Self-Healing MXene- and Graphene-Based Composites: Properties and Applications. *Nano-Micro Letters, 15*(1), 100. DOI: 10.1007/s40820-023-01074-w PMID: 37052734

Zhang, H., Yang, Y., Ren, D., Wang, L., & He, X. (2021). Graphite as anode materials: Fundamental mechanism, recent progress and advances. *Energy Storage Materials, 36*, 147–170. DOI: 10.1016/j.ensm.2020.12.027

Zhang, R., Long, Z., Long, G., Wang, J., Wang, X., Zhang, X., & Jiang, Y. (2024). Mechanism of graphene oxide concrete macro-micro properties evolution under large temperature difference freeze-thaw action. *Construction & Building Materials, 415*, 135019. DOI: 10.1016/j.conbuildmat.2024.135019

Chapter 5

Nano-Engineered Surface Coatings for Corrosion Resistance in UAV-Based Fog Dispersal Systems

S. Khalid
https://orcid.org/0009-0005-6894-0087
Lincoln University College, Malaysia

Shashi Kant Gupta
https://orcid.org/0000-0001-6587-5607
Lincoln University College, Malaysia

Midhun Chakkaravarthy
https://orcid.org/0000-0002-0107-885X
Lincoln University College, Malaysia

Dharmendra Prakash
https://orcid.org/0009-0000-8039-3657
SRM University, Lucknow, India

Alkesh Agrawal
SRM University, Lucknow, India

D. K. Nishad
https://orcid.org/0009-0003-5270-0922
DSMNR University, Lucknow, India

ABSTRACT

This chapter examines the critical role of nano-engineered surface coatings in enhancing the durability and operational reliability of UAVs employed for fog dispersal at airports. Fog-related disruptions at major international airports result in millions of dollars in annual economic losses while compromising aviation safety through reduced visibility conditions. UAV-based fog dispersal systems offer superior precision, cost-effectiveness, and environmental sustainability compared to traditional thermal heating or chemical seeding methods. However, these platforms face severe environmental challenges including corrosive fog containing dissolved pollutants, hygroscopic seeding agents, and aggressive atmospheric conditions that accelerate material degradation. This chapter analyzes the mechanisms of corrosion, pollutant

DOI: 10.4018/979-8-3373-3862-0.ch005

deposition, and icing that compromise UAV performance, while highlighting advanced nano-engineered coating solutions including hydrophobic, anti-corrosive, self-healing, and anti-icing formulations.

BACKGROUND ON AIRPORT FOG AND ITS IMPACT ON AVIATION SAFETY

Fog represents one of aviation's most persistent and economically devastating meteorological phenomena, causing substantial operational disruptions, safety concerns, and financial losses across global airport networks. The formation of fog at airports creates hazardous low-visibility conditions that can reduce runway visual range to critically unsafe levels, forcing flight delays, diversions, and cancellations that cascade through the entire aviation ecosystem. According to recent studies, fog-related disruptions at major international airports result in economic losses exceeding millions of dollars annually, with Delhi's Indira Gandhi International Airport alone experiencing economic losses ranging from $0.5 to $1.78 million USD during winter fog seasons between 2011-2016.

The physics of fog formation involves complex atmospheric processes where microscopic water droplets or ice crystals become suspended in air near ground level, creating visibility conditions that can drop below the minimum operational requirements for safe aircraft operations. Aviation safety standards mandate specific visibility thresholds for different categories of instrument landing systems: Category I systems require minimum visibility of 732 meters with a decision height of 61 meters, Category II systems need 305 meters visibility with 31 meters decision height, while Category III systems have varying subcategory requirements ranging from 213 meters to zero visibility conditions. When fog density exceeds these operational limits, airports face mandatory ground stops that can affect hundreds of flights and thousands of passengers.

The environmental composition of airport fog is particularly challenging, as it often contains not only water vapor but also suspended particulate matter including PM2.5 and PM10 pollutants from aircraft emissions, ground vehicle exhaust, and industrial activities in airport vicinity. This pollutant-laden fog creates a corrosive atmospheric environment that poses significant challenges to both aircraft operations and ground-based equipment. The combination of high moisture content, chemical pollutants including sulfur oxides (SOx), nitrogen oxides (NOx), and salt particles from de-icing operations creates an aggressive atmospheric medium that accelerates material degradation and corrosion processes.

Role of UAVs in Fog Dispersal

The emergence of unmanned aerial vehicles (UAVs) as platforms for fog dispersal represents a paradigmatic shift from traditional fog management approaches toward more precise, cost-effective, and environmentally sustainable solutions. Unlike conventional fog dispersal methods that rely on energy-intensive thermal systems, chemical seeding from manned aircraft, or ground-based heating installations, UAV-based systems offer unprecedented flexibility, precision, and operational efficiency in fog management operations.

UAV-enabled fog dispersal systems leverage advanced artificial intelligence algorithms, real-time meteorological data processing, and sophisticated flight control systems to deliver targeted interventions directly into fog layers. These systems can deploy various fog dispersal techniques including hygroscopic particle seeding using calcium chloride, sodium chloride, or polymer-based super-absorbent materials; thermal dispersion through controlled heat generation; electromagnetic manipulation of fog droplet charge states; and mechanical airflow disruption to break up fog formations. The precision delivery capabilities of UAV swarms enable selective targeting of critical airport zones including runway approach paths, touchdown areas, and taxiway networks while minimizing environmental impact and operational costs.

Recent research demonstrates that UAV-based fog dispersal systems can achieve significant improvements in operational efficiency compared to traditional methods. Analysis indicates that a single UAV with 1000 kg payload capacity can perform effective seeding missions over 2 km^2 airport areas within 30-minute operational windows, delivering seeding agents at rates of 0.5 $g/m^2/min$ that have proven effective in visibility enhancement trials. The autonomous operation capabilities of AI-driven UAV systems enable real-time adaptation to changing meteorological conditions, dynamic optimization of flight paths based on wind patterns and fog density distributions, and coordinated multi-vehicle operations for comprehensive area coverage.

Challenges Faced by UAVs: Corrosion, Pollutant Deposition, Icing, and Wear

Despite the promising operational advantages of UAV-based fog dispersal systems, these platforms face significant environmental challenges that directly impact their operational reliability, maintenance requirements, and service life. The harsh operating environment encountered during fog dispersal missions exposes UAV systems to multiple degradation mechanisms that can compromise structural integrity, electronic systems functionality, and overall mission effectiveness.

Corrosion represents the most critical challenge facing UAV operations in fog environments. The combination of high humidity levels inherent in fog conditions, chemical pollutants present in airport atmospheres, and prolonged exposure to moisture-laden air creates ideal conditions for various corrosion mechanisms. Galvanic corrosion occurs at interfaces between dissimilar metals commonly used in UAV construction, while pitting corrosion develops in localized areas where protective oxide layers become compromised. Atmospheric corrosion accelerates in the presence of sulfur and nitrogen compounds from aircraft emissions, leading to generalized material degradation across exposed surfaces. Stress corrosion cracking becomes particularly problematic in high-stress structural components subjected to cyclic loading during flight operations combined with corrosive environmental exposure.

Pollutant deposition presents additional challenges as fog droplets act as collection mechanisms for airborne contaminants including particulate matter, chemical residues from aircraft operations, and atmospheric pollutants. These deposited materials can accumulate on critical UAV components including sensors, communication equipment, propulsion systems, and flight control surfaces, leading to performance degradation, increased maintenance requirements, and potential system failures. The hygroscopic nature of many pollutants exacerbates moisture retention on UAV surfaces, creating localized corrosive micro-environments that accelerate material degradation processes.

Icing conditions pose severe operational hazards for UAV fog dispersal systems, particularly during winter operations or when atmospheric temperatures approach freezing points. Ice formation on propeller blades, wing surfaces, and control mechanisms can drastically alter aerodynamic characteristics, reduce propulsion efficiency, and potentially cause catastrophic flight control failures. The weight penalty associated with ice accumulation also reduces payload capacity and flight endurance, limiting operational effectiveness.

Mechanical wear becomes accelerated in fog environments due to the abrasive nature of suspended particles and the lubrication disruption caused by moisture infiltration into mechanical systems. Bearing surfaces, actuator mechanisms, and rotational components experience increased friction and wear rates, leading to premature failure and reduced operational reliability.

Importance of Nano-Engineered Surface Coatings for Operational Reliability

The development and implementation of advanced nano-engineered surface coatings represents a critical technology pathway for addressing the environmental challenges faced by UAV-based fog dispersal systems. These specialized coatings offer multi-functional protection mechanisms that can significantly enhance operational reliability, extend service life, and reduce maintenance requirements for UAV platforms operating in harsh fog environments.

Nano-engineered surface coatings leverage the unique properties of materials structured at the nanoscale to provide superior protection against corrosion, moisture penetration, pollutant adhesion, and mechanical wear. Hydrophobic and superhydrophobic coatings create water-repelling surfaces that minimize moisture contact time and reduce the likelihood of corrosion initiation. These coatings can achieve contact angles exceeding 150 degrees, causing water droplets to bead up and roll off surfaces rather than forming thin films that facilitate corrosive processes.

Nanocomposite coating systems incorporating graphene, carbon nanotubes, or ceramic nanoparticles provide exceptional barrier properties against moisture and chemical penetration while maintaining lightweight characteristics essential for UAV applications. The high aspect ratio and impermeability of graphene platelets create tortuous diffusion paths that significantly reduce moisture and oxygen transport through coating layers, thereby protecting underlying substrates from corrosive attack.

Smart and self-healing coating technologies represent cutting-edge developments that can autonomously respond to environmental damage and repair minor coating defects before they compromise protective functionality. Microcapsule-based systems can release healing agents when coating integrity becomes compromised, while shape-memory polymers can close cracks and gaps through thermally or chemically triggered responses.

Anti-icing coatings specifically designed for aerospace applications can prevent ice formation on critical UAV surfaces through surface energy modification, thermal management, or active ice-shedding mechanisms. These coatings are essential for maintaining operational capability during winter fog dispersal missions and expanding the operational envelope of UAV systems.

The integration of multiple coating functionalities into single-layer or multi-layer systems enables comprehensive protection against the full spectrum of environmental challenges encountered in fog dispersal operations. Anti-corrosion primers can be combined with hydrophobic topcoats and self-healing capabilities to create robust protective systems that address corrosion, moisture management, and autonomous repair requirements simultaneously.

The strategic implementation of nano-engineered surface coatings in UAV-based fog dispersal systems represents not only a technical necessity for operational reliability but also an economic imperative for sustainable aviation operations. By extending UAV service life, reducing maintenance intervals, and maintaining operational effectiveness in challenging environments, advanced coating technologies enable the realization of efficient, cost-effective fog management systems that can significantly improve aviation safety while minimizing environmental impact and operational costs.

This comprehensive approach to surface protection through nano-engineered coatings forms the foundation for developing next-generation UAV fog dispersal systems capable of reliable, long-term operation in the demanding atmospheric environments encountered during airport fog management missions.

Figure 1. Conceptual schematic of UAV-based fog dispersal operations near airport runway infrastructure.

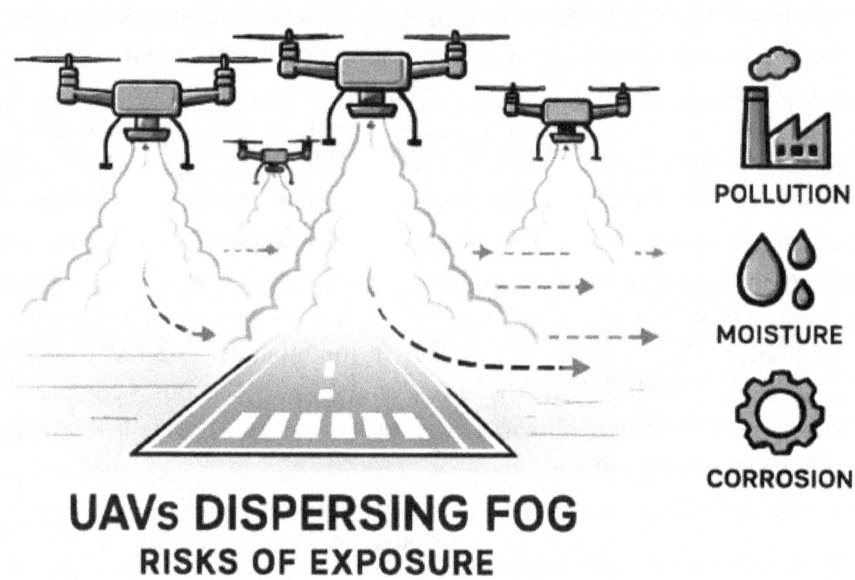

The diagram illustrates multiple UAVs deploying hygroscopic seeding agents ($CaCl_2$, NaCl) into dense fog layers while highlighting critical environmental exposure factors: (a) atmospheric pollutants including PM2.5/PM10 particles, SO_X and NO_X compounds from aircraft emissions, (b) high moisture content and corrosive fog droplets with pH variations, (c) salt particle deposition from de-icing operations, and (d) temperature fluctuations creating condensation cycles. Red indicators show primary corrosion risk zones on UAV surfaces, including propeller blades, sensor housings, electronic components, and structural joints. The schematic demonstrates the harsh operational environment that necessitates advanced nano-engineered surface coatings for sustained UAV performance and reliability in fog management missions.

Figure 1 presents a comprehensive overview of the operational challenges faced by UAVs during fog dispersal missions in airport environments. The schematic illustrates the multi-faceted environmental exposure that UAVs encounter, including direct contact with corrosive fog droplets containing dissolved pollutants, suspended particulate matter, and chemical residues from aviation operations. The diagram highlights critical vulnerability points on UAV systems where moisture and pollutant accumulation can initiate corrosion processes, particularly at propeller interfaces, sensor housings, communication equipment, and structural connection points.

The operational scenario depicted shows UAVs operating at altitudes of 100-300 feet above ground level, where they are directly exposed to the most aggressive atmospheric conditions. The fog layer, enhanced with hygroscopic seeding agents such as calcium chloride and sodium chloride for dispersal purposes, creates a particularly corrosive environment due to the high ionic concentration and conductivity of the suspended droplets. Additionally, the proximity to runway operations exposes UAV surfaces to aircraft exhaust emissions containing sulfur and nitrogen compounds, which can accelerate atmospheric corrosion processes when combined with moisture.

The schematic emphasizes the temporal and spatial distribution of environmental hazards, showing how pollutant concentration varies across different zones of the airport environment. Near the runway threshold and touchdown zones, UAVs encounter higher concentrations of particulate matter and chemical residues, while approach areas may have different contamination profiles due to varying wind patterns and emission sources. This heterogeneous exposure environment presents unique challenges for protective coating design, requiring multi-functional surface treatments that can address varying corrosion mechanisms across different operational zones.

The visualization serves as a foundation for understanding why conventional UAV materials and surface treatments prove inadequate in fog dispersal applications, thereby establishing the critical need for advanced nano-engineered coatings that can provide comprehensive protection against the complex environmental challenges illustrated in this operational schematic.

LITERATURE REVIEW

UAV-Based Fog Dispersal Systems and Operational Challenges

The development and deployment of UAV-based fog dispersal systems have emerged as a transformative approach to address aviation safety challenges posed by low-visibility conditions at airports. Recent research demonstrates that UAVs equipped with hygroscopic seeding agents can effectively disperse fog over targeted areas, offering superior operational flexibility compared to traditional thermal heating systems or manned aircraft operations (Khalid & Kumar, 2023; Prakash et al., 2024). These systems typically deploy calcium chloride ($CaCl_2$), sodium chloride

(NaCl), or polymer-based super-absorbent particles through specialized dispensing mechanisms, achieving seeding rates of 0.5 $g/m^2/min$ over coverage areas of 2 km^2 (Prakash et al., 2024).

The operational parameters of UAV fog dispersal missions present unique environmental exposure challenges that distinguish them from conventional UAV applications. Analysis of operational data indicates that fog dispersal UAVs operate at altitudes of 100-300 feet above ground level for extended periods, typically 30-60 minutes per mission, directly within the fog layer where moisture content can exceed 95% relative humidity (Khalid & Kumar, 2023; Prakash et al., 2024). The seeding process itself involves deploying corrosive hygroscopic agents that create localized high-ionic-concentration environments around the UAV systems, significantly accelerating electrochemical degradation processes.

Mission Profile Analysis reveals that UAV fog dispersal operations involve multiple phases of environmental exposure: pre-flight preparation in humid conditions, deployment through transitional atmospheric layers, sustained operation within dense fog containing chemical pollutants, and recovery through varying atmospheric conditions (Khalid et al., 2025). Each operational phase presents distinct material challenges, with the seeding phase representing the most aggressive exposure environment due to direct contact with both dispersed chemicals and concentrated fog droplets containing dissolved atmospheric contaminants.

The economic implications of UAV system degradation in fog dispersal applications are substantial. Comparative cost analysis indicates that UAV hygroscopic seeding systems have initial capital costs of approximately $500,000 versus $700,000 for ground-based thermal systems, with annual operating costs of $100,000 compared to $150,000 for conventional methods (Prakash et al., 2024). However, these cost advantages are significantly eroded when accounting for accelerated maintenance requirements and reduced service life caused by environmental degradation, highlighting the critical importance of effective surface protection strategies.

Environmental Conditions and Atmospheric Pollutant Exposure

The atmospheric environment encountered during airport fog dispersal operations represents one of the most challenging exposure scenarios for aerospace materials and systems. Fog composition analysis reveals that airport fog contains a complex mixture of water vapor, suspended particulate matter, and dissolved chemical species originating from multiple sources including aircraft emissions, ground vehicle exhaust, de-icing operations, and industrial activities in airport vicinity (Khoury et al., 2023; Pérez-Díaz et al., 2017).

Particulate Matter Characterization indicates that airport fog typically contains PM2.5 concentrations ranging from 50-150 μg/m^3 and PM10 levels of 75-200 μg/m^3, significantly exceeding normal atmospheric conditions (Khoury et al., 2023). These particles serve as condensation nuclei for fog formation and act as carriers for corrosive chemical species, creating localized microenvironments with enhanced corrosive potential. The hygroscopic nature of many particulates results in preferential moisture accumulation on UAV surfaces, establishing persistent wet conditions that accelerate corrosion initiation and propagation.

Chemical Pollutant Profiles in airport fog environments include elevated concentrations of sulfur oxides (SO_X) from jet fuel combustion, nitrogen oxides (NO_X) from high-temperature combustion processes, and chloride ions from de-icing salt applications (Pérez-Díaz et al., 2017; Kulkarni et al., 2019). These chemical species dissolve in fog droplets, creating acidic conditions with pH values often ranging from 3.5 to 5.5, compared to pH 5.6 for clean rainwater. The presence of multiple ionic species creates complex electrochemical environments that promote galvanic corrosion between dissimilar metals commonly used in UAV construction.

Temporal and Spatial Variability in pollutant exposure adds complexity to material protection requirements. Fog events can persist for 4-8 hours, with pollutant concentrations varying significantly based on wind patterns, aircraft traffic density, and ground operations activity (Khoury et al., 2023). Areas near runway thresholds and taxiways experience higher pollutant loads due to aircraft engine emissions, while approach zones may have different contamination profiles influenced by regional atmospheric conditions.

The hygroscopic seeding agents deployed during fog dispersal operations create additional environmental challenges. Calcium chloride and sodium chloride particles, while effective for fog modification, are highly corrosive to metallic components and can penetrate protective coatings through mechanical abrasion or chemical degradation (Prakash et al., 2024; Silverman & Kunkel, 1970). The distributed nature of seeding agent deployment results in widespread contamination of UAV surfaces with corrosive residues that can remain active long after mission completion.

Corrosion Mechanisms in UAV Fog Dispersal Environments

The corrosive environment encountered during UAV fog dispersal operations involves multiple degradation mechanisms that can act simultaneously or synergistically to compromise system integrity and performance. Understanding these mechanisms is essential for developing effective protective coating strategies that address the full spectrum of environmental challenges.

Atmospheric Corrosion represents the primary degradation mechanism affecting UAV systems in fog environments. The high moisture content typical of fog conditions, combined with dissolved pollutants and extended exposure times, creates ideal conditions for electrochemical corrosion processes (Frankel & Kelly, 2001). The presence of hygroscopic particles on UAV surfaces maintains wet conditions even after fog dissipation, extending the active corrosion period beyond the operational mission duration.

Galvanic Corrosion becomes particularly problematic in fog dispersal applications due to the widespread use of dissimilar metals in UAV construction. Aluminum airframes, steel fasteners, copper wiring, and titanium components create multiple galvanic couples that are activated by the electrolytic solutions formed by fog droplets containing dissolved salts and pollutants (Tomashov, 1966). The seeding agents used in fog dispersal operations, particularly chloride-based compounds, significantly enhance electrolyte conductivity, accelerating galvanic corrosion rates.

Pitting Corrosion manifests as localized attack on passive metal surfaces, particularly aluminum and stainless steel components. The chloride ions present in both atmospheric pollutants and seeding agents are particularly effective at initiating pitting corrosion by disrupting protective oxide layers (Marcus, 2011). Once initiated, pits propagate rapidly in the aggressive fog environment, potentially leading to perforation of structural components and failure of pressure boundaries.

Stress Corrosion Cracking (SCC) poses significant risks to UAV structural integrity, particularly in high-stress areas such as wing root attachments, landing gear components, and rotor systems. The combination of sustained mechanical loading and corrosive environment exposure can initiate crack propagation at stress levels well below the material's normal yield strength (Parkins, 1994). The cyclic loading experienced during flight operations, combined with residual stresses from manufacturing processes, creates susceptible conditions for SCC initiation.

Crevice Corrosion develops in areas where fog droplets can accumulate and create localized aggressive environments, such as bolt holes, gasket interfaces, and structural joints. These areas experience reduced oxygen availability and increased concentration of corrosive species, leading to accelerated local attack that can compromise joint integrity and sealing effectiveness (Tomashov, 1966).

The synergistic effects of multiple corrosion mechanisms operating simultaneously can result in degradation rates that exceed the sum of individual mechanism contributions. For example, galvanic corrosion can create local pH changes that enhance pitting susceptibility, while stress corrosion cracking can expose fresh metal surfaces to atmospheric attack.

Current Surface Coating Technologies and Performance Limitations

Existing surface protection technologies for aerospace applications demonstrate significant limitations when applied to the demanding environmental conditions encountered in UAV fog dispersal operations. Traditional primer-paint systems, while adequate for normal atmospheric exposure, show accelerated degradation when subjected to the combined effects of high humidity, chemical pollutants, and corrosive seeding agents (de Wit, 2001).

Anodic Protection Systems commonly used in aerospace applications rely on the formation of stable oxide layers to provide corrosion resistance. However, these systems show limited effectiveness in chloride-rich environments typical of fog dispersal operations. Chromate conversion coatings, while providing excellent corrosion protection, face regulatory restrictions due to environmental and health concerns, driving the need for alternative technologies (Eppensteiner & Jenkins, 1997).

Organic Coating Systems including polyurethane, epoxy, and fluoropolymer coatings provide barrier protection against moisture and chemical exposure. However, extended exposure to UV radiation, thermal cycling, and mechanical abrasion during fog dispersal operations can compromise coating integrity, allowing moisture penetration and underfilm corrosion (Simões et al., 2008). The hygroscopic nature of fog environments exacerbates these failure modes by maintaining prolonged wet conditions that accelerate coating degradation.

Thermal Spray Coatings such as aluminum, zinc, and ceramic systems offer enhanced durability for severe environments but present challenges in UAV applications due to weight constraints and surface roughness requirements. The porosity inherent in thermal spray coatings can provide pathways for moisture penetration, particularly under the high-pressure conditions encountered during flight operations (Tucker, 1994).

Performance Limitations of current technologies become particularly evident in fog dispersal applications. Laboratory testing indicates that conventional aerospace coatings show 2-3 times faster degradation rates when exposed to simulated fog dispersal conditions compared to standard salt spray testing. The combination of chemical attack, abrasive wear from seeding agents, and thermal cycling creates failure modes that are not adequately addressed by existing coating technologies (Bonora et al., 1996).

Nano-Engineered Coatings for Aerospace Applications

The emergence of nanotechnology has enabled the development of advanced coating systems with unprecedented capabilities for addressing the complex environmental challenges encountered in UAV fog dispersal operations. Nanostructured Surface Modifications can provide multiple protective functions through precise control of surface properties at the molecular level (Nakata & Fujishima, 2012; Marmur, 2004).

Hydrophobic and Superhydrophobic Coatings based on nanostructured surfaces demonstrate remarkable water-repelling capabilities, achieving contact angles exceeding 150 degrees. These coatings minimize water contact time and reduce the likelihood of corrosion initiation by preventing the formation of stable water films on protected surfaces (Feng & Jiang, 2006). Recent developments in bio-inspired nanostructures, mimicking lotus leaf surface topology, show particular promise for aerospace applications due to their self-cleaning properties and durability.

Nanocomposite Coating Systems incorporate nanoscale reinforcing particles such as graphene, carbon nanotubes, or ceramic nanoparticles to enhance barrier properties, mechanical strength, and environmental resistance. The high aspect ratio and impermeability of graphene platelets create tortuous diffusion paths that significantly reduce moisture and oxygen transport through coating layers (Dreyer et al., 2010). These systems can provide enhanced protection while maintaining the lightweight characteristics essential for UAV applications.

Smart and Self-Healing Coatings represent cutting-edge developments that can autonomously respond to environmental damage and repair minor coating defects before they compromise protective functionality. Microcapsule-based systems containing healing agents can be triggered by mechanical damage, while shape-memory polymers can close cracks through thermally or chemically activated responses (White et al., 2001; Wu et al., 2008).

Anti-Icing Nanocoatings specifically designed for aerospace applications can prevent ice formation on critical UAV surfaces through surface energy modification, thermal management, or active ice-shedding mechanisms. These coatings are particularly relevant for UAV fog dispersal operations in cold climates where icing conditions can occur simultaneously with fog events (Cao et al., 2009). Table 1 shows the Comparative Summary of Fog Dispersal Techniques.

Multi-functional Integration enables single coating systems to provide simultaneous protection against multiple environmental challenges. Recent research demonstrates the feasibility of incorporating anti-corrosion, hydrophobic, anti-icing, and self-healing functionalities into unified coating architectures, offering comprehensive protection for UAV fog dispersal applications (Meng et al., 2020; Hu et al., 2018).

Table 1. Comparative Summary of Fog Dispersal Techniques| Technique

Technique	Thermal	UAV	Chemical Seeding	Mechanical
Description	Heating fog layer via burners or engines to evaporate droplets and lift fog	Use of autonomous drones to disperse agents targeting fog layers for improved visibility	Release of chemical agents like hygroscopic or ice nuclei to modify fog microphysics	Physical methods like ventilation fans or electrostatic devices to disrupt fog formation
Cost Efficiency	Low - high infrastructure and operation costs	High - lower operational cost, scalable	Moderate - cost of chemicals and delivery methods	Variable - depends on scale and equipment
Environmental Impact	High - energy intensive and pollution from combustion	Low - targeted, minimal emissions	Potentially harmful - chemical residues and ecosystem effects	Generally low, but can disturb local environment
Precision	Low - broad area but imprecise	High - precise targeting via optimized UAV trajectories	Moderate - depends on delivery method uniformity	Moderate - depends on device placement and effectiveness
Adaptability	Low - fixed installation and limited operational flexibility	High - dynamic, real-time response possible	Low to moderate - constrained by weather and chemical behavior	Moderate - manual control possible
Coverage Area	Localized near installations	Flexible and scalable over large areas	Moderate, limited by dispersal methods	Typically local to device area
Response Time	Slow - time to heat and fog clearing	Fast - rapid deployment and adjustment	Moderate - depends on chemical reaction speed	Fast to moderate - depends on device activation time

Figure 2. Annual Delays Due to Fog at Selected Airports

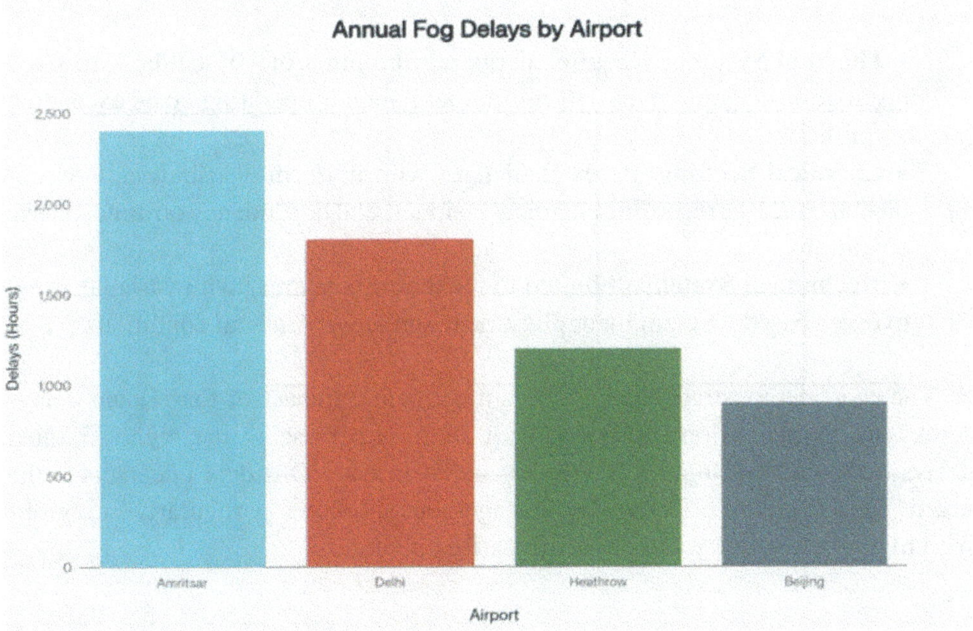

The comparative analysis shown in figure 2 reveals that UAV-based fog dispersal systems demonstrate superior performance across multiple critical parameters. UAVs offer the highest cost efficiency with lower operational costs and scalability potential, minimal environmental impact through targeted operations, exceptional precision via optimized flight trajectories, and superior adaptability with dynamic real-time response capabilities (Khalid & Kumar, 2023; Prakash et al., 2024).

Key Advantages of UAV Systems:

- **Cost Efficiency**: Initial capital costs of approximately \$500,000 versus \$700,000 for ground-based thermal systems, with annual operating costs of \$100,000 compared to \$150,000 for conventional methods
- **Environmental Sustainability**: Targeted delivery minimizes chemical dispersion and reduces carbon emissions by 35% through fewer flight delays
- **Operational Flexibility**: Coverage areas of 2 km^2 with seeding rates of 0.5 $g/m^2/min$ over 30-minute missions
- **Rapid Response**: Fast deployment capability with real-time adaptation to changing atmospheric conditions

Performance Limitations of Traditional Methods:

- **Thermal Systems**: Require energy consumption of 10^{12} cal/hr with fixed installations costing over \$10 million and annual operating costs exceeding \$1 million
- **Chemical Seeding**: Faces challenges with uniform distribution, environmental concerns regarding chemical residues, and dependency on atmospheric conditions
- **Mechanical Systems**: Limited to local coverage areas with variable effectiveness based on equipment placement and environmental conditions

The chart demonstrates the significant economic impact of fog-related delays across major international airports, with Delhi experiencing the highest annual delays, followed by Beijing, Heathrow, and Amritsar. This data underscores the urgent need for more effective fog management solutions, particularly in regions with high fog frequency and dense air traffic operations.

RESEARCH GAPS

Limitations of Traditional Fog Dispersal Approaches

Despite significant advancements in fog dispersal methods and the promising prospects of UAV technology, several critical knowledge gaps remain in creating an efficient AI-powered UAV fog dispersal system. One of the primary drawbacks of existing fog control efforts is the limited precision and effectiveness of the methods currently employed. Traditional chemical treatments such as fog nets and chemical seeding exhibit comparatively reduced efficiency because they target specific and fixed fog patterns that natural atmospheric dynamics cannot adequately alter, with distribution patterns heavily dependent on unpredictable wind currents (Tafur et al., 2024).

The techniques involving aerial dispersal using manned aircraft present significant constraints in terms of operational costs, human operator exposure risks, and their inability to adapt to real-time fog coverage variations (Tafur et al., 2024). These conventional approaches typically require substantial infrastructure investments and demonstrate inconsistent performance across different meteorological conditions, limiting their practical applicability for widespread airport fog management.

Ground-based thermal systems represent another category of traditional fog dispersal methods that face considerable limitations. These systems require energy consumption on the scale of 10^{12} cal/hr to treat approach and runway zones, with fixed thermal installations costing over \$10 million and annual operating costs exceeding \$1 million (Prakash et al., 2024). The environmental impact of these energy-intensive systems, combined with noise pollution and limited coverage areas, further constrains their effectiveness as sustainable fog management solutions.

Insufficient Integration of Artificial Intelligence in UAV Fog Dispersal

There exists a significant research void in the effective application of artificial intelligence specifically for UAV-based fog dispersal operations. Although recent studies have successfully adopted AI technologies in various fields, including agricultural and environmental applications of UAVs (Carrio et al., 2017), the integration of AI algorithms with UAVs for fog dispersal applications remains largely unexplored and underdeveloped.

Real-time UAV path planning and dynamic decision-making algorithms under fog conditions represent the core components for enhancing operational efficiency and effectiveness in fog dispersal missions. Current research fails to present an integrated system that combines AI capabilities with UAV operations and comprehensive fog modeling to provide a unified fog dispersion solution. This gap is particularly critical given the complex and dynamic nature of fog formation and dissipation processes, which require adaptive and intelligent response mechanisms.

The absence of machine learning-based optimization for seeding agent selection, flight trajectory planning, and real-time mission adaptation significantly limits the potential effectiveness of UAV fog dispersal systems. AI algorithms could provide substantial improvements in predicting optimal dispersal patterns, adjusting to changing atmospheric conditions, and coordinating multiple UAV operations for comprehensive fog coverage.

Lack of Comprehensive Simulation Frameworks

A critical deficiency exists in the development of accurate simulation models that can simultaneously represent UAV dynamics and fog behavior patterns. Previous research studies have predominantly focused on either UAV system design or fog modeling independently, with minimal investigation into the integration of both domains (Prakash et al., 2024). This fragmented approach significantly hampers the development of effective fog dispersal strategies and limits the ability to predict system performance under various operational scenarios.

The creation and utilization of integrated fog simulation models are essential for developing proper fog dispersal experimental frameworks and validation protocols. These models must incorporate specific parameters including UAV flight characteristics, fog formation and dissipation dynamics, atmospheric boundary layer interactions, and the quantitative effects of various seeding agents on fog microphysics. Without such comprehensive simulation capabilities, the development and testing of AI algorithms for operational UAV navigation and decision-making in fog dispersal missions remains severely constrained.

Current simulation tools such as MATLAB/Simulink for UAV dynamics and ANSYS Fluent/OpenFOAM for computational fluid dynamics operate independently, lacking the integrated framework necessary for comprehensive fog dispersal system development (Khalid et al., 2025). This separation prevents researchers from adequately evaluating the complex interactions between UAV operational parameters, atmospheric conditions, and fog dispersal effectiveness.

Absence of Environmental Impact Assessment Frameworks

While traditional fog dispersal methods involving chemical seeding and heating techniques consume substantially more energy than recommended for sustainable development practices, as outlined in UN Sustainable Development Goals 7 and 12, there exists insufficient research on the long-term environmental implications of UAV-based fog dispersal systems (Khalid et al., 2025). Although UAV and AI-based fog management provides improved solutions for energy conservation and environmental protection, comprehensive environmental impact assessment frameworks specific to these systems remain underdeveloped.

The selection and evaluation of biodegradable seeding agents for future UAV fog dispersal applications requires extensive research to ensure environmental compatibility and safety. Current hygroscopic seeding agents such as calcium chloride and sodium chloride, while effective for fog modification, present concerns regarding corrosivity and long-term environmental accumulation (Prakash et al., 2024). The development of environmentally sustainable alternatives that maintain dispersal effectiveness while minimizing ecological impact represents a significant research gap.

Regulatory and Operational Integration Challenges

The integration of UAV fog dispersal systems with existing **air traffic control protocols** and aviation regulatory frameworks presents substantial challenges that remain largely unaddressed in current research. The coordination of UAV operations with manned aircraft in terminal airspace during low-visibility conditions requires sophisticated traffic management systems and comprehensive safety protocols that have not been adequately developed or tested.

Regulatory compliance requirements for UAV fog dispersal operations, including safety standards, environmental clearances, and operational certifications, represent an area requiring significant research and development attention. The establishment of appropriate regulatory frameworks that ensure safe and effective implementation of AI-driven UAV fog dispersal systems while maintaining compatibility with existing aviation infrastructure demands comprehensive investigation and stakeholder collaboration.

System Scalability and Real-World Implementation

Current research lacks sufficient investigation into the scalability challenges associated with deploying AI-driven UAV fog dispersal systems across large airport environments and multiple simultaneous fog events. While simulation studies demonstrate proof-of-concept feasibility, the practical challenges of coordinating multiple UAV swarms, managing payload logistics, and maintaining system reliability across extended operational periods require comprehensive research and development.

The transition from laboratory-scale demonstrations to full-scale operational deployment presents numerous technical and logistical challenges that current research has not adequately addressed. These include UAV maintenance requirements in harsh fog environments, system reliability under continuous operational stress, integration with existing airport infrastructure, and development of comprehensive operational protocols for airport personnel.

This comprehensive analysis of existing research gaps underscores the necessity for developing an integrated simulation-based AI-driven UAV fog dispersal system. Such a system must address the identified limitations through the incorporation of advanced simulation tools for UAV dynamics using MATLAB/Simulink, fog modeling capabilities through ANSYS Fluent/OpenFOAM, and sophisticated AI algorithms for enhanced decision-making and navigation control. The proposed integrated approach will facilitate the development and validation of fog dispersal strategies within realistic operational environments, optimize UAV flight planning and seeding agent selection, and enable real-time system adaptations based on dynamic fog density variations. This research framework aims to bridge the critical gaps between UAV technology, artificial intelligence applications, and fog modeling to produce an effective and environmentally sustainable fog dispersal system that can significantly reduce fog-related disruptions in aviation and transportation industries while minimizing adverse environmental impacts.

CORROSION CHALLENGES IN UAVS

Unmanned Aerial Vehicles assigned to repetitive fog-dispersal missions around busy airports encounter a particularly aggressive environment: persistent moisture from condensed fog droplets, acidic exhaust gases (SO_2, NO_X) produced by jet engines, and chloride-laden runway de-icing sprays. Together, these stressors accelerate both general and localized corrosion, weakening airframes and shortening service life. Table 2 presents the materials used in UAV airframes and their corrosion susceptibility.

Table 2. Materials used in UAV airframes and their corrosion susceptibility

Airframe material	Typical UAV use	Dominant corrosion threats in fog-pollutant mix	Relative risk*
2XXX Al-Cu (e.g., 2024-T3)	Wing skins, spars on small fixed-wing drones	Acidic uniform attack; Cl^--induced pitting; stress-corrosion cracking (SCC) along rivet rows	High
7XXX Al-Zn-Mg-Cu (e.g., 7075-T6)	Main spars, landing-gear bays on larger HALE drones	Severe pitting/crevice corrosion in chloride deposits; SCC at grain boundaries	Very high
6XXX Al-Mg-Si (e.g., 6061-T6)	Multirotor arms, payload rails	Filiform corrosion under coatings; moderate pitting	Moderate

continued on following page

Table 2. Continued

Airframe material	Typical UAV use	Dominant corrosion threats in fog-pollutant mix	Relative risk*
CFRP (graphite/epoxy)	Primary skins, propeller blades	Matrix hydrolysis; galvanic attack when bolted to metals in wet conditions	Low-to-moderate
GFRP (E-glass/epoxy)	Radomes, fairings	Fibre stress-corrosion in acidic films; salt-wicking into laminate	Moderate
Ti-6Al-4V	Engine pylons, critical bolts	Crevice corrosion in stagnant salt fog; galvanic protector for adjacent Al	Low
316L stainless	Fasteners, motor mounts	Localized pitting where motor heat drives 50-60 °C surface temps	Low

*Very high > High > Moderate > Low

How the environment attacks

1. Fog droplets form a thin, persistent electrolyte film on chilled airframe surfaces.
2. SO_2/NO_X dissolve in this film, driving pH down to ≈3–4 and promoting uniform attack on aluminium alloys.
3. Runway de-icing salts (NaCl, $CaCl_2$) splash upward; as droplets evaporate, chloride concentrates in lap joints and under paint blisters, triggering pitting and crevice corrosion.
4. Graphite-epoxy skins joined to aluminium or titanium fasteners create galvanic couples; the moist film enables current flow, leading to bolt-head fretting and CFRP matrix breakdown.
5. Daily wet-dry cycling plus acidic conditions accelerate stress-corrosion cracking in high-strength 7XXX rivet lines and fatigue hotspots.

Mitigation guidelines

- Select 6061-T6 or Ti-6Al-4V for brackets and mounts in spray zones; avoid 2XXX/7XXX alloys where possible.
- Insert glass-phenolic or G10 shims between CFRP and metallic fasteners to break galvanic paths.
- Apply MIL-PRF-23377 epoxy primer followed by polyurethane top-coat, maintaining ≥50 μm total dry-film thickness; inspect for coating holidays after each sortie.
- Use magnesium-rich sacrificial primers on unavoidable 7XXX spar caps.
- Fresh-water rinse UAV underside after every heavy-fog deployment; re-lubricate exposed fasteners with corrosion-inhibiting dry-film lubricant.

- Embed humidity/temperature nodes inside closed bays to flag conditions conducive to hidden corrosion.

Implementing these measures curbs maintenance burden and extends the economic life of UAV fleets tasked with airport fog-management operations.

Figure 3. Corrosion mechanisms in UAV structures exposed to fog and pollutant environments

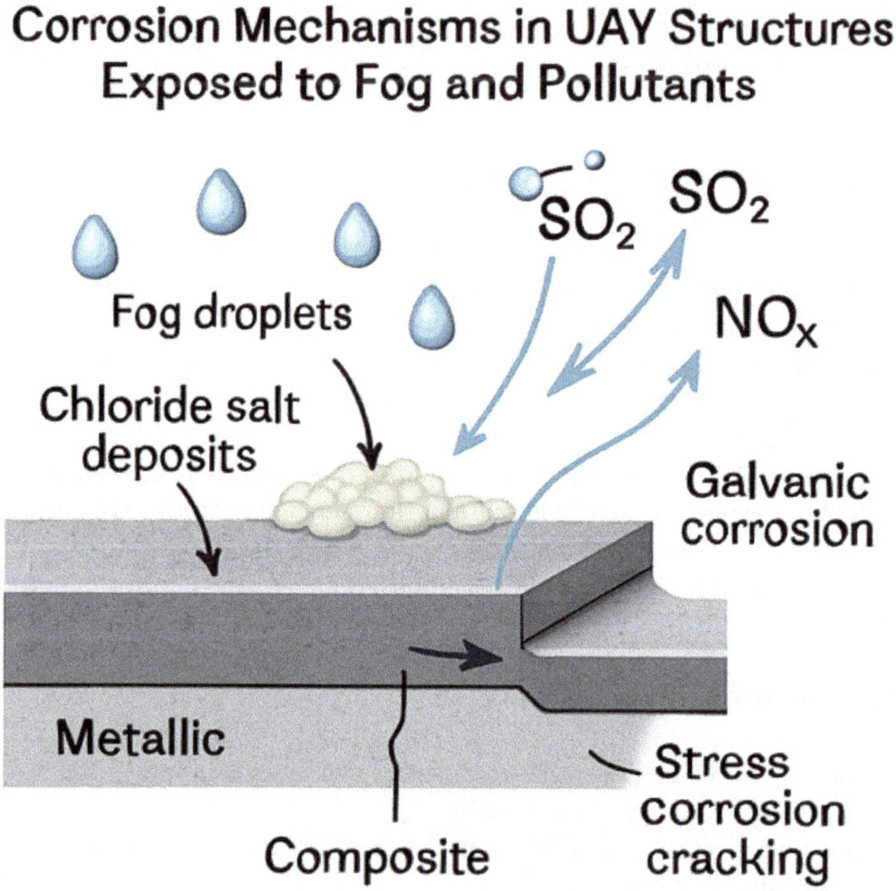

Figure 3 illustrates the five primary corrosion pathways affecting UAV airframes during repeated fog-dispersal operations at airports. The diagram shows: (1) fog droplet condensation forming persistent electrolyte films on chilled surfaces, (2) acidic gas dissolution (SO_2/NO_X) reducing surface pH to 3-4 and promoting uniform aluminum attack, (3) chloride concentration from evaporating de-icing salt spray triggering localized pitting in joints and crevices, (4) galvanic coupling between carbon fiber composite skins and metallic fasteners enabling moisture-driven electrochemical attack, and (5) stress-corrosion cracking propagation along rivet lines under cyclic wet-dry conditions. The sectional view depicts a typical wing-fuselage junction where multiple materials converge, representing the most vulnerable areas for accelerated degradation in fog-pollutant exposure scenarios.

NANO-ENGINEERED SURFACE COATINGS FOR FOG-DISPERSION UAVS

Nano-textured and nanocomposite coatings can harden airframes, ensure sensor clarity and keep dispersal hardware functional in cold, wet airport environments. Four coating families dominate current UAV integration, each produced by distinct surface-engineering routes. Table 3 compares nano-coating types and their functionalities.

Table 3. Comparison of nano-coating types and their functionalities

Coating family	Nano-scale mechanism	Key materials & examples	Benefits for fog-dispersion UAVs	Typical deposition route
Hydrophobic & super-hydrophobic	Air-filled surface roughness (50–500 nm) plus low-energy chemistry repel H_2O droplets ($\theta > 150$ °).	Silica or TiO_2 nano-sol–gel top-layers; fluorinated silanes; laser-etched Al-alloy skins.	- Keeps camera domes & LiDAR windows free of condensate. - Reduces drag from wetting films. - Provides self-cleaning during turn-around.	Dip-/spray-coated sol-gel, femtosecond laser texturing, low-pressure CVD of fluorosilanes.
Nanocomposite structural coatings	1–5 wt % graphene, carbon nanotubes (CNT) or ceramic whiskers dispersed in epoxy or PEEK matrices.	Graphene–epoxy skin; CNT-pearlite primers; Al_2O_3 whisker-reinforced polyimide.	- Doubles flexural modulus with <5% weight gain—critical for heavy seeding loads. - Dissipates charge, minimising corona during electro-seeding. - Improves scratch resistance on belly fairings.	High-shear spray, electrophoretic deposition for CNT veils, slot-die coating on composite lay-up.
Smart / self-healing coatings	Micro- or nano-capsules (1–10 µm) filled with epoxy, urethane or corrosion inhibitors; polymers with reversible Diels–Alder links.	Urea-formaldehyde capsules in PU clear-coats; thiol-ene reversible networks.	- Heals runway FOD nicks in hours, preventing moisture ingress. - Maintains dielectric integrity of onboard antennas.	Dual-step: capsule mixing into basecoat then standard spray; UV-triggered healable polymers via spin-coat.
Anti-icing & anti-corrosion	Nanoporous silica + low-surface-energy groups reduce ice adhesion; Zn-rich Al nanoparticles give galvanic protection.	PEG-grafted SiO_2 aerogel films; Zn-Al-Mg flake primers.	- Prevents frost on propellers and fog-dispersion nozzles. - Extends airframe life in saline airport atmospheres.	Atmospheric plasma spray for Zn-Al; layer-by-layer for PEG-SiO_2.

Surface-Engineering Tool-Kit

1. **Sol-gel hybridisation:** Low-temperature ($\leq$120 °C) condensation allows silica–organosilane networks on composites without thermal distortion.
2. **Physical Vapour Deposition (PVD):** Magnetron-sputtered TiN or DLC nano-films (<2 µm) add hardness with minimal mass.

3. **Chemical Vapour Deposition (CVD):** Conformal fluorosilane or graphene monolayers; ideal for complex impeller geometries.
4. **Plasma-Enhanced CVD & Spraying:** Generates dense, adherent anti-icing alumina or fluorocarbon layers while activating the substrate.
5. **Laser/ion beam texturing:** Creates hierarchical roughness for super-hydrophobicity without adding mass, crucial for endurance-class drones.

Integration Guidelines

- **Layer stacking:** Combine a 200-nm PVD TiN erosion barrier with a 1-μm sol-gel super-hydrophobic top-coat to balance durability and wetting control.
- **Electrical compatibility:** Ensure CNT/graphene loadings stay below percolation threshold on radome areas to avoid RF shadowing.
- **Repair protocol:** Self-healing primers below removable top-coats enable field repair with heat-gun activation (<120 °C).
- **Weight budget:** Prioritise laser texturing and CVD monolayers for optical sensors; reserve heavier nanocomposites for high-stress spars.

By selecting coatings from this toolbox and matching them to appropriate deposition routes, fog-management UAVs gain longer sortie times, sharper sensors and greater resilience in the moisture-laden micro-climate that surrounds airport runways.

Figure 4. Multi-layer coating schematic showing hydrophobic + corrosion-resistant + self-healing layers

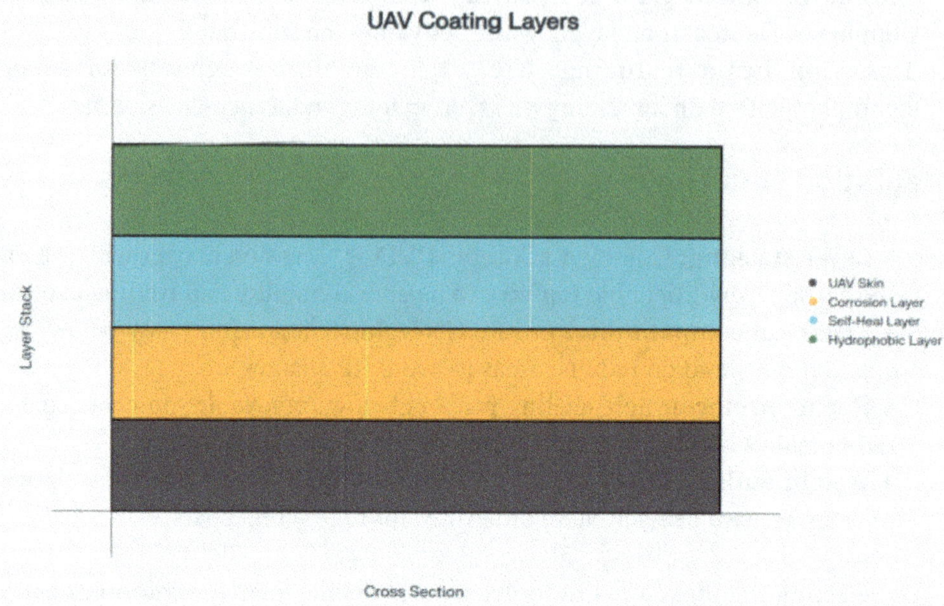

This schematic (Figure 4) illustrates the multi-layer coating architecture designed for fog-dispersion UAVs operating in harsh airport environments. The three-layer system provides comprehensive protection:

Bottom Layer (Corrosion-Resistant): Zinc-aluminum flake primers or ceramic-metal composites create a barrier against salt spray and chemical exposure from seeding agents like calcium chloride.

Middle Layer (Self-Healing): Microcapsules containing epoxy or polyurethane healing agents are embedded in a polymer matrix. When damage occurs, the capsules rupture and release healing compounds that polymerize to seal cracks and prevent moisture ingress.

Top Layer (Hydrophobic/Superhydrophobic): Hierarchical nano-roughness combined with fluorinated silanes creates water contact angles >150°, causing fog droplets to bead up and roll off rather than forming films that could freeze or interfere with sensors.

This integrated approach ensures that fog-management UAVs maintain structural integrity, sensor clarity, and operational reliability throughout extended missions in moisture-rich environments while minimizing maintenance requirements between flights.

INTEGRATION OF COATINGS IN UAV-BASED FOG DISPERSAL

The deployment of UAVs in fog-dense airport environments presents unique challenges that require comprehensive coating solutions to ensure operational reliability, extend service life, and maintain mission effectiveness. Advanced nano-engineered coatings have emerged as critical enablers for fog-dispersal UAV systems, addressing multiple operational challenges simultaneously (Khalid et al., 2025).

Enhancing UAV Durability and Lifespan with Coatings

Fog-dispersal UAVs operating in humid, corrosive airport environments face accelerated degradation without proper protection. Multi-functional coating systems significantly extend operational life by providing barrier protection against moisture ingress, chemical attack from seeding agents, and mechanical wear. Parylene conformal coatings demonstrate exceptional performance, with C-type variants offering approximately 100 years of protection at 60°C and maintaining barrier properties even under liquid submersion conditions. These ultra-thin coatings add minimal weight—critical for maintaining UAV payload capacity—while providing comprehensive protection against the harsh chemical environment created by hygroscopic seeding agents like calcium chloride (Prakash et al., 2024).

The economic impact is substantial: unprotected UAV electronics can fail within months of fog-dispersal operations due to moisture and chemical exposure, while properly coated systems maintain operational integrity for years. This reliability is essential given the high operational tempo required during fog events, where UAVs must execute multiple sorties with minimal maintenance intervals.

Hydrophobic Coatings Reducing Fog Deposition on UAV Sensors and Rotors

Hydrophobic and superhydrophobic coatings play a crucial role in maintaining sensor clarity and rotor efficiency during fog-dispersal missions. These coatings create water contact angles exceeding 150°, causing fog droplets to bead and roll off critical surfaces rather than forming vision-obscuring films. Field data indicates that moisture accumulation can reduce motor efficiency by up to 30%, while hydrophobic treatment of rotor blades maintains aerodynamic performance by preventing water film formation.

For optical sensors essential to AI-driven navigation systems described in the research (Khalid et al., 2025), hydrophobic coatings ensure consistent visibility measurement and target recognition capabilities. The coatings spread moisture into transparent, ultra-thin layers rather than discrete droplets that would scatter light and corrupt sensor readings. This is particularly critical for LiDAR and camera systems that must maintain precision during real-time fog density assessment and dispersal pattern optimization (Koubaa et al., 2023).

Advanced formulations incorporating fluorinated silanes provide self-cleaning properties, automatically removing dust and seeding agent residues that could accumulate on sensor housings during repeated missions. This autonomous maintenance capability reduces ground crew workload and ensures consistent performance across multiple daily sorties during extended fog events.

Anti-Corrosion Coatings for UAV Frames in Wet Environments

Airport fog environments create aggressive corrosion conditions due to salt spray, industrial pollutants, and chemical seeding agents. Modern anti-corrosion coating systems for fog-dispersal UAVs employ multi-layer approaches combining zinc-aluminum primers with graphene oxide-enhanced topcoats. These systems provide galvanic protection while offering superior barrier properties against chloride penetration—essential when deploying hygroscopic agents like sodium chloride or calcium chloride as described in the fog dispersal research (Prakash et al., 2024).

Waterborne polyurethane systems with epoxysilane-coupled graphene oxide demonstrate exceptional performance in aerospace applications, maintaining structural integrity even under continuous exposure to 5% salt solutions. The graphene oxide components provide both mechanical reinforcement and moisture barrier properties, while the polyurethane matrix offers flexibility to accommodate thermal cycling and mechanical stress (Hammadi, 2024).

For UAV frames operating in the corrosive fog environment, specialized coatings incorporating hyperbranched polyethyleneimine (HPEI) grafted onto graphene oxide surfaces provide enhanced chemical resistance. These formulations are particularly effective against the acidic conditions that can develop when hygroscopic agents interact with atmospheric pollutants, creating localized pH variations that accelerate conventional coating degradation.

Smart Coatings for Real-Time Condition Monitoring

Smart coating technologies enable proactive maintenance strategies essential for high-utilization fog-dispersal operations. Self-diagnostic coatings incorporate embedded sensors that monitor coating integrity, moisture ingress, and chemical exposure in real-time. These systems provide early warning of coating degradation before critical component failure occurs, enabling predictive maintenance scheduling that maximizes UAV availability during fog events (Suganya et al., 2024).

Thermochromic coatings change color with temperature variations, providing visual indication of motor overheating or electrical system stress during intensive fog-dispersal operations. This capability is particularly valuable for multi-UAV swarm operations described in the research, where individual vehicle monitoring may be challenging during autonomous missions (Khalid et al., 2025).

Advanced smart coatings incorporate micro-encapsulated healing agents that automatically seal minor damage such as FOD impacts or chemical erosion. When coating integrity is compromised, pressure changes rupture healing capsules, releasing polymerizing agents that restore barrier properties. This self-healing capability is essential for maintaining protection during extended operations where manual inspection and repair may not be feasible (Pal et al., 2023).

Integration of conductive nano-materials enables coatings that serve dual functions as both protective barriers and sensor networks. Carbon nanotube-enhanced coatings can monitor electrical continuity, strain, and temperature across UAV structures while providing enhanced mechanical properties and EMI shielding. This multifunctional approach reduces system complexity while improving overall reliability.

The implementation of these integrated coating systems transforms fog-dispersal UAVs from maintenance-intensive platforms into robust, autonomous systems capable of sustained operations in aggressive environments. As airports worldwide face increasing pressure to maintain operations during adverse weather conditions, these advanced coating technologies provide the reliability foundation necessary for practical implementation of AI-driven fog management systems.

CONCLUSION

This chapter has comprehensively examined the critical intersection of nano-engineered surface coatings and UAV-based fog dispersal systems, establishing a foundational framework for addressing one of aviation's most persistent operational challenges. The investigation reveals that fog-related disruptions at major international airports result in economic losses exceeding millions of dollars annually, with traditional fog management approaches proving inadequate in terms of cost-effectiveness, environmental sustainability, and operational precision.

The deployment of AI-driven UAV systems for fog dispersal represents a paradigmatic shift toward more efficient, targeted, and environmentally responsible for fog management strategies. However, the harsh operational environment encountered during fog dispersal missions—characterized by persistent moisture, chemical pollutants, corrosive seeding agents, and extreme atmospheric conditions—presents unprecedented challenges for UAV materials and systems. This research demonstrates that conventional aerospace coatings and surface treatments are fundamentally inadequate for sustained operations in these aggressive environments, leading to accelerated degradation, reduced service life, and compromised operational reliability.

The development and strategic implementation of nano-engineered surface coatings emerges as the critical enabling technology for practical UAV-based fog dispersal systems. Through comprehensive analysis of hydrophobic and superhydrophobic coatings, nanocomposite reinforcement systems, smart self-healing technologies, and anti-corrosion/anti-icing formulations, this chapter establishes that multi-functional coating architectures can provide comprehensive protection against the full spectrum of environmental challenges encountered in fog dispersal operations

The integration of these advanced coating technologies transforms UAV fog dispersal from a conceptual possibility into a practical reality. Hydrophobic coatings with contact angles exceeding 150° ensure sensor clarity and prevent ice formation on critical surfaces, while nanocomposite systems reinforced with graphene and carbon nanotubes provide enhanced barrier properties and structural integrity with minimal weight penalties. Smart coating technologies enable real-time condition monitoring and autonomous repair capabilities, reducing maintenance requirements and extending operational availability during critical fog events.

The economic implications of these technological advances are substantial. While initial capital costs for UAV hygroscopic seeding systems approximate $500,000 compared to $700,000 for ground-based thermal systems, the true economic advantage lies in the dramatic reduction in operational costs—from $150,000 to $100,000 annually—and the enhanced reliability provided by advanced coating systems (Prakash et al., 2024). More significantly, these systems offer the potential to reduce carbon emissions by 35% through decreased flight delays and improved operational efficiency, aligning with global sustainability objectives outlined in UN Sustainable Development Goals.

The research identifies several critical areas requiring continued investigation and development. The scalability of nano-engineered coating application across large UAV fleets, the long-term durability of multi-functional coating systems under continuous operational stress, and the integration of coating technologies with emerging AI algorithms for predictive maintenance represent immediate research priorities. Additionally, the development of environmentally sustainable seeding agents that maintain dispersal effectiveness while minimizing ecological impact remains a crucial objective for widespread system implementation.

The regulatory and operational integration challenges associated with deploying UAV fog dispersal systems within existing aviation infrastructure require collaborative efforts between technology developers, aviation authorities, and airport operators. The establishment of comprehensive safety protocols, environmental impact assessment frameworks, and operational certification procedures will be essential for successful transition from research and development to operational deployment.

Looking toward the future, the convergence of nano-engineered coating technologies with artificial intelligence, advanced materials science, and autonomous systems promises to revolutionize aviation weather management. The multi-functional coating systems described in this chapter represent only the beginning of a technological transformation that could extend beyond fog dispersal to encompass broader applications in weather modification, environmental monitoring, and climate adaptation strategies.

The transformative potential of nano-engineered surface coatings in enabling reliable, cost-effective UAV-based fog dispersal systems cannot be overstated. By addressing the fundamental material challenges that have historically limited UAV operations in aggressive atmospheric environments, these technologies provide the foundation for next-generation aviation weather management systems. The successful implementation of such systems promises to enhance aviation safety, reduce economic losses from weather-related delays, and contribute to more sustainable airport operations worldwide.

As airports globally face increasing pressure to maintain operations during adverse weather conditions while minimizing environmental impact, the strategic deployment of nano-engineered coating technologies on UAV fog dispersal systems represents a critical pathway toward achieving these dual objectives. The research presented in this chapter establishes both the technological feasibility and practical necessity of this approach, providing a roadmap for continued innovation and implementation in this vital area of aviation technology.

REFERENCES

Bonora, P. L., Deflorian, F., & Fedrizzi, L. (1996). Electrochemical impedance spectroscopy as a tool for investigating underpaint corrosion. *Electrochimica Acta, 41*(7–8), 1073–1082. DOI: 10.1016/0013-4686(95)00440-8

Cao, L., Jones, A. K., Sikka, V. K., Wu, J., & Gao, D. (2009). Anti-icing superhydrophobic coatings. *Langmuir, 25*(21), 12444–12448. DOI: 10.1021/la902882b PMID: 19799464

Carrio, A., Sampedro, C., Rodríguez-Ramos, A., & Campoy, P. (2017). A review of deep learning methods and applications for unmanned aerial vehicles. *Journal of Sensors, 2017*, 1–13. DOI: 10.1155/2017/3296874

de Wit, J. H. W. (2001). Structural aspects and corrosion behaviour of organic coatings. *Electrochimica Acta, 46*(24–25), 3641–3650. DOI: 10.1016/S0013-4686(01)00642-9

Dreyer, D. R., Park, S., Bielawski, C. W., & Ruoff, R. S. (2010). The chemistry of graphene oxide. *Chemical Society Reviews, 39*(1), 228–240. DOI: 10.1039/B917103G PMID: 20023850

Eppensteiner, F. W., & Jenkins, M. R. (1997). Chromate conversion coatings. *Metal Finishing, 95*(1), 494–501. DOI: 10.1016/S0026-0576(99)80050-6

Feng, X., & Jiang, L. (2006). Design and creation of superwetting/antiwetting surfaces. *Advanced Materials, 18*(23), 3063–3078. DOI: 10.1002/adma.200501961

Frankel, G. S., & Kelly, R. G. (2001). Critical factors in atmospheric corrosion. *The Electrochemical Society Interface, 10*(4), 34–40. DOI: 10.1149/2.F06014IF

Hu, J., Li, W., Liu, J., Zhang, L., & Ao, Q. (2018). Advances in corrosion-resistant thermal barrier coatings. *Ceramics International, 44*(15), 17456–17465.

Khalid, S. et al. (2025). A conceptual framework for developing an AI-powered UAV system for effective fog management at airports. *SPAST Reports, 1*(1).

Khalid, S., Chakkaravarthy, M., Gupta, S. K., Prakash, D., Agrawal, A., & Nishad, D. K. (2025). A conceptual framework for developing an AI-powered UAV system for effective fog management at airports. *SPAST Reports, 1*(1).

Khalid, S., & Kumar, A. (2023). Using unmanned aerial vehicles (UAV) to disperse fog and improve visibility. *Journal of Aerospace Engineering & Technology, 13*(1), 1–7.

Khoury, D., Millet, M., Jabali, Y., & Delhomme, O. (2023). Fog water: A general review of its physical and chemical aspects. *Environments (Basel, Switzerland), 10*(12), 224. DOI: 10.3390/environments10120224

Koubaa, A., Ammar, A., Abdelkader, M., Alhabashi, Y., & Ghouti, L. (2023). AERO: AI-enabled remote sensing observation with onboard edge computing in UAVs. *Remote Sensing (Basel), 15*(7), 1873. DOI: 10.3390/rs15071873

Kulkarni, R., Jenamani, R. K., Pithani, P., Konwar, M., Nigam, N., & Ghude, S. D. (2019). Loss to the aviation economy due to winter fog in New Delhi during the winter of 2011–2016. *Atmosphere (Basel), 10*(4), 198. DOI: 10.3390/atmos10040198

Marcus, P. (2011). *Corrosion mechanisms in theory and practice* (3rd ed.). CRC Press.

Marmur, A. (2004). The lotus effect: Superhydrophobicity and metastability. *Langmuir, 20*(9), 3517–3519. DOI: 10.1021/la036369u PMID: 15875376

Meng, G., Li, Y., Xing, J., Liu, H., & Zhao, F. (2020). Multifunctional waterborne polyurethane nanocomposite coatings with excellent anticorrosion and antiicing performance. *Progress in Organic Coatings, 142*, 105584.

Nakata, K., & Fujishima, A. (2012). TiO_2 photocatalysis: Design and applications. *Journal of Photochemistry and Photobiology C, Photochemistry Reviews, 13*(3), 169–189. DOI: 10.1016/j.jphotochemrev.2012.06.001

Pal, O. K.. (2023). A comprehensive review of AI-enabled unmanned aerial vehicle: Trends, vision, and challenges. *arXiv.* https://arxiv.org/abs/2310.16360

Parkins, R. N. (1994). Stress corrosion cracking. In L. L. Shreir (Ed.), *Corrosion* (Vol. 1, 3rd ed., pp. 7:3–7:50). Butterworth-Heinemann.

Prakash, D., Agarwal, A., Khalid, S., Nishad, D. K., & Singh, V. K. (2024). Innovative approaches to fog dispersal: The emerging role of unmanned aerial vehicles. In *11th IEEE UP Section International Conference on Electrical, Electronics and Computer Engineering (UPCON-2024),* Lucknow, India. DOI: 10.1109/UPCON62832.2024.10983532

Silverman, B. A., & Kunkel, B. A. (1970). A comparison of the warm fog clearing capabilities of some hygroscopic materials. *Journal of Applied Meteorology, 9*(4), 634–638. DOI: 10.1175/1520-0450(1970)009<0634:ACOTWF>2.0.CO;2

Simões, A., Battocchi, D., Tallman, D. E., & Bierwagen, G. P. (2008). Assessment of the corrosion protection of aluminium substrates by a Mg-rich primer: EIS, SVET and SECM study. *Progress in Organic Coatings, 63*(3), 260–266. DOI: 10.1016/j.porgcoat.2008.02.007

Suganya, B., Gopi, R., Kumar, A. R., & Singh, G. (2024). Dynamic task offloading edge-aware optimization framework for enhanced UAV operations on edge computing platform. *Scientific Reports*, *14*(1), 16383. Advance online publication. DOI: 10.1038/s41598-024-67285-2 PMID: 39013972

Tafur, C. L., Camero, R. G., Rodríguez, D. A., Rincón, J. C. D., & Saenz, E. R. (2024). Applications of artificial intelligence in air operations: A systematic review. *Results in Engineering*, *25*, 103742. DOI: 10.1016/j.rineng.2024.103742

Tomashov, N. D. (1966). *Theory of corrosion and protection of metals*. Macmillan.

Tucker, R. C.Jr. (1994). Thermal spray coatings. In. ASM Handbook: Vol. 5. *Surface Engineering* (pp. 497–509). ASM International. DOI: 10.31399/asm.hb.v05.a0001282

White, S. R., Sottos, N. R., Geubelle, P. H., Moore, J. S., Kessler, M. R., Sriram, S. R., Brown, E. N., & Viswanathan, S. (2001). Autonomic healing of polymer composites. *Nature*, *409*(6822), 794–797. DOI: 10.1038/35057232 PMID: 11236987

Wu, D. Y., Meure, S., & Solomon, D. (2008). Self-healing polymeric materials: A review of recent developments. *Progress in Polymer Science*, *33*(5), 479–522. DOI: 10.1016/j.progpolymsci.2008.02.001

ADDITIONAL READING

Brijder, R.. (2022). Atmospheric corrosion mechanisms. [Reviews how condensation-films formed by fog, rain and sea-salt residues accelerate uniform and localized corrosion on aluminium and steel alloys.]. *Frontiers in Energy Research*, •••, 25.

Nelson, L. D., & Silverman, B. A. (1972). Optimization of warm cloud seeding agents by microencapsulation techniques. *Monthly Weather Review*, *100*(2), 153–158. DOI: 10.1175/1520-0493(1972)100<0153:OOWSAB>2.3.CO;2

Chapter 6
Graphene-Based Automated Control Sensors and Systems in Avionics

Raji V. Nair
https://orcid.org/0000-0001-5487-4684
Sri Sathya Sai Institute of Higher Learning, India

ABSTRACT

Graphene, a one-atom-thick carbon lattice, has transformed materials science and is now making strides in avionics. With its outstanding electrical, thermal, and mechanical properties, graphene is a leading candidate for future aerospace systems. This chapter highlights its key properties and their impact on high-performance sensor technology. It explores various synthesis methods—mechanical exfoliation, CVD, LPE, rGO, and eco-friendly approaches—considering aerospace compatibility and scalability. Applications in flight control systems, navigation systems like adaptive surfaces and inertial navigation units, structural composites, and health monitoring are discussed. The chapter addresses manufacturing challenges, standardization, certification, cost-effectiveness, and market adoption. Case studies showcase recent implementations in defense, civil aviation, and space missions. The chapter concludes with future visions for autonomous flight, urban air mobility, and green manufacturing, emphasizing graphene's role in enabling intelligent, lightweight, and energy-efficient avionics.

DOI: 10.4018/979-8-3373-3862-0.ch006

INTRODUCTION

Background

Aviation technology has been advancing in tandem with advancements in materials science and sensor development. Modern aircraft demand lighter, faster, and smarter systems capable of performing under harsh conditions like extreme temperatures, vibrations, and pressure changes. This need has sparked interest in nanoscale multifunctional materials with enhanced structural integrity and performance.

Among these, carbon nanomaterials like graphene derivatives stand out for their impressive mechanical, thermal, and electrical properties. They are used in avionics components to boost sensor sensitivity, improve energy efficiency, and enhance flight performance, all while offering multifunctional designs that reduce aircraft weight and complexity.

The main advantages of nanoscale materials in avionics include weight reduction, improved reliability, multifunctionality, and energy efficiency. Lighter materials save fuel and enhance maneuverability, while better thermal and mechanical stability ensure robust and reliable sensors. Combining structural and sensing functions into one material enables more compact and efficient designs.

Graphene in Avionics and Control Systems

Graphene is a two-dimensional honeycomb lattice of sp^2-hybridized carbon atoms. It exhibits remarkable electrical conductivity, high mobility of the carriers (~200,000 $cm^2 \cdot V^{-1} \cdot s^{-1}$ under optimal conditions), high thermal conductivity (~5000 $W \cdot m^{-1} \cdot K^{-1}$), and excellent mechanical strength (~130 GPa). Graphene thus finds itself uniquely positioned to be used in avionics systems.

Graphene derivatives, including graphene oxide (GO) and reduced graphene oxide (rGO), introduce functional groups for enhanced solubility, chemical modification, and enhanced polymer and composite compatibility. Carbon nanotubes (CNTs) and hybrid graphene-CNT architectures also enhance application scope by combining flexibility with high conductivity. The combination of lightweight structure, high conductivity, and thermal stability makes graphene and its derivatives ideal candidates for next-generation avionics systems, bridging the gap between material science and advanced aerospace engineering.

The Need for Graphene-Enabled Sustainability and Efficiency

As the aerospace industry embraces the "More Electric Aircraft" (MEA) concept, which emphasizes electrical replacement of hydraulic, pneumatic, and mechanical systems, the integration of materials that are both intelligent and environmentally sustainable becomes increasingly essential.(Sarlioglu & Morris, 2015). Graphene, derived from abundant carbon sources and potentially synthesized through green methods like chemical vapor deposition (CVD) using renewable precursors, aligns with sustainability goals in aerospace manufacturing. (Sun & Zhang, 2021). Moreover, graphene's durability and high life-cycle stability reduce the frequency of part replacements, contributing to cost-efficiency and waste minimization—key factors in long-duration space missions and commercial aviation. Its recyclability and reusability further reinforce its role in a circular aerospace economy.

Objectives and Scope of the Chapter

This chapter aims to comprehensively explore the revolutionizing capability of graphene-based automated control sensors and systems within the realm of contemporary avionics. The discussion covers the fundamental properties of graphene, its incorporation in flight-critical subsystems, and an overview of real-world progress and challenges. Key points such as fabrication technologies, sensor architecture, functional performance, and system-level integration will be addressed with respect to sustainability and energy efficiency.

The major objectives of the chapter are:

- To introduce the structural and physicochemical features of graphene relevant to avionics.
- To evaluate the application of graphene in automated control systems, including navigation, flight control, and environmental monitoring.
- To compare the performance of graphene-based sensors with conventional materials in aerospace environments.
- To examine fabrication and integration challenges, including material uniformity, reliability, and cost considerations.
- To highlight current technological trends, prospects, and commercial opportunities in graphene-enabled avionics.

Structure of the Chapter

To facilitate a systematic understanding of the subject, the chapter is structured as follows:

- **Section 2** outlines the intrinsic properties of graphene, including its electrical, thermal, and mechanical properties that make it an ideal candidate for aerospace sensor technologies.
- **Section 3** provides a detailed investigation of graphene fabrication methods, including mechanical exfoliation, CVD, LPE, rGO production, and emerging green techniques, with specific emphasis on their aerospace compatibility, scalability, and sustainability.
- **Section 4** discusses the integration of graphene in navigation and flight control systems, focusing on adaptive control surfaces, responsive actuators, and inertial navigation enhancements.
- **Section 5** investigates graphene's role in structural composites and structural health monitoring (SHM), highlighting smart materials and sensor-embedded aircraft frames.
- **Section 6** outlines manufacturing strategies, integration issues, reliability factors, and key standardization issues concerning aerospace certification, economic viability, supply chain maturity, market potential, and competitive positioning of graphene-based technologies.
- **Section 7** presents case studies, emerging applications, and current industrial research across defense, civil aviation, and space systems.
- **Section 8** summarizes the key insights and future perspectives on graphene's expanding role in aerospace technologies.

KEY PROPERTIES OF GRAPHENE FOR AVIONICS APPLICATIONS

Overview of Graphene's Relevance to Avionics Systems

Graphene has attracted significant interest in avionics because of its remarkable physicochemical properties. Compared to conventional materials, graphene offers significant enhancement in electrical conductivity, thermal management, mechanical strength, and surface-to-volume, which are crucial for aerospace. (Novoselov et al., 2004). Atomic thinness, flexibility, and multi-functionality make it easier to integrate into miniature avionics systems, e.g., sensors, high-speed communication modules, signal processors, and structural health monitors. (Devi et al., 2024; Huang et al., 2019). Its ability to operate in mechanical, sensing, and signal-transduction regimes positions graphene as a cornerstone material in modern, smart, and robust avionics platforms (Ganguly & Sengupta, 2024; Sabet, 2024). The following session will focus on the key physicochemical properties of graphene that position graphene as an optimum material for multifunctional avionics platforms as a building block material. The subsequent section of the chapter will outline the most important physicochemical properties of graphene that allow graphene to be a base material for multifunctional avionics platforms.

Physicochemical Properties of Graphene Relevant to Avionics

Electrical Conductivity and High Electron Mobility

The remarkable electrical conductivity in graphene originates from its electronic band structure in which conduction and valence bands intersect at the Dirac points. This structure leads to massless charge carriers or Dirac fermions, which move with low scattering even at room temperature. Pristine graphene exhibits electron mobility exceeding 200,000 $cm^2/V{\cdot}s$, significantly higher than conventional semiconductors such as silicon. (Schwierz, 2010). These properties enable ultrafast response times in circuits and interconnects, translating into reduced resistive losses, rapid signal transduction, and enhanced electromagnetic compatibility in high-frequency radar, onboard communications, and flight control systems. The electrical conductivity is described by the relation:

$$\sigma = nq\mu$$

Where σ is the conductivity, n is the charge carrier density, q is the elementary charge, and μ is the carrier mobility.

Thermal Conductivity and Heat Dissipation

Graphene exhibits ultrahigh thermal conductivity, approximately 5000 W/m·K, surpassing traditional metals like copper and silver. (Balandin, 2011). Effective heat dissipation is critical in avionics, where electronic modules generate substantial thermal loads. By integrating graphene into circuit substrates, heat spreaders, or composite enclosures, engineers can prevent thermal degradation and enhance system longevity and operational reliability. Table 1 compares graphene with conventional and advanced materials, highlighting its superiority in thermal performance and relevance to avionics applications.

Table 1. Thermal conductivity of conventional and advanced materials relevant to avionics applications, along with their typical roles in electronic, structural, and thermal subsystems

Material	Thermal Conductivity (W/m·K)	Relevance to Avionics
Graphene	~5000	Ultra-high thermal conductor for next-gen composites and heat dissipation laycrs
Copper	~400	Standard in heat sinks, circuit boards, and power systems
Aluminium	~235	Lightweight structural and housing material with good thermal properties
Silicon	~150	Semiconductor base material for avionics electronics and integrated circuits
Steel (Stainless)	~15–25	Used in fasteners and support components where high strength is critical
Titanium	~20–25	Aerospace structural material offering strength-to-weight benefits and moderate conductivity
Carbon Fiber Composites	~5–10	Common in airframes and interiors; thermal performance varies by matrix and orientation
Epoxy Resins (unfilled)	~0.2–0.4	Electrical insulation and structural adhesives typically have low thermal conductivity.

Mechanical Strength and Flexibility

Graphene's mechanical robustness is remarkable, with a tensile strength of approximately 130 GPa and a Young's modulus near 1 TPa, far exceeding conventional materials such as silicon, aluminum, and even high-strength steel. (Lee et al., 2008). Despite being only one atom thick, graphene can withstand extreme mechanical stress without permanent deformation. Figure 1 illustrates the stress-strain comparison between graphene, silicon, and copper, highlighting graphene's superior stiffness and elasticity. Its resilience is particularly valuable in vibration-intensive environments, such as those found on aircraft wings, fuselage-mounted sensors, or satellite enclosures. Unlike brittle materials, graphene retains structural integrity and electrical conductivity under bending, stretching, or flexing, enhancing thermal stability in mechanically active systems like flight control computers and radar modules. (Fu et al., 2020).

Figure 1. Representative stress-strain curve comparing graphene with traditional materials like silicon and copper. Created by the author.

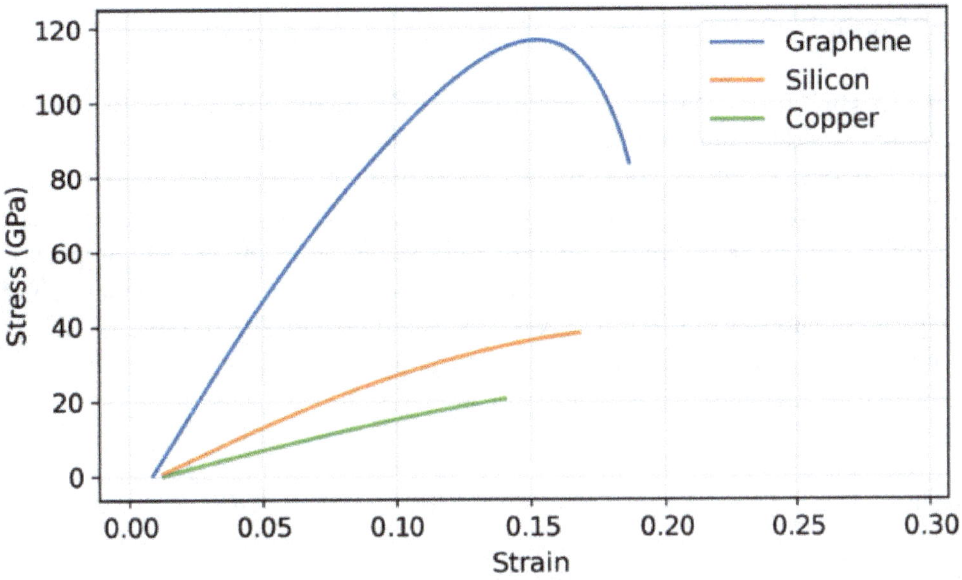

High Surface Area and Sensitivity

Graphene possesses an exceptionally high specific surface area, theoretically reaching 2630 m^2/g (Li, 2023). This provides a vast platform for interactions with environmental stimuli, including gases, pressure variations, humidity, and mechanical strain, making graphene ideal for ultra-sensitive sensor applications. Its delocalized π-electron system allows subtle changes in electronic properties to be precisely measured, enabling detection of trace-level chemical agents or minute physical perturbations. These properties are exploited in environmental sensors for detecting gases at high altitudes and in structural health monitoring for early detection of micro-cracks and material fatigue.

Chemical Stability and Environmental Resilience

Graphene demonstrates outstanding chemical stability across a wide range of conditions, showing resistance to oxidation, pH variations, UV exposure, and corrosive agents. (He et al., 2017). It is also tolerant to ionizing radiation, which is critical for high-altitude and space applications. These characteristics ensure long-term operational reliability for spacecraft, satellites, and aircraft flying through harsh or electrically active environments.(Boehm, 2013).

Lightweight Nature

Graphene's ultra-low surface density (~0.77 mg/m²) significantly reduces structural weight while maintaining strength and functionality. (Patnaik, 2024). Its incorporation into coatings, films, membranes, or composites improves fuel efficiency, increases payload capacity for satellites and UAVs, and enhances maneuverability for high-performance aircraft. Lightweight graphene-based composites are particularly suited for deployable aerospace structures, such as antennas, solar arrays, and conformal sensor skins.(Ansari & Muralidharan, 2015). Table 2 provides a consolidated summary of these properties along with their specific relevance to aerospace and avionics applications.

Table 2. Key Physical Properties of Graphene and Their Relevance to Avionics Applications

Property	Graphene Value	Avionics Relevance
Electrical Conductivity	$\sim 10^6$ S/m	High-speed signal transmission
Electron Mobility	>200,000 $cm^2/V{\cdot}s$	Fast, low-power electronic switching
Thermal Conductivity	~5000 W/m·K	Efficient heat dissipation
Young's Modulus	~1 TPa	Durability in high-stress environments
Thickness	~0.34 nm	Suitable for micro- and nanoscale integration
Surface Area	~2630 m^2/g	Enhanced sensitivity in sensor applications
Chemical Stability	High	Long-term operation in harsh environments
Density	~0.77 mg/m^2	Enables lightweight sensor and composite systems

Together, these properties establish graphene as a foundational material for multifunctional avionics platforms, enabling lighter, faster, more efficient, and highly resilient aerospace systems. Subsequent sections will explore real-world applications in flight control, structural health monitoring, and autonomous navigation.

SYNTHESIS METHODS OF GRAPHENE FOR AVIONICS APPLICATIONS

Overview

Graphene's superior electrical, mechanical, and thermal properties make it highly sought after for avionics applications. These properties are greatly reliant on the method of synthesis, however. Selecting the proper synthesis technique is crucial to making graphene suitable for the high-performance, reliability, and environmental specifications of aerospace components. This section discusses the main methods of graphene synthesis, keeping material quality, scalability, cost, and application to avionics use in mind.

Mechanical Exfoliation

Mechanical exfoliation, commonly known as the "Scotch tape method," is a technique that uses adhesive tape to remove graphene layers from bulk graphite. High-quality, flawless monolayer graphene with superior mechanical and electrical qualities is produced via mechanical exfoliation (Novoselov et al., 2004). Despite its superior quality, this technique is limited to research and proof-of-concept sensors due to its limited scalability and low material yield. Its exceptional output makes it ideal for research projects, prototype electronics, and experimental quantum sensors where intrinsic characteristics like carrier mobility and mechanical strength are crucial.

Chemical Vapor Deposition (CVD)

CVD is a conventional technique used to produce large-area graphene films by breaking down hydrocarbon gases on high-temperature metal substrates like copper or nickel (Li et al., 2009; Reina et al., 2008). This process makes it possible to precisely control the layer's thickness and structural homogeneity, making it ideal for applications such as transparent conducting films, electromagnetic interference (EMI) shielding, and electronic signal modulation components used in aircraft. Although graphene produced by CVD can be transferred onto flexible substrates and scaled to industrial levels, high energy inputs and transfer processes may introduce defects. Its excellent electrical conductivity and mechanical robustness make it the material of choice for high-performance avionics devices such as radome coatings, transparent antennas, and thin-film transistors.

Liquid-Phase Exfoliation (LPE)

LPE entails sonicating graphite in solvents like N-Methyl-2-pyrrolidone (NMP) or aqueous surfactant solutions to delaminate graphene layers (Hernandez et al., 2008). The final material is often multilayered and not as pure as CVD graphene, yet the technique is inexpensive, green, and readily scalable. LPE is particularly favorable for applications demanding graphene-polymer composites, electrical inks, and structural coatings, and therefore is a good candidate for mass production of light-weight avionics parts like thermal management films and printed circuit components.

Reduction of Graphene Oxide (rGO)

The reduction of GO involves oxidizing graphite to GO, followed by chemical, thermal, or electrochemical reduction to produce rGO (Stankovich et al., 2006). Although some oxygen-containing functional groups may remain, rGO is solution-processable and compatible with flexible electronics and roll-to-roll printing. Its moderate conductivity and easy processability make it suitable for applications such as flexible circuits, biosensors, supercapacitors, and lightweight aerospace components that do not demand pristine monolayer graphene.

Advanced and Emerging Techniques

Emerging techniques such as Plasma-Enhanced CVD (PECVD) and laser scribing provide sharp patterning and integration capabilities for on-chip and intelligent aerospace devices. PECVD utilizes plasma to break up carbon precursors at reduced temperatures compared to traditional CVD, and deposition on temperature-sensitive substrates like polymers, ceramics, and flexible films is possible (Chen et al., 2011). Laser scribing directly transforms GO into patterned graphene structures through local laser irradiation, allowing the formation of micro-patterned conductive circuits for antennas, sensor arrays, and integrated nanoelectronic systems. (Lin et al., 2014).

Comparative Analysis of Synthesis Techniques

Selection of the synthesis method is important for matching material properties to performance needs in avionics, including thermal control, signal transmission, and structural incorporation. Table 3 gives a comparison analysis of prevalent graphene synthesis methods according to material quality, scalability, cost, and appropriateness for diverse avionics-related applications.

Table 3. Comparative Overview of Graphene Synthesis Techniques for Avionics Applications

Method	Quality	Scalability	Cost	Avionics Applications
Mechanical Exfoliation	Very High	Very Low	High	Research-grade sensors, quantum electronics
Chemical Vapor Deposition	High	High	Moderate	Transparent electrodes, EMI shielding, signal conductors
Liquid-Phase Exfoliation	Medium	High	Low	Composite coatings, printed circuits, thermal films
Reduction of Graphene Oxide	Medium	High	Low	Flexible electronics, energy storage, smart skins
PECVD / Laser Scribing	High (Patterned)	Medium	High	Nanoelectronics systems, smart interfaces, embedded sensors

Considerations for Aerospace Integration

Environmental Compatibility

Environmental Compatibility is a concern. Space materials are generally required to be light, temperature-sensitive, and non-metallic substrate-compatible. Methods like PECVD and LPE are highly beneficial in such scenarios, as they come with low processing temperatures and can easily accommodate non-stiff substrates. (Choucair et al., 2009; Price, 2017). PECVD makes it possible to directly deposit graphene on polymer substrates like polyimide and PET, and hence is a good candidate for flexible electronics, wearables, sensors, and curved interior parts. LPE enables flexible processes like spray coating, roll-to-roll processing, and inkjet printing on thermoplastics, carbon fiber-reinforced polymers, and other aerospace materials. This process compatibility ensures long-term adhesion, thermal stability, and seamless mechanical integration in aircraft structures.

Performance Demands

Performance requirements are crucial. High-frequency signal transmission, structural strength, or electromagnetic interference (EMI) shielding demand premium graphene quality. CVD is ideal for producing monolayer or bilayer graphene, providing enhanced electrical conductivity, high carrier mobility, and mechanical strength (Du et al., 2022). These traits are essential for applications like transparent antennas, radome coatings, and avionics signal guiding layers. The high homogeneity and purity of CVD-grown graphene ensure signal uniformity and long-term reliability in precise devices like navigation systems and onboard diagnostics. Thus, synthesis methods that provide better structural and electrical integrity for use in applications are always desirable when high-performance needs are fulfilled.

Cost and Scalability

Cost and scalability become therefore major drivers for aerospace systems that demand large-scale deployment, e.g., structural health monitoring networks or EMI cabin panel shielding, e.g., structural health monitoring networks or EMI shielding of cabin panels. Methods that generate high-quality material at the expense of production quantity are thus preferred. Liquid-Phase Exfoliation (LPE) provides a low-cost, scalable technique susceptible to mass production of sensors and conductive coatings. (Chavalekvirat et al., 2024). Likewise, the reduction of rGO provides a cost-effective method of producing semi-conductive films suitable for energy storage modules, environmental sensors, or disposable biosensors. (Saleh & Hassan, 2023). These techniques provide designers with flexibility, particularly where minor variation in layer homogeneity or material perfection is tolerable.

Sustainability Considerations

Sustainability is driving the choice of means to produce graphene. Green synthesis processes are striving to minimize the environmental impact with guaranteed strong performance. Bio-exfoliation using natural surfactants or enzymes avoids the use of toxic solvents. (Taha et al., 2020). Microwave-assisted synthesis offers an energy-efficient alternative to energy- and labor-intensive thermal reduction at high temperatures. (Xie et al., 2019). Electrochemical exfoliation reduces chemical waste even more and allows control over flake dimensions and functional group distribution. (Liu et al., 2019). These methods support the aerospace industry's goals of carbon-neutral manufacturing, minimization of wasteful damage, and compliance with green procurement standards as outlined by regulatory agencies.

In conclusion, a synthesis route for graphene needs to be selectively tailored depending on the functional, economic, and environmental limits of the intended aerospace application. Performance specification, cost-effectiveness, environmental compatibility, scalability, and sustainability are some of the aspects that lead to the choice of an ideal synthesis route. Table 4 summarizes some of the most critical factors that influence the choice of graphene synthesis routes and indicates their relevance for avionics integration.

Table 4. Synthesis Method Selection Matrix for Graphene Integration in Avionics

Consideration	Preferred Synthesis Techniques	Typical Applications in Avionics
Environmental Compatibility	PECVD, LPE	Flexible electronics, interior sensors, polymer integration
Performance Demands	CVD	EMI shielding, transparent electrodes, signal processing
Cost and Scale	LPE, rGO	Structural coatings, disposable sensors, printed electronics
Sustainability	Bio-exfoliation, Microwave-assisted, Electrochemical	Eco-friendly production, lightweight green composites

INTEGRATION OF GRAPHENE IN NAVIGATION AND FLIGHT CONTROL SYSTEMS

Graphene's exceptional electrical conductivity, mechanical robustness, flexibility, and atomic thinness have ushered in new avenues in the development of next-generation avionics with a promise of light, energy-efficient, and highly responsive control. (Panghal et al., 2025). Being multimodal, it is well-suited for sensing, actuation, and signal transmission, enabling adaptive control surfaces, responsive actuators, and high-end inertial navigation units (INUs). All these technologies have the potential to transform flight control platform architecture in the future.

Recent research exhibits how graphene can enhance the efficiency of major flight components. As can be seen in Figure 2, graphene is being integrated more and more into antenna arrays, navigation modules, and communications systems. Its greater conductivity and tunable electronic properties facilitate high-frequency signal transmission, low-noise amplification, and device miniaturization. In navigation modules, graphene-based circuits enhance sensitivity and processing rate, while in antennas and communication systems, the material facilitates flexible designs, low electromagnetic interference, and wider operational bandwidth, altogether promoting contemporary flight control and communication architecture.

Figure 2. Illustration of graphene integration in key avionics subsystems, including antenna arrays, navigation modules, and communication systems. Created by the author.

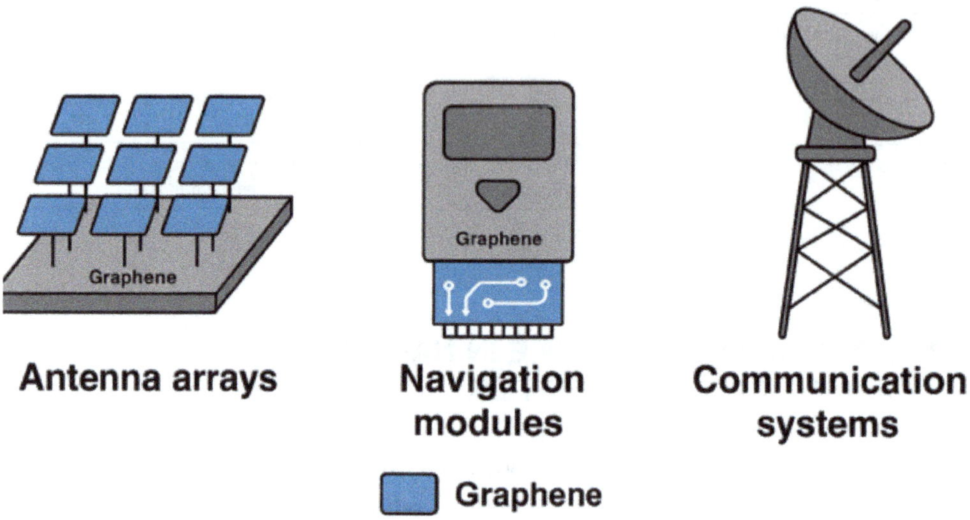

Role of Graphene in Avionics Control Architectures

Modern aircraft require real-time control systems capable of adapting to dynamic flight conditions. These systems rely on high-bandwidth data exchange, accurate sensor feedback, and robust actuators that must perform reliably under extreme temperatures, pressures, and mechanical stresses. (Geim & Novoselov, 2007). Traditional materials often impose limitations due to weight, rigidity, and reduced sensitivity in harsh environments.

Graphene-based components overcome many of these challenges. Their combination of high-frequency signal response, mechanical flexibility, ultralight weight, fatigue resistance, and miniaturization potential makes them ideal for compact, multifunctional avionics architectures. Graphene enables embedded sensor networks and actuator systems capable of precise, high-speed operation under extreme conditions.

Graphene-Enhanced Adaptive Control Surfaces

To optimize aerodynamic performance, adaptive control surfaces like rudders, ailerons, and flaps are crucial. Conventional designs make use of stiff, mechanically or hydraulically driven structures that are ill-suited to quickly changing conditions, prone to wear, and have a slow response time. Graphene enables flexible composites that integrate graphene-based sensors and electroactive materials, enabling a new generation of "smart skin" technologies that may dynamically alter the control surface's geometry. (Pina et al., 2015). These surfaces are equipped with high-resolution strain sensors that can continuously provide real-time feedback by detecting submicron deformations. This capability is further enhanced by electroactive graphene composites, such as graphene PVDF, which flex or stretch in response to electrical inputs. (Maji et al., 2024). This enables accurate airflow control, which lowers drag, increases agility, and maximizes energy efficiency. High-frequency reaction times, low weight and power needs, integrated self-sensing capabilities, high sensitivity and durability, and operational resilience under harsh conditions are some of the main benefits. (Zitoun et al., 2022).

Graphene-Enhanced Responsive Actuators

The weight and power consumption, and wear of conventional aerospace actuators like piezoelectric, shape memory alloy, or hydraulic, are limited, particularly in high-speed or space-constrained settings. (Huang et al., 2012). Because of their remarkable thermal stability, electrical conductivity, and flexibility, graphene-based actuators, which use graphene polymer composites, GO, or rGO, offer quick, accurate, and energy-efficient actuation. (Lee, 2016).

Electroactive graphene-polymer composites (e.g., graphene-PVDF) deform via ion migration or heat-induced shape changes, showing faster response than pure PVDF. (Wang et al., 2015). GO films respond to humidity or electrostatic stimuli, with asymmetric designs enabling significant bending (~1800°) for moisture-responsive applications. (Qiu et al., 2018). These actuators can control micro-flaps, wing-warping surfaces, and morphing fins in UAVs and space vehicles, enhancing maneuverability, lift-to-drag ratios, and aerodynamic adaptation.

Graphene-based actuators demonstrate ultra-fast response times (milliseconds), high energy efficiency, minimal weight addition, long-term durability, and scalable fabrication for flexible electronics and printed systems. Table 5 compares conventional actuators with graphene-based alternatives, highlighting graphene's superior responsiveness, miniaturization potential, and multifunctional sensing.

Table 5. Comparison of Conventional vs. Graphene-Based Actuators for Flight Control Applications

Parameter	Conventional Actuators	Graphene-Based Actuators
Response Time	Moderate (10–100 ms)	Fast (1–10 ms)
Weight	Relatively heavy (due to metal/ mechanical parts)	Ultra-lightweight (few mg–g range)
Integration Complexity	High (requires mechanical linkages, control units)	Low (flexible, can be printed or laminated)
Power Requirement	High (requires hydraulic or high-voltage systems)	Low (operates at ≤10 V)
Mechanical Endurance	Moderate (susceptible to wear/fatigue)	High (resilient under cyclic stress)
Scalability	Limited	High (suitable for nano/micro-scale applications)
Temperature Tolerance	Moderate	High (graphene stable up to ~3000 K in inert atmosphere)
Self-sensing Capability	Rare (external sensors needed)	Inherent (strain-sensitive properties enable feedback)

Graphene in Inertial Navigation Units (INUs)

INUs enable precise positional awareness without relying on GPS, critical in GPS-denied environments such as tunnels, military zones, or deep space (Goossens et al., 2017). Traditional MEMS-based accelerometers and gyroscopes face limitations in scaling, power efficiency, and signal drift (Lu et al., 2021).

Graphene-based nanoelectromechanical systems (NEMS) provide a high-performance solution, offering low mass, high mechanical strength, and enhanced Young's modulus (~1 TPa) for ultra-sensitive angular velocity detection. (Wu et al., 2021) (Geng & Li, 2024). These NEMS devices minimize drift, reduce noise, and enable compact, CMOS-compatible sensors suitable for UAVs, nanosatellites, and hypersonic platforms. (Wang et al., 2020) (Ryan et al., 2024).

Figure 3. Conceptual diagram of a graphene-based quantum inertial navigation sensor utilizing gyroscope and accelerometer systems with quantum-enhanced precision. Created by the author.

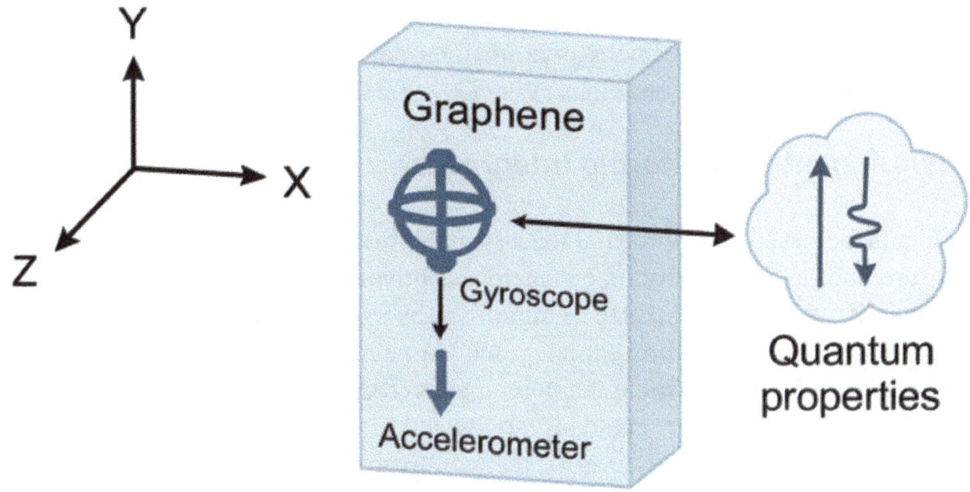

Thermal and EMI Shielding in Flight Control Electronics

Flight control electronics are sensitive to thermal fluctuations and electromagnetic interference (EMI), particularly in high-altitude, hypersonic, or stealth operations. (Han et al., 2016). Graphene, with its superior electrical and thermal conductivity and lightweight flexibility, provides multifunctional solutions. Graphene films embedded in epoxy or printed as thin coatings improve heat dissipation (5000 W/m·K) (Balandin et al., 2008; Gao et al., 2021), while multilayer graphene and rGO films offer >90 dB EMI shielding in the X-band (8–12 GHz), reducing crosstalk and protecting avionics (Oliveira et al., 2022). Nonetheless, in the case of very thin coatings (<10 μm) that apply to lightweight aerospace electronics, shielding effectiveness values are usually between 40–70 dB over the C- to Ku-band (4–18 GHz)—still adequate to suppress crosstalk, radar interference, and electromagnetic leakage in avionics enclosures and signal processors. Graphene coatings also dissipate static charges and enhance fire safety, ensuring compliance with aerospace standards.

The EMI shielding action of graphene is based on a synergistic reflection-absorption combination, where the delocalized π-electrons are involved in high electrical conductivity to allow effective reflection, and the layered structure and defect sites in rGO increase absorption through repeated internal reflections.

Additionally, graphene coatings may dissipate electrostatic charges, improve fire retardancy, and comply with FAA and EASA material safety requirements, further guaranteeing aerospace compliance.

Applications include printed graphene layers on processor housings for thermal and EMI management and graphene-enhanced thermal interface materials (TIMs) replacing conventional thermal pastes in flight data recorders, providing reliable operation under extreme conditions.(Wu et al., 2023).

Environmental Monitoring Integration

Graphene-based environmental sensors enable real-time monitoring of humidity, pressure, and gas concentrations, enhancing adaptive control and flight safety. (Kitadai et al., 2021; Wang et al., 2016). Humidity sensors on control surfaces can detect moisture accumulation to prevent icing, while graphene-based capacitive pressure sensors dynamically adjust engine components to optimize thrust. Gas detection via chemiresistors monitors cabin air quality and external hazards.

A key advantage of these sensors is their ability to detect trace levels of gases and humidity in real-time, even in the dynamic and high-altitude environments characteristic of modern aviation. Figure 4 illustrates a conceptual layout of a graphene-based environmental sensor array, designed for the detection of gas and humidity levels within aircraft interiors. These sensors can be integrated into ventilation systems, cockpit instrumentation, and cargo holds, enabling continuous monitoring of environmental parameters that are crucial for health and safety.

Figure 4. Conceptual illustration of graphene-based environmental sensors integrated in an aircraft cabin for real-time detection of gases (e.g., CO_2, NO_X) and humidity levels, enhancing air quality monitoring and passenger safety. Created by the author.

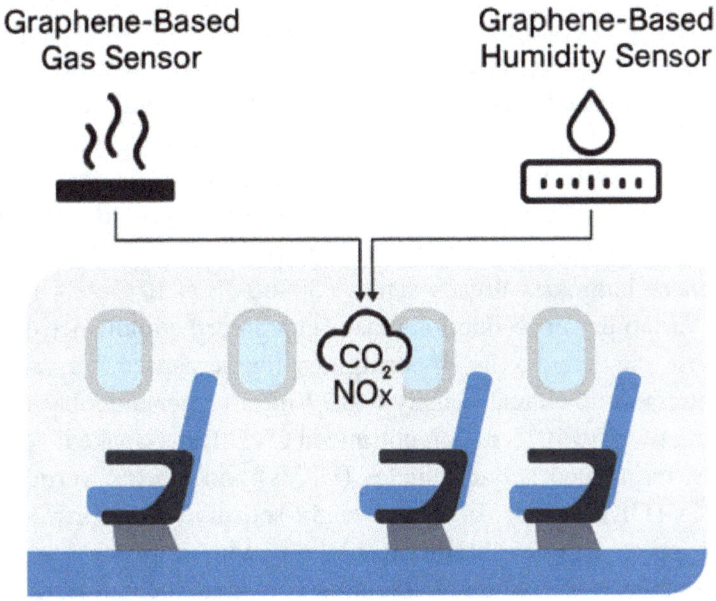

Challenges and Opportunities

Although graphene integration into flight control and avionics systems holds transformative potential, several technical and certification-related barriers still limit large-scale adoption in regulated aerospace environments.

One persistent challenge is material uniformity and scalability. Chemical vapor deposition (CVD) has enabled the production of wafer-scale and even meter-scale graphene sheets, yet maintaining consistent quality across these areas remains difficult. Variations in layer thickness (often ±1–2 layers), point defect density ($>10^{11}$ cm^{-2} in some batches), and oxygen content can alter electrical conductivity and mechanical response. Such inconsistencies complicate sensor calibration and actuator reliability, sometimes requiring post-growth annealing or plasma treatments to reach aerospace qualification standards. (Deng et al., 2019); (Zhong et al., 2015).

Mechanical durability under flight conditions is another concern. Laboratory studies show pristine monolayer graphene has a tensile strength near 130 GPa and a fatigue limit exceeding 1 GPa at 10^7 cycles; however, polycrystalline CVD films exhibit a 20–40% drop in fatigue strength after 10^6 vibration cycles at 200–250 Hz and temperature swings between −50 °C and +150 °C (Bunch et al., 2007; Nagaraj, 2023). Since avionics hardware may experience more than 10^9 low-amplitude cycles during its service life, long-term crack propagation and delamination at grain boundaries remain active research areas before certification can be achieved.

System compatibility also poses significant engineering hurdles. Legacy flight control architectures were not designed for ultrathin, high-mobility carbon materials. Integrating graphene-based sensors and EMI shields often requires redesigning connectors, adapting digital signal protocols, and passing electromagnetic compatibility (EMC) tests such as DO-160 Section 21. New graphene composites must demonstrate shielding effectiveness of at least 40 dB from 1 MHz to 6 GHz to match or exceed aluminum foils traditionally used in avionics (Oliveira et al., 2022). High-quality few-layer graphene laminates already achieve 50–60 dB at 10 GHz with <0.5 mm thickness, but variability in production makes guaranteed compliance difficult.

Despite these challenges, the opportunities for aerospace are considerable. Graphene's extremely low areal density (~0.77 mg m^{-2} per monolayer) and high electrical conductivity (>10^6 S m^{-1} in optimized CVD films) make it attractive for next-generation unmanned aerial vehicles (UAVs) and electric vertical take-off and landing (eVTOL) aircraft. In these weight-sensitive platforms, substituting conventional copper wiring or aluminum shielding with graphene composites can cut wiring mass by 30–40% while maintaining comparable conductivity and EMI protection, directly translating to longer flight endurance and reduced battery load. (Liang et al., 2025).

Graphene's flexibility and ability to form thin conformal coatings also support modular retrofits of legacy aircraft. Instead of replacing entire control systems, avionics suppliers can integrate lightweight graphene-based sensors, strain gauges, or EMI-shielding skins onto existing fuselage and wing structures. This incremental approach reduces downtime and certification costs while extending the lifespan and capability of current fleets (Geim & Novoselov, 2007).

Equally important are sustainability benefits. Traditional avionics shielding and sensor substrates often rely on rare-earth or heavy-metal components. Recent life-cycle analyses show that microwave-assisted or electrochemical graphene production can cut energy use by 35–50% compared with high-temperature reduction methods, and bio-exfoliation using natural surfactants avoids toxic solvents entirely. (Favour, 2025). Replacing aluminum wiring harnesses with graphene composites could lower aircraft manufacturing emissions by ~15% and reduce operational fuel consumption due to weight savings, aligning with the aerospace sector's 2050 carbon-neutral targets.

By systematically addressing this material, mechanical, and certification barriers, while exploiting graphene's low weight, multifunctionality, and environmentally responsible production routes, the aerospace industry can move toward safe, scalable, and commercially viable deployment of graphene-enabled avionics and flight control systems.

GRAPHENE IN AEROSPACE STRUCTURAL COMPOSITES AND STRUCTURAL HEALTH MONITORING (SHM)

The integration of graphene into aerospace structural components is transforming material design, merging exceptional mechanical properties with intelligent sensing capabilities. Graphene-enhanced materials not only offer superior strength and reduced weight but also enable in-situ monitoring of stress, strain, fatigue, and structural damage, forming the foundation for smart, self-aware structures. This innovation reduces the frequency of manual inspections, improves aircraft safety, and prolongs service life.

Graphene-Reinforced Structural Composites

Modern aerospace relies heavily on composites for their high strength-to-weight ratios, corrosion resistance, and design flexibility. Incorporating graphene nanoplatelets (GNPs) or functionalized GO into conventional composites, such as carbon fiber-reinforced polymers (CFRPs), results in next-generation materials with remarkable improvements in mechanical, thermal, and electrical properties (Asyraf et al., 2022).

Mechanical Enhancements

Graphene's exceptional tensile strength (~130 GPa) and Young's modulus (~1 TPa) enhance load transfer at the fiber-matrix interface. Even a 0.1–1 wt% graphene addition can improve tensile strength by 25–40% due to improved stress distribution. (Mirabedini et al., 2020). Graphene nanosheets act as crack deflectors, dissipating energy and delaying fracture, which increases toughness and impact resistance. (Iqbal et al., 2024). Under cyclic loading, graphene bridges microcracks and inhibits propagation, extending fatigue life—critical for wings, fuselage skins, and rotor components. Experimental studies indicate fatigue life extensions of 20–35% under repeated stress cycles typical of commercial aircraft (10^5–10^6 cycles at 30–50 MPa stress range) (Florea et al., 2024).

Thermal and Electrical Conductivity

Graphene's high thermal conductivity (>2000 W/m·K) efficiently dissipates heat generated by avionics or aerodynamic friction (Balandin, 2011). Electrically, graphene forms conductive pathways for lightning strike dissipation and acts as electromagnetic interference (EMI) shielding for sensitive electronics. Frequency-dependent studies demonstrate >90 dB shielding in the 30 MHz–10 GHz range for graphene-infused composites, outperforming conventional aluminum coatings in both weight and performance (Song et al., 2021).

Interface Chemistry and Functionalization

Graphene's integration depends critically on interfacial bonding. Functionalization with carboxyl, hydroxyl, or amine groups enhances compatibility with polymer matrices, forming covalent or hydrogen bonds that strengthen the matrix-fiber interface. Chemical functionalization of graphene with certain chemical groups—like carboxyl (-COOH), hydroxyl (-OH), or amine (-NH_2) moieties—is crucial to enhancing interfacial compatibility. This chemical coupling improves load transfer efficiency and long-term durability, particularly under thermal cycling and mechanical fatigue. (Iqbal et al., 2024). Experimental investigations show that even low loadings of graphene (0.1–1 wt%) can enhance tensile strength by 25–40%, while fatigue life can be extended by 20–35% under repeated cyclic loading typical of aerospace applications (10^5–10^6 cycles at 30–50 MPa)(Florea et al., 2024).

Graphene-Based Sensors for Structural Health Monitoring (SHM)

Structural Health Monitoring (SHM) is a critical component of aerospace safety and maintenance. Conventional SHM systems, such as piezoelectric sensors and fiber Bragg gratings, have supported aircraft reliability for decades, but they come with limitations: fragility, added weight, complex wiring requirements, and restricted sensing coverage. (Jiang et al., 2024). Graphene-based sensors offer an attractive alternative. Their flexible, lightweight, and ultrathin design enables them to be printed or embedded right on the surface of aircraft so that distributed, real-time sensing over broad areas can be achieved without adding weight or complexity. (Das et al., 2024).

The design of multimodal SHM systems for strain, stress, temperature gradient, acoustic noise, and delamination sensing in composite structures is strengthened by graphene's exceptional electrical conductivity, thermal stability, and mechanical strength. Unlike conventional rigid devices, graphene sensors can conform to curved aerodynamic surfaces, retain sensitivity under large strain cycles, and operate across extreme aerospace environments. Furthermore, their compatibility with wireless and autonomous data acquisition systems makes them especially relevant for next-generation smart aircraft.

Recent investigations highlight the use of graphene strain sensors as superior alternatives to traditional rigid strain gauges. With gauge factors of more than 100, as opposed to 2–5 in conventional metal sensors, graphene devices provide both flexibility and ultra-lightweight integration, and are therefore well-suited for wings, flaps, and fuselage joints. (Nagaraj, 2023). Figure 5 is a conceptual plan view of an embedded graphene strain sensor in an aircraft flap and how it can monitor deformation in real time during flight and give an early indication of fatigue, stress buildup, or failure.

Figure 5. Schematic representation of a graphene-based strain sensor integrated within an aircraft flap structure. Created by the author.

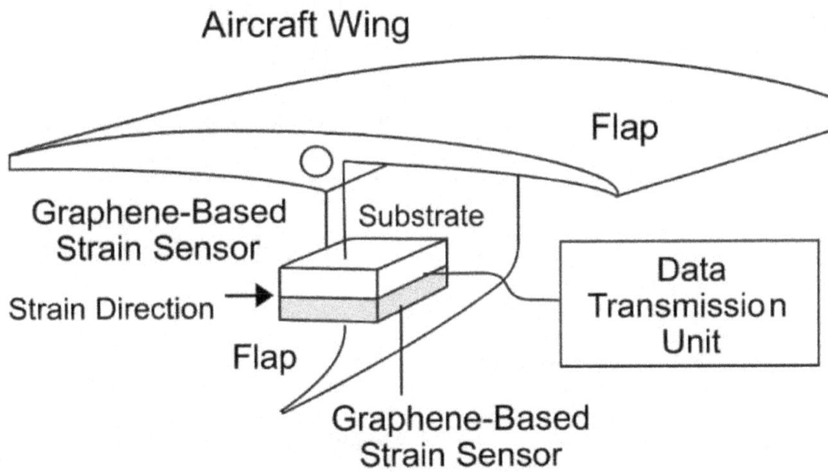

Graphene Sensing Mechanisms

The versatility of graphene sensors arises from several underlying transduction mechanisms. In piezoresistive sensing, resistance varies predictably with applied strain because lattice deformation alters charge carrier mobility (Cao et al., 2020).

This impact allows for the sensitivity to detect micro-strains as low as 0.01%, which is vital for detecting stress evolution in high-load structures (Reghat et al., 2021). Capacitive sensing, often employed by multilayer graphene–polymer composites, presents sensitivity to displacement or delamination of fuselage skins and sandwich panels. Wirelessly monitoring such capacitance changes at low power can support autonomous inspection regimes (Yang et al., 2019). Furthermore, graphene is very sensitive to acoustic stress waves and heat variations. The sensitivities allow sensors to sense transient events, like hail or bird impacts, to provide instantaneous diagnostics prior to detectable surface damage (Blaikie, 2020).

The advantages of graphene-based SHM devices in performance over conventional sensors are listed in Table 6. It stands a chance to be a future aero system because it is more sensitive, has a higher operating temperature range, and is insensitive to cycling loading.

Table 6. Comparison of Conventional vs. Graphene-Based SHM Sensors for Aerospace Applications

Parameter	Conventional Sensors	Graphene-Based Sensors
Flexibility	Rigid	Flexible/stretchable
Weight	Moderate to high	Ultra-lightweight
Sensitivity (Gauge Factor)	2–5 (strain gauges)	>100 (graphene strain sensors)
Temperature Range	–50 to 150 °C	–150 to 300 °C
Durability (Cycles)	$<10^4$	$>10^6$
Integration Ease	Discrete placement	Layered or printed onto surfaces

Applications in Aerospace SHM

In practical use, graphene-based SHM sensors offer multiple functionalities. Incorporated layers in composite skins can sense abrupt resistance variations upon impacts and detect bird strikes, hail damage, or runway debris events with high accuracy. (Shalabi, 2022). Printed sensor arrays over fuselage joins and wing spars monitor strain accumulation continuously and alert against fatigue long before critical levels are approached. (Kim et al., 2021). Just as important is their capacity to identify subsurface delamination in honeycomb or carbon fiber laminates, where standard visual inspection is not enough (Steinke et al., 2021).

The integration of graphene is also catalyzing a transition from passive to adaptive aerospace structures. Graphene-enhanced composites can simultaneously sense strain and enable actuation, forming the basis for self-sensing and morphing systems. (Carvalho et al., 2022). For instance, shape memory polymers doped with graphene exhibit enhanced conductivity, allowing control surfaces to adjust dynamically in response to aerodynamic loads. These materials demonstrate rapid recovery and resilience, offering new pathways for morphing wing and rudder designs. (Sudan, 2024).

Graphene-based self-healing coatings represent another innovation. Polymer matrices containing graphene oxide or reduced graphene oxide have demonstrated the ability to autonomously repair microcracks upon thermal or mechanical activation, restoring up to 95% of mechanical strength after damage. (Idumah & Odera, 2020). Such coatings reduce maintenance frequency and extend component service life.

Finally, graphene "smart skin" arrays emulate the tactile sensitivity of biological skin. Stretchable graphene sensors printed onto aircraft surfaces can detect pressure, temperature, and strain in real time. (Chen et al., 2018). Laser scribing and inkjet printing techniques, which can scale up to large structures, can integrate the arrays. (Ji et al., 2023). Collectively, these smart skins support improved aerodynamic control, impact detection, and early warning of failure conditions.

Integration Challenges and Certification Barriers

In spite of these promising developments, many integration challenges persist. Uniform dispersion of graphene into polymer matrices remains a limitation to the reproducibility of electrical and mechanical properties. Aggregation of nanosheets contributes to localized defects that diminish sensor reliability, necessitating optimization of dispersion and curing. (Kim et al., 2010); (Macosko, 2010). Environmental stability is also a challenge, as graphene oxide derivatives break down upon extended humidity and UV exposure. (Pei & Cheng, 2012). Protective encapsulation or hybridization strategies are hence essential in order to guarantee long-term performance.

Certification and standardization are also key roadblocks. Complimentary organizations like the FAA and EASA need painstaking verification, such as cyclic fatigue testing, EMI compatibility tests, and long-term simulations. (Asyraf et al., 2022; del Cuvillo Mezquita, 2023). Compliance with these needs enforces coordinated test and compliance frameworks, which have only just started to materialize. Without codified standards, mass commercial deployment of graphene SHM systems will continue to be constrained.

Future Outlook and Industry Trends

In the future, graphene-based SHM systems will merge with digital aerospace technologies to form more clever and responsive maintenance environments. Digital twins—virtual models of physical aircraft—are being coupled with graphene sensor inputs to simulate structural behavior and predict degradation with high fidelity (Santos-Ceballos et al., 2024). Machine learning and artificial intelligence models are further augmenting this ability by processing big sensor datasets to predict crack development and remaining service life and identify anomalies. (Mardanshahi et al., 2025; Sabato et al., 2023).

Along with this, multifunctional composites of graphene are being engineered to integrate indigenous sensing with mechanical strength, particularly for UAVs and conventional aircraft, where material efficiency is extremely important. (del Bosque et al., 2025). At the same time, networked graphene sensors in composite form over large areas are becoming feasible due to additive manufacturing techniques, like inkjet and 3D printing. (Wang et al., 2019; Wu et al., 2023).

Together, these advancements herald a revolution: SHM systems will be designed in as integrated, multi-faceted elements of aerospace design, not as isolated add-ons. The marriage of the properties of graphene with predictive analysis from AI, digital twins, and advanced manufacturing technology has positioned it as the material of choice for future aircraft that are more efficient, safer, and responsive.

SCALABLE MANUFACTURING STRATEGIES, INTEGRATION CHALLENGES, RELIABILITY, AND CERTIFICATION PATHWAYS FOR GRAPHENE-BASED AVIONICS

Graphene-based avionics technologies hold out the potential for revolutionary performance gains through the integration of ultra-low weight, superior electrical and thermal conductivity, and mechanical strength. These properties can substantially enhance flight control, navigation, and structural health monitoring while minimizing the total system mass and power consumption. However, translating these laboratory-scale demonstrations into airworthy products remains a significant challenge. The challenges include the scalability of graphene manufacturing, its integration with standard aerospace electronic architectures, long-term reliability under harsh flight conditions, and compliance with stringent certification and regulatory standards. This section covers these areas, combining the latest developments and industry views.

Scalable Fabrication Strategies for Graphene Sensors and Components

For graphene-based avionics hardware to be economically viable, the manufacturers need to make defect-controlled, reproducible graphene layers with consistent electrical performance at an industrial production scale. Several fabrication strategies have emerged, each with trade-offs in cost, quality, and throughput.

Chemical Vapor Deposition (CVD) produces high-quality monolayer and few-layer graphene with carrier mobilities exceeding 10,000 $cm^2 V^{-1} s^{-1}$—ideal for high-frequency electronics and electromagnetic interference (EMI) shielding. However, its slow growth rates, substrate dependency, and expensive vacuum systems limit mass adoption.

Liquid Phase Exfoliation (LPE) is a cost-effective, high-throughput process capable of generating graphene inks for printable sensors and composites. Although suitable for large-area coatings, variability in flake thickness and lower conductivity compared to CVD-grown films can restrict high-speed data applications.

Laser or Plasma Reduction of GO enables maskless patterning on flexible substrates with relatively low cost and environmental compatibility. The challenge remains in achieving uniform reduction and controlling oxygen content to maintain long-term conductivity.

Roll-to-Roll (R2R) Transfer is emerging as the most promising industrial-scale approach. It supports meter-scale continuous films and is compatible with flexible polyimide or polyethylene terephthalate (PET) substrates. Recent R2R lines report throughput >50 m h^{-1} and sheet resistances below 30 Ω sq^{-1}, but wrinkles and microcracks introduced during transfer still reduce yield.

Inkjet and aerosol jet printing have shown potential for low-cost, patternable avionics interconnects and sensor arrays. For example, Li et al. demonstrated fully printed graphene micro-supercapacitors with areal capacitance up to 11 mF cm^{-2} and 95% retention after 10,000 cycles, highlighting reliability under repetitive load conditions. (Li et al., 2017).

Table 7 summarizes the advantages and limitations of prominent graphene fabrication methods relevant to avionics and aerospace applications. Techniques such as CVD offer high-quality graphene layers ideal for advanced electronics, while scalable methods like Liquid Phase Exfoliation and Roll-to-Roll Transfer enable cost-effective integration into composites, coatings, and printed circuits. Each method presents trade-offs in terms of scalability, quality, and application suitability.

Table 7. Comparative Overview of Graphene Fabrication Techniques for Aerospace Integration

Method	Advantages	Limitations
Chemical Vapor Deposition (CVD)	High-quality monolayers, good control	Expensive, substrate-limited, slow throughput
Liquid Phase Exfoliation (LPE)	Scalable, low-cost, adaptable to ink formulations	Thickness variations, lower conductivity
Laser/Plasma Reduction of GO	Maskable patterning, suitable for flexible devices	Uniformity challenges
Roll-to-Roll Transfer	Industrial compatibility, high-throughput	Substrate-induced defects, wrinkles

Integration with Aerospace Control Systems

Graphene-based avionics solutions offer revolutionary performance benefits through the merging of extremely low weight, superior electrical and thermal conductivity, and mechanical strength. These capabilities have the potential to substantially enhance flight control, navigation, and structural health monitoring while minimizing overall system mass and power consumption. However, these laboratory demonstrations need to be transferred into airworthy products. The primary obstacles include the scalability of graphene production, its smooth integration into current aerospace electronic systems, long-term dependability under extreme flight conditions, and adherence to strict certification and regulative requirements. These are the areas covered under this section, incorporating the latest developments and industry viewpoints.

The seamless integration of graphene-based sensors into existing avionics frameworks, such as flight control systems, environmental monitoring units, inertial navigation units, and emerging digital twin platforms, is crucial to the successful deployment of these sensors in future aircraft. This is necessary for these sensors to be used in future aircraft. Integration models should focus on lowering weight, size, energy use, and real-time data accuracy in order to work well in harsh aerospace environments.

Flexible Printed Electronics (FPE)

FPE makes it possible to directly deposit graphene interconnects and sensor circuits onto curved aircraft skins, wings, and control panels. PET and polyimide substrates can tolerate temperatures ranging from -55 °C to +150 °C, which satisfies standard aerospace environmental requirements (Tong, 2022). In recent FPE implementations, antenna arrays have been printed directly onto fuselage surfaces, reducing weight by over 85% over coaxial cabling without sacrificing radio frequency performance. (Riaz et al., 2023). For electric propulsion systems and fuel economy, this lightweighting is essential.

Hybrid Sensor Packages

Combining graphene's ultrahigh sensitivity with silicon-based CMOS readouts has enabled hybrid MEMS–graphene packages with multiplexed sensing capability. (Ramesh et al., 2020); (Jiang et al., 2021). These architectures integrate graphene field-effect transistors (GFETs) as the front-end sensing layer and CMOS for signal amplification, signal conditioning, and wireless transmission. Polyimide or ceramic encapsulation of coating is applied to improve UV and humidity resistance, adding in excess of 20,000 thermal cycles to service life. Hybridization reduces risk by leveraging proven avionics electronics certification pathways while improving performance.

Wireless Sensor Networks (WSNs)

Wireless, dispersed SHM networks are made possible by graphene's low sheet resistance ($<30\ \Omega\ sq^{-1}$) and energy harvesting compatibility. These devices use low-power radio frequency protocols that comply with aviation standards to exchange information and capture energy from vibration or heat gradients (Sadeghi et al., 2024). Prototypes of graphene-based air quality sensors demonstrated dependable operation for 2,000 hours at cabin altitudes of 35,000 feet with no signal drift greater than 3% (Darwish et al., 2024), with potential use in next-generation self-powered avionics diagnostics.

Reliability and Environmental Stability of Graphene-Based Systems in Avionics

Material reliability and environmental strength are of the utmost importance in aerospace engineering because avionics are subjected to extremely dynamic and, in most instances, extreme conditions throughout the service lifetime of aircraft—from ground operation, take-off to long-duration high-altitude flight. Because of its extremely high mechanical strength, thermal stability, and electrical conductivity, graphene is a desirable material for avionics in the future. To ensure safe, long-term, and certified use, its long-term behavior must be determined under stressors unique to aviation.

Environmental Stress Factors in Aerospace Applications

Graphene-enabled avionics are continuously exposed to a combination of severe environmental challenges. Aircraft components frequently undergo sharp thermal gradients, with temperatures shifting from sub-zero levels near −60 °C at cruising altitudes to as high as +120 °C near engines or under intense solar heating during ascent and descent. (Yavuz & Ozkol, 2024). In parallel, continuous vibration and cyclic mechanical fatigue, generated by aerodynamic forces, propulsion systems, and onboard actuators, impose oscillatory stresses on sensors and electronics that may alter calibration and structural integrity over time. (Lalanne, 2014). Prolonged ultraviolet (UV) exposure further threatens materials mounted on the aircraft skin or other externally facing surfaces, potentially degrading structural and electrical performance. (Liu et al., 2024). Rapid variations in cabin pressure and the presence of unpressurized compartments can lead to cyclic condensation and humidity fluctuations, which compromise unsealed electronics and sensitive interfaces. (Jones & Foreman, 2015). Additionally, avionics must tolerate significant electromagnetic interference (EMI) generated by radar, high-power communication modules, and other onboard systems, as well as resist radiation at high altitudes or in near-space environments, where cosmic rays and ionizing radiation can impair sensor longevity and stability. (Mishra et al., 2018; Sihver et al., 2015)

Comparative Performance Under Operational Stress

Experimental research has recently compared the performance of graphene-based materials with traditional aerospace substrates under simulated real-world flight conditions. The performance in terms of fatigue resistance, thermal cycling stability, and EMI shielding. These studies indicate that typical polymer-based traces micro-crack and delaminate following repeated exposure to −60 to +120 °C cycles, while graphene films preserve stable conductivity with no detectable delamination. When subjected to high-frequency mechanical fatigue (10^6 Hz), typical conductors show greater than 20% signal drift after 100 hours, while graphene devices show less than 5% degradation even after 500 hours. (Meng et al., 2024). In 30 MHz–10 GHz EMI testing, standard aluminum or copper shielding achieves 50–70 dB attenuation, while graphene-based thin coatings have already achieved over 90 dB shielding in layers less than 1 mm thick. (L. Xu et al., 2021). The findings underline graphene's excellent fatigue resistance, dimensional stability, and electromagnetic shielding, and emphasize its application in next-generation avionics and flight-critical systems.

Enhancing Durability Through Advanced Material Engineering

Graphene is commonly incorporated into sophisticated composite structures that improve adhesion, environmental stability, and structural toughness in order to ensure stable long-term performance. An efficient method is to encapsulate graphene in high-performance polymer matrices. For example, sensors used near fuels, lubricants, or hydraulic fluids can benefit from graphene–polyvinylidene fluoride (PVDF) composites because of their good piezoelectric response and moderate chemical resistance (Zhang et al., 2022). Likewise, graphene–epoxy laminates exhibit functional stability with long-term vibration and flexural loading owing to their high mechanical stiffness and strong interfacial adhesion (Guo et al., 2023). The composites were discovered to impart a robust protective coating against moisture, oxidation, and corrosive species, vibration damping, and tolerate aggressive thermal cycling without delamination (TabkhPaz et al., 2018).

Case studies depict these benefits as well. Following over 500 hours of mechanical fatigue testing mimicking a turbine, graphene strain sensors have been proven to maintain consistent signals. (Meng et al., 2024). Thin graphene-integrated coatings have reduced the overall shielding weight by over 30% when compared to conventional copper or aluminum solutions, while achieving an EMI shielding effectiveness of over 90 dB (L. Xu et al., 2021). Such findings support the hypothesis that tailored composite engineering can mitigate the mechanical and environmental risks that are typically inherent in next-generation in-flight technology.

Certification and Regulatory Considerations

Beyond their technical effectiveness, graphene-based solutions need to comply with stringent certification and safety criteria established by regulatory bodies such as the European Aviation Safety Agency (EASA) and the Federal Aviation Administration (FAA). Complex devices, including graphene-enhanced sensor circuits or control modules that cannot be fully validated through testing alone, require structured design, verification, and documentation processes to demonstrate predictable and fail-safe performance (DO, 2000a, 2000b). At the system level, the SAE ARP4754B standard establishes a rigorous, safety-driven development framework for civil aircraft, complemented by ARP4761, which defines safety assessment processes for integrated hardware and software. (Darrah, 2022; Voros, 2021).

Material qualification also remains critical. Regulations such as FAA 14 CFR Part 25 require extensive testing of mechanical, thermal, and chemical resistance to confirm that new materials, including graphene composites, can withstand the severe operating environment of aviation. (Administration, 2023). EASA's parallel guidelines emphasize not only performance and safety but also production traceability and process control to ensure consistent quality (Databank). Electromagnetic compatibility (EMC) is also essential; graphene's high conductivity enables effective EMI shielding, with studies showing >80–90 dB effectiveness in layers under one millimeter thick. (Nawaz et al., 2021; J. Xu et al., 2021).

A current gap in the regulatory landscape is the absence of graphene-specific testing and qualification protocols. Standardized methods to assess flake dispersion, layer uniformity, fatigue life, long-term adhesion, and EMI performance are still under development. (Allan et al., 2019; Barkan et al., 2023). The establishment of these guidelines will be essential for streamlining airworthiness approval and accelerating the safe integration of graphene technologies into commercial and defense aerospace fleets.

Economic and Cost–Benefit Analysis of Graphene Integration

The integration of graphene components into the aerospace system brings both initial costs and extended operational advantages. Fabrication and processing of high-quality graphene continue to be pricier than traditional materials, but the lifecycle cost can be offset through reduced weight, increased fuel efficiency, and reduced maintenance requirements. For instance, substituting conventional EMI shielding components or structural composites with their graphene-enhanced counterparts decreases aircraft weight by a few kilograms per module, which corresponds directly to quantifiable fuel savings throughout the lifespan of the aircraft. Graphene also has superior thermal management characteristics that can decrease cooling system requirements, which again lowers operational costs. A structured cost–benefit analysis shows that while investment is higher upfront, the net economic benefit is realized in mid-term operating cycles, especially for commercial and military fleets (Kim et al., 2021).

Supply Chain Maturity and Qualification Issues

In addition to regulatory adherence, reliable material supply and manufacturing consistency continue to be essential for aerospace uptake of graphene. Existing supply chains for high-purity graphene flakes, films, and composites are narrow, with variability in flake size, layer uniformity, and defect density that could impact system performance. Qualification of such materials involves not only typical aerospace testing but also tight traceability and process control to guarantee batch-to-batch consistency. New industrial platforms are starting to overcome these issues using standard production procedures, supplier certification, and incorporating quality control at every level. (Barkan et al., 2023). Closing the performance gap between laboratory-scale synthesis and bulk production will be crucial for fulfilling both performance requirements and regulatory demands in actual aerospace applications.

Market Potential and Adoption Timeline

A staged progression is predicted in the adoption of graphene to markets in the aerospace market. Near-term uses (1–5 years) target lightweight EMI shielding, thermal management coatings, and sensor enhancement for UAVs and satellite systems. Mid-term deployment (5–10 years) could extend into integration with commercial aircraft composites, multifunctional coatings, and avionics modules. Adoption for the long term (>10 years) may witness graphene as a regular material in major structural elements and future-generation flight control systems, subject to additional cost reduction, supply chain maturity, and approval of large-scale manufacturing. Increased need for light, efficient, and multi-functional systems makes graphene a strategic material in defense as well as civilian aerospace markets, with market expectations predicting steep growth in the coming decade. (Darrah, 2022).

Comparative Evaluation and Competitor Analysis

Graphene is competing with other advanced materials, including CNTs, MXenes, and engineered metamaterials, in aerospace applications. Whereas CNTs offer high strength-to-weight ratios and MXenes provide superior electrical conductivity, graphene alone integrates ultra-high electrical and thermal conductivity, mechanical flexibility, and thin-film processability. Metamaterials are capable of engineered electromagnetic responses, but usually at the expense of intricate fabrication and compromised structural integrity. Comparative analysis suggests that graphene provides a trade-off among balanced performance, multifunctionality, and integration viability, especially in sensors, EMI shielding, and structural health monitoring systems. Yet, cost, standardization, and fabrication at large areas are domains where incumbent materials still have advantages until now, and hence the necessity for ongoing R&D and industrial development.

CASE STUDIES, EMERGING APPLICATIONS, AND INDUSTRIAL RESEARCH

The application of graphene-based automated control systems in avionics is increasingly moving from laboratory proofs to initial-stage operational deployment. With increasingly data-intensive, light, and multifunctional aerospace systems, the combination of high mechanical strength, superior thermal management, and extraordinary electrical conductivity of graphene has elevated it to become a prime contender for flight electronics in future aircraft. In this section, representative case studies and research efforts conducted in industries are presented, categorized into the defense, civil aviation, and space industries, to demonstrate the direction of the technology toward real-world certification and implementation.

Defense Applications: Smart Combat Aircraft and UAV Systems

Defense aerospace has long led the way in terms of adopting new materials, based on demanding requirements for survivability, weight reduction, and resistance in hostile environments. Graphene's high strength with low weight, combined with its inherent electromagnetic interference (EMI) shielding, makes it exceptionally attractive to unmanned combat aerial vehicles (UCAVs) and new generation fighter aircraft.

One prominent example is from the U.S. Defense Advanced Research Projects Agency (DARPA) under its "NextGen Structural Materials Program." Advanced UCAV prototypes were fabricated with graphene–polymer composite skins containing embedded strain sensors and shape-memory actuators (Kausar, 2022). These "adaptive skins" dynamically adjusted control surface angles and wing camber to counter environmental turbulence and high-g maneuvers. The result was improved agility and survivability, while the graphene layers also reduced the radar cross-section and shielded critical avionics from electronic warfare (EW) attacks. (Amiet et al., 2010).

Additional research within hypersonic drone platforms has focused on graphene-based strain gauges capable of tracking real-time deformation under extreme aerodynamic loading and temperatures exceeding 800 °C. In parallel, self-sensing armor panels integrated into next-generation fighters have demonstrated the ability to transmit stress and impact data directly to cockpit displays, supporting predictive threat response and maintenance planning.

Civil Aviation: Predictive Maintenance and In-flight Monitoring

Commercial aviation is increasingly adopting condition-based maintenance (CBM) and digital twin platforms to reduce unscheduled downtime and operating costs. Graphene's durability and piezoresistive properties make it well-suited for distributed sensing networks embedded in primary structures.

A major initiative is the Airbus Smart Maintenance Program, conducted in partnership with a European nanotechnology consortium. In this pilot, graphene-based strain sensors were embedded in wing spars and landing gear beams. (Florea et al., 2024). These sensors were linked to a digital twin model, enabling early detection of micro-cracks and fatigue accumulation. Over a two-year validation period, the system reduced unplanned aircraft groundings by more than 30% and optimized maintenance intervals, resulting in significant cost savings and enhanced fleet availability.

New trends in civil aviation also involve incorporating graphene-based CO_2 and humidity sensors in cabin environmental control systems for real-time monitoring of air quality. Additionally, graphene thermal interface materials (TIMs) are also under consideration for application to avionics modules to facilitate thermal dissipation in long-endurance and high-power electronics, potentially extending hardware life cycles and reducing cooling system loads.

Space Systems: Radiation-Hardened Sensors and Structural Health Monitoring

Graphene's atomic thinness, ultra-high tensile strength, and radiation resistance offer special benefits for spacecraft and satellite platforms, where weight and long-term durability are paramount.

A collaborative pilot with the European Space Agency (ESA) and NASA successfully illustrated the deployment of graphene-based nanoelectromechanical system (NEMS) strain sensors within the CubeSat structural. (Camps et al., 2023). These sensors successfully tracked micrometeoroid impacts, temperature-induced dimensional changes, and vibration patterns in low-Earth orbit while maintaining operational stability under high-energy particle radiation. Graphene-enhanced coatings also mitigated radiation-induced degradation in sensitive electronics, outperforming traditional aluminum shielding. (Hodson et al., 2013).

Looking ahead, graphene–polymer composites are being investigated as lightweight shields for flight computers, offering improved protection against cosmic rays and ultraviolet degradation. Additionally, graphene-integrated thruster control systems are being explored to enable ultra-precise orientation in nano- and micro-satellite platforms.

Industrial Research and Prototyping Initiatives

Several major aerospace corporations and academic–industry consortia are accelerating the transition of graphene technologies from prototype to certified use. Research activities cover EMI shielding, thermal management, structural sensing, and navigation systems. Table 8 summarizes representative projects.

Table 8. Ongoing Industrial and Institutional Research on Graphene in Avionics.

Organization	Focus Area	Status	Approx. TRL (1–9)	Reference
Lockheed Martin	Lightweight EMI-shielding graphene skins for UAVs	Lab-stage prototypes tested	4-5	(Kargar et al., 2018)
NASA Ames Research	Graphene coatings for thermal management in hypersonic flight	Wind tunnel testing is ongoing	5-6	(Mason, 2022)
Boeing + MIT Labs	Graphene composites for next-gen aircraft panels	Field trials in progress (2025–26)	6-7	(Bello, 2023)
Airbus	Predictive maintenance using embedded graphene sensors	Limited in-service deployment	7-8	(Florea et al., 2024)
ISRO & IIT Kanpur	Miniaturized graphene-based inertial sensors for satellites	Low-orbit prototype launched	6-7	(Somanath & Spaceport)

This global activity reflects growing confidence in graphene's readiness for Technology Readiness Level (TRL) 6–7 applications, where systems move beyond laboratory validation to operational environment testing.

Strategic Implications and Future Outlook

The convergence of graphene-enabled intelligent sensing, adaptive actuation, EMI shielding, and lightweight structural integration is redefining avionics architectures. Military programs emphasize survivability, stealth, and resilience under extreme operating conditions, while commercial and space platforms are leveraging graphene for sustainability, predictive safety, and autonomous health monitoring.

However, successful large-scale deployment depends on more than just material performance. Regulatory certification remains a key barrier, requiring compliance with standards such as RTCA DO-254 for electronic hardware and SAE ARP4754B/ARP4761 for system safety assurance. Moreover, aerospace supply chains must adapt to ensure reproducible, cost-competitive, and traceable production of high-quality graphene materials that meet FAA and EASA material qualification standards.

Another critical factor is market positioning relative to competing advanced materials such as carbon nanotubes (CNTs) and MXenes. While CNTs offer comparable conductivity, graphene provides superior planar coverage for EMI shielding at lower weight, and MXenes excel in thermal conductivity but face oxidation stability challenges. As the cost per square meter of graphene films continues to decline due to scalable chemical vapor deposition (CVD) and roll-to-roll production, graphene is becoming increasingly competitive for both military and civil aviation.

Looking forward, strategic success will require cross-disciplinary collaboration among materials scientists, avionics engineers, and certification authorities, as well as sustained public–private partnerships to bridge the technology readiness gap. Developing standardized testing protocols and technical glossaries for non-specialists will also help interdisciplinary teams navigate the complexity of graphene integration. Ultimately, the material's multifunctionality—if paired with robust reliability validation and cost-benefit justification—could make graphene a cornerstone of the next generation of safe, intelligent, and energy-efficient aerospace systems.

CONCLUSION AND FUTURE PERSPECTIVES

Summary of Key Insights

This chapter has discussed the revolutionary influence of graphene in transforming the design and functionality of aerospace control systems. From adaptive sensing and actuation to real-time structural health monitoring (SHM), graphene has emerged less as an evolutionary step in mainstream materials but more as the foundation for a novel avionics and flight architecture. Its supreme electrical conductivity, thermal stability, mechanical strength, and electromagnetic interference (EMI) shielding capabilities offer designers the capability to craft smarter, lighter, and more resilient aircraft and spacecraft.

Several of the important insights can be found in this review. Graphene's characteristic properties, including high electrical conductivity with low weight, enabled the production of adaptive control surfaces and responsive actuators that enhance flight safety, maneuverability, energy efficiency, and overall responsiveness (Su et al., 2025). Its application in multifunctional systems like inertial navigation and environmental sensing systems further establishes its multipurpose capabilities. Application of graphene smart composites in the frameworks of aircraft provides the potential for real-time monitoring of structural health, together with predictive maintenance, which further reduces life-cycle costs while enhancing safety margins. Concurrently, advances in scalable manufacturing methods and other avenues of acquiring certification have started to meet long-standing concerns over material variation, cost savings, and aerospace qualification. This flowed from positive engagements from the world's major space agencies and airborne aerospace companies such as NASA, ESA, Boeing, Lockheed Martin, and Airbus. Their considerable investments in graphene systems and their early pilot projects provide evidence of the technology's immediate potential (Barkan et al., 2023; Prater et al., 2020).

Future Perspectives

The use of graphene in aerospace technology and space applications will grow exponentially as materials science, digital engineering, and green production converge. Among the major advances will be the integration of graphene sensor networks with digital twins and AI technologies. These platforms will drive self-awareness, autonomous flight systems capable of adapting to turbulence, structural wear and tear, or mission-dependent specifications without human intervention. (Chugh et al., 2024).

In aerospace applications, graphene's ultrahigh tolerance for radiation, thermal cycling, and mechanical stress positions it ideally to be utilized in deep space exploration and satellite electronics. It can be utilized to provide effective radiation shielding when integrated in polymer matrices without losing lightweight and structural properties, thus extending the operational lifespan of spacecraft control systems and sensors in harsh cosmic conditions. (Laurenzi et al., 2024).

The electrification of aviation and the growth of urban air mobility (UAM) further increase the use of lightweight multifunctional materials. Graphene has the ability to serve as a structure, EMI shield, thermal management layer, and distributed sensing network simultaneously, which is optimal for electric vertical take-off and landing (eVTOL) aircraft, drones, and hybrid-electric aircraft. (Hu et al., 2025; Souza, 2024). Another near-future frontier is the production of self-healing and reconfigurable graphene composites. These will heal their aerodynamic shape during flight, which can allow for morphing airframes that are survivable as well as adaptable. (Mohapatra et al., 2023; Sabet, 2024).

Sustainability issues are also increasingly a consideration in aerospace materials planning. Renewable biomass or recycled carbon-based graphene represents a path towards decreasing aircraft manufacturing's environmental impact without sacrificing performance. (Oluremi, 2025; Safian et al., 2020). The integration of green production into aerospace technology can make graphene one of the first high-performance, sustainable materials to have a broad impact on flight systems.

Strategic Roadmap

In order to fully appreciate these advantages, the aerospace sector must have clear and established paths from laboratory research to approved application. Promoting the evolution of standards specific to the aerospace sector for the production of graphene, testing, and measurement of performance will be essential. Regulatory standards such as those managed by the Federal Aviation Administration (FAA) and the European Union Aviation Safety Agency (EASA) should include graphene testing in current certification procedures, for example, standards RTCA DO-254 for electronic hardware and SAE ARP4754B/ARP4761 for system-level safety assurance. (Kausar et al., 2023). Industry–academia collaboration will be vital for bridging the gap between experimental innovation and commercial readiness. Public–private partnerships can help scale up production techniques such as roll-to-roll graphene synthesis, advanced 3D printing, and ink-based deposition systems to meet cost and quality requirements for mass production. At the same time, the seamless integration of graphene-enabled sensors with Internet of Things (IoT) platforms, AI-driven analytics, and cloud-based digital twins will accelerate predictive maintenance and autonomous flight capabilities.

It will also be important to provide interdisciplinary teams with technical glossaries and standardized test protocols to help materials scientists, avionics engineers, and regulatory bodies work within a common framework. Such harmonization will be essential for meeting safety standards while enabling innovation.

Concluding Remarks

Graphene is moving beyond the stage of novel laboratory material to become a foundational enabler of next-generation aerospace systems. Its exceptional mechanical, electrical, and thermal properties—combined with its potential for multifunctional integration—are redefining avionics, navigation, and flight control design. (Novoselov et al., 2012). Already, prototypes in adaptive control surfaces, ultra-sensitive inertial sensors, EMI-shielding skins, and self-monitoring structural components demonstrate that graphene can support lighter, safer, and more intelligent aircraft capable of operating reliably under the extreme conditions of aviation and space.

Significant challenges remain, particularly in achieving reproducible large-scale manufacturing, long-term environmental stability, and full regulatory certification. Yet the trajectory is clearly forward-looking. Ongoing pilot programs by NASA, ESA, Airbus, Boeing, and Lockheed Martin indicate growing confidence in graphene's readiness for operational systems. (Kausar et al., 2023; Scalia et al., 2023; Valorosi et al., 2020). Simultaneously, the rise of electrified aviation, unmanned aerial systems, UAM platforms, and deep-space exploration offers fertile ground for deploying graphene-based technologies that combine structural performance with real-time intelligence.

Ultimately, the emergence of graphene-enabled aerospace systems represents more than incremental improvement; it signals a paradigm shift toward materials that can sense, adapt, repair, and communicate while reducing weight and environmental impact. As manufacturing technologies mature and certification pathways solidify, graphene is poised to become a cornerstone of modern aviation and space exploration—ushering in an era of safer, smarter, and more sustainable flight.

REFERENCES

Administration, F. A. (2023). *Pilot's Handbook of Aeronautical Knowledge (2025): FAA-H-8083-25C*. Simon and Schuster.

Allan, J. E., Ferreri, A., Grande, S., Giannantonio, R., & Matteucci, F. (2019). *Technology transfer in nanotechnology*. Publications Office of the European Union.

Amiet, A., Edwards, D., Ibrahim, M., Mart, P., McAdam, G., St John, N., & Truong, V. (2010). Exploitation of Smart Materials and Sensors as Disruptive Technologies.

Ansari, S., & Muralidharan, M. (2015). Electronic applications of polyurethane and its composites. In *Flexible and stretchable electronic composites* (pp. 87–134). Springer.

Asyraf, M., Ilyas, R., Sapuan, S., Harussani, M., Hariz, H., Aiman, J.,... Norkhairunnisa, M. (2022). Advanced composite in aerospace applications: opportunities, challenges, and future perspective. Advanced Composites in Aerospace Engineering Applications, 471–498.

Balandin, A. A. (2011). Thermal properties of graphene and nanostructured carbon materials. *Nature Materials, 10*(8), 569–581. DOI: 10.1038/nmat3064 PMID: 21778997

Balandin, A. A., Ghosh, S., Bao, W., Calizo, I., Teweldebrhan, D., Miao, F., & Lau, C. N. (2008). Superior thermal conductivity of single-layer graphene. *Nano Letters, 8*(3), 902–907. DOI: 10.1021/nl0731872 PMID: 18284217

Barkan, T., Walthall, R., Dixit, S., David, A., Webb, P., & Fletcher, S. (2023). Aerospace Advancements.

Bello, A. (2023). Review of Composite Structures in Aeronautic Applications Part.

Blaikie, A. (2020). *Graphene Electromechanical Resonators and Their Use in Thermal Detectors*. University of Oregon.

Boehm, F. J. (2013). Nanomedical device and systems design in space applications. Nanomedical device and systems design: challenges, possibilities, visions. CRC Press, Taylor and Francis Group, Boca Raton, FL, 585–652.

Bunch, J. S., Van Der Zande, A. M., Verbridge, S. S., Frank, I. W., Tanenbaum, D. M., Parpia, J. M.,... McEuen, P. L. (2007). Electromechanical resonators from graphene sheets. science, 315(5811), 490-493.

Camps, A. J., Ruiz-de-Azua, J. A., Perez-Portero, A., Fernandez, L., & Munoz-Martin, J. F. (2023). Evolving capabilities and limitations of future CubeSat missions. In *Next Generation CubeSats and SmallSats* (pp. 79–120). Elsevier. DOI: 10.1016/B978-0-12-824541-5.00027-3

Cao, X., Zhang, J., Chen, S., Varley, R. J., & Pan, K. (2020). 1D/2D nanomaterials synergistic, compressible, and response rapidly 3D graphene aerogel for piezoresistive sensor. *Advanced Functional Materials*, *30*(35), 2003618. DOI: 10.1002/adfm.202003618

Carvalho, A. F., Kulyk, B., Fernandes, A. J., Fortunato, E., & Costa, F. M. (2022). A review of the applications of graphene in mechanical transduction. *Advanced Materials*, *34*(8), 2101326. DOI: 10.1002/adma.202101326 PMID: 34288155

Chavalekvirat, P., Hirunpinyopas, W., Deshsorn, K., Jitapunkul, K., & Iamprasertkun, P. (2024). Liquid phase exfoliation of 2D materials and its electrochemical applications in the data-driven future. *Precision Chemistry*, *2*(7), 300–329. DOI: 10.1021/prechem.3c00119 PMID: 39473901

Chen, S., Brown, L., Levendorf, M., Cai, W., Ju, S.-Y., Edgeworth, J., & Piner, R. D. (2011). Oxidation resistance of graphene-coated Cu and Cu/Ni alloy. *ACS Nano*, *5*(2), 1321–1327. DOI: 10.1021/nn103028d PMID: 21275384

Chen, S., Jiang, K., Lou, Z., Chen, D., & Shen, G. (2018). Recent developments in graphene-based tactile sensors and E-skins. *Advanced Materials Technologies*, *3*(2), 1700248. DOI: 10.1002/admt.201700248

Choucair, M., Thordarson, P., & Stride, J. A. (2009). Gram-scale production of graphene based on solvothermal synthesis and sonication. *Nature Nanotechnology*, *4*(1), 30–33. DOI: 10.1038/nnano.2008.365 PMID: 19119279

Chugh, V., Basu, A., Kaushik, A., Bhansali, S., & Basu, A. K. (2024). Employing nano-enabled artificial intelligence (AI)-based smart technologies for prediction, screening, and detection of cancer. *Nanoscale*, *16*(11), 5458–5486. DOI: 10.1039/D3NR05648A PMID: 38391246

Darrah, P. D. (2022). Standard Approach to Identifying and Defining Functions for Systems Development and Safety Assessments. SAE International Journal of Advances and Current Practices in Mobility, 4(2022-01-0008), 989–993.

Darwish, M. A., Abd-Elaziem, W., Elsheikh, A., & Zayed, A. A. (2024). Advancements in nanomaterials for nanosensors: A comprehensive review. *Nanoscale Advances*, *6*(16), 4015–4046. DOI: 10.1039/D4NA00214H PMID: 39114135

Das, G. S., Tripathi, V. K., Dwivedi, J., Jangir, L. K., & Tripathi, K. M. (2024). Nanocarbon-based sensors for the structural health monitoring of smart biocomposites. *Nanoscale, 16*(4), 1490–1525. DOI: 10.1039/D3NR05522A PMID: 38186362

Databank, E. I. A. E. E. (2019). European Union Aviation Safety Agency. In.

Del Bosque, A., Vergara, D., & Fernández-Arias, P. (2025). An overview of smart composites for the aerospace sector. *Applied Sciences (Basel, Switzerland), 15*(6), 2986. DOI: 10.3390/app15062986

Del Cuvillo Mezquita, R. (2023). Aeronautic structures reinforced with graphene subjected to dynamic loadings. Analysis and modeling Universidad Carlos III de Madrid.

Deng, B., Liu, Z., & Peng, H. (2019). Toward mass production of CVD graphene films. *Advanced Materials, 31*(9), 1800996. DOI: 10.1002/adma.201800996 PMID: 30277604

Devi, N., Kumar, R., Singh, S., & Singh, R. K. (2024). Recent development of graphene-based composite for multifunctional applications: Energy, environmental, and biomedical sciences. *Critical Reviews in Solid State and Material Sciences, 49*(1), 72–140. DOI: 10.1080/10408436.2022.2132910

DO. R. (2000a). 254 Design Assurance Guidance for Airborne Electronic Hardware»,© RTCA. Inc., Washington, DC, USA.

Do, R. (2000b). *Design Assurance Guidance for Airborne Electronic Hardware. Design Assurance Guidance for Airborne Electronic Hardware*. RTCA Inc.

Du, J., Tong, B., Yuan, S., Dai, N., Liu, R., Zhang, D., Cheng, H.-M., & Ren, W. (2022). Advances in flexible optoelectronics based on chemical vapor deposition-grown graphene. *Advanced Functional Materials, 32*(42), 2203115. DOI: 10.1002/adfm.202203115

Favour, A. A. (2025). Economic and Environmental Benefits of Waste-Derived Graphene Nanosheets for Carbon Dioxide Mitigation.

Florea, M., Constantin, V.-S., Bucur, A.-C., Faur, R., Predu, D., & Cazacu, A. (2024). Materials and Structures Used in Aeronautics: Present and Future Perspectives. INCAS Bulletin, 16(4).

Fu, Y., Hansson, J., Liu, Y., Chen, S., Zehri, A., Samani, M. K.,...Zhang, Z.-B. (2020). Graphene-related materials for thermal management. 2D Materials, 7(1), 012001.

Ganguly, S., & Sengupta, J. (2024). Graphene-based nanotechnology in the Internet of Things: A mini review. *Discover Nano*, *19*(1), 110. DOI: 10.1186/s11671-024-04054-0 PMID: 38954113

Gao, J., Yan, Q., Lv, L., Tan, X., Ying, J., Yang, K., & Xiang, R. (2021). Lightweight thermal interface materials based on hierarchically structured graphene paper with superior through-plane thermal conductivity. *Chemical Engineering Journal*, *419*, 129609. DOI: 10.1016/j.cej.2021.129609

Geim, A. K., & Novoselov, K. S. (2007). The rise of graphene. *Nature Materials*, *6*(3), 183–191. DOI: 10.1038/nmat1849 PMID: 17330084

Geng, J., & Li, Y. (2024). Higher-order anti-PT-symmetric self-injection locking micro-resonator optical gyroscope. *Physica Scripta*, *99*(7), 075502. DOI: 10.1088/1402-4896/ad4fc6

Goossens, S., Navickaite, G., Monasterio, C., Gupta, S., Piqueras, J. J., Pérez, R., & Galán, T. (2017). Broadband image sensor array based on graphene–CMOS integration. *Nature Photonics*, *11*(6), 366–371. DOI: 10.1038/nphoton.2017.75

Guo, L., Chen, Z., Han, H., Liu, G., Luo, M., Cui, N., Dong, H., & Li, M.-Z. (2023). Advances and outlook in modified graphene oxide (GO)/epoxy composites for mechanical applications. *Applied Nanoscience*, *13*(5), 3273–3287. DOI: 10.1007/s13204-022-02653-w

Han, D., Zhao, Y.-H., Bai, S.-L., & Ping, W. C. (2016). High shielding effectiveness of multilayer graphene oxide aerogel film/polymer composites. *RSC Advances*, *6*(95), 92168–92174. DOI: 10.1039/C6RA20976A

He, K., Chen, G., Zeng, G., Peng, M., Huang, Z., Shi, J., & Huang, T. (2017). Stability, transport, and ecosystem effects of graphene in water and soil environments. *Nanoscale*, *9*(17), 5370–5388. DOI: 10.1039/C6NR09931A PMID: 28406500

Hernandez, Y., Nicolosi, V., Lotya, M., Blighe, F. M., Sun, Z., De, S., McGovern, I. T., Holland, B., Byrne, M., Gun'Ko, Y. K., Boland, J. J., Niraj, P., Duesberg, G., Krishnamurthy, S., Goodhue, R., Hutchison, J., Scardaci, V., Ferrari, A. C., & Coleman, J. N. (2008). High-yield production of graphene by liquid-phase exfoliation of graphite. *Nature Nanotechnology*, *3*(9), 563–568. DOI: 10.1038/nnano.2008.215 PMID: 18772919

Hodson, R., McCabe, M., Paulick, P., Ruffner, T., Some, R., Chen, Y.,…Redifer, M. (2013). Flight Avionics Hardware Roadmap.

Hu, L., Yan, X., & Yuan, Y. (2025). Development and challenges of autonomous electric vertical take-off and landing aircraft. *Heliyon, 11*(1), e41055. DOI: 10.1016/j.heliyon.2024.e41055 PMID: 39758371

Huang, H., Su, S., Wu, N., Wan, H., Wan, S., Bi, H., & Sun, L. (2019). Graphene-based sensors for human health monitoring. *Frontiers in Chemistry, 7*, 399. DOI: 10.3389/fchem.2019.00399 PMID: 31245352

Huang, Y., Liang, J., & Chen, Y. (2012). The application of graphene-based materials for actuators. *Journal of Materials Chemistry, 22*(9), 3671–3679. DOI: 10.1039/c2jm15536b

Idumah, C. I., & Odera, S. (2020). Recent advancement in self-healing graphene polymer nanocomposites, shape memory, and coating materials. *Polymer-Plastics Technology and Materials, 59*(11), 1167–1190. DOI: 10.1080/25740881.2020.1725816

Iqbal, A. A., Iqbal, A. P., & Nuruzzaman, D. M. (2024). A review on fatigue fracture characteristics in graphene reinforced nanocomposites. *Ain Shams Engineering Journal, 15*(6), 102730. DOI: 10.1016/j.asej.2024.102730

Ji, J., Zhao, W., Wang, Y., Li, Q., & Wang, G. (2023). Templated laser-induced graphene-based tactile sensors enable wearable health monitoring and texture recognition via a deep neural network. *ACS Nano, 17*(20), 20153–20166. DOI: 10.1021/acsnano.3c05838 PMID: 37801407

Jiang, B., Wang, S., Sun, J., & Liu, Z. (2021). Controllable synthesis of wafer-scale graphene films: Challenges, status, and perspectives. *Small, 17*(37), 2008017. DOI: 10.1002/smll.202008017 PMID: 34106524

Jiang, X., Lu, D., Yin, B., & Leng, Z. (2024). Advancing carbon nanomaterials-engineered self-sensing cement composites for structural health monitoring: A state-of-the-art review. *Journal of Building Engineering, 87*, 109129. DOI: 10.1016/j.jobe.2024.109129

Jones, F., & Foreman, J. (2015). The response of aerospace composites to temperature and humidity. In *Polymer composites in the aerospace industry* (pp. 335–369). Elsevier. DOI: 10.1016/B978-0-85709-523-7.00012-8

Kargar, F., Barani, Z., Balinskiy, M. G., Magana, A. S., Lewis, J. S., & Balandin, A. A. (2018). Graphene Composites with Dual Functionality: Electromagnetic Shielding and Thermal Management. arXiv preprint arXiv:1808.03401.

Kausar, A. (2022). Shape memory polymer/graphene nanocomposites: State-of-the-art. e-Polymers, 22(1), 165-181.

Kausar, A., Ahmad, I., Eisa, M., & Maaza, M. (2023). Graphene nanocomposites in the Space Sector—fundamentals and advancements. C, 9(1), 29.

Kim, D. S., Jeong, J.-M., Park, H. J., Kim, Y. K., Lee, K. G., & Choi, B. G. (2021). Highly concentrated, conductive, defect-free graphene ink for screen-printed sensor application. *Nano-Micro Letters*, *13*(1), 1–14. DOI: 10.1007/s40820-021-00617-3 PMID: 34138339

Kim, H., Abdala, A. A., & Macosko, C. W. (2010). Graphene/polymer nanocomposites. *Macromolecules*, *43*(16), 6515–6530. DOI: 10.1021/ma100572e

Kitadai, H., Yuan, M., Ma, Y., & Ling, X. (2021). Graphene-based environmental sensors: Electrical and optical devices. *Molecules (Basel, Switzerland)*, *26*(8), 2165. DOI: 10.3390/molecules26082165 PMID: 33918751

Lalanne, C. (2014). *Mechanical vibration and shock analysis, fatigue damage* (Vol. 4). John Wiley & Sons.

Laurenzi, S., Zaccardi, F., Toto, E., Santonicola, M. G., Botti, S., & Scalia, T. (2024). Fused Filament Fabrication of Polyethylene/Graphene Composites for In-Space Manufacturing. *Materials (Basel)*, *17*(8), 1888. DOI: 10.3390/ma17081888 PMID: 38673245

Lee, C., Wei, X., Kysar, J. W., & Hone, J. (2008). Measurement of the elastic properties and intrinsic strength of monolayer graphene. science, 321(5887), 385-388.

Lee, J. S. (2016). Fabrication of Zinc Oxide Nanostructure/Poly (vinylidene fluoride) Hybrid Thin Films and Their Application for Pressure Sensors and Acoustic Actuators].

Li, J., Sollami Delekta, S., Zhang, P., Yang, S., Lohe, M. R., Zhuang, X., Feng, X., & Ostling, M. (2017). Scalable fabrication and integration of graphene microsupercapacitors through full inkjet printing. *ACS Nano*, *11*(8), 8249–8256. DOI: 10.1021/acsnano.7b03354 PMID: 28682595

Li, S. (2023). Applications of Graphene: Supercapacitors and Gas Sensors. MATEC Web of Conferences.

Liang, S., Wu, X., Wu, Z., & Liu, W. (2025). Progress and Challenges on Materials Used to Optimize Flight Efficiency, Improve Safety and Reliability, and Reduce Cost of Electric Vertical Take-Off and Landing Aircraft. *Journal of Applied Polymer Science*, *142*(24), 57029. DOI: 10.1002/app.57029

Lin, J., Peng, Z., Liu, Y., Ruiz-Zepeda, F., Ye, R., Samuel, E. L., Yacaman, M. J., Yakobson, B. I., & Tour, J. M. (2014). Laser-induced porous graphene films from commercial polymers. *Nature Communications*, *5*(1), 5714. DOI: 10.1038/ncomms6714 PMID: 25493446

Liu, F., Wang, C., Sui, X., Riaz, M. A., Xu, M., Wei, L., & Chen, Y. (2019). Synthesis of graphene materials by electrochemical exfoliation: Recent progress and future potential. *Carbon Energy*, *1*(2), 173–199. DOI: 10.1002/cey2.14

Liu, S., Li, Z., Feng, Y., Zhang, T., Ma, B., & Wang, W. (2024). Effect of ultraviolet radiation on the low-velocity impact and compression after impact performances of aerospace composite laminates. *Polymer Composites*, *45*(4), 3284–3295. DOI: 10.1002/pc.27990

Lu, Y., Guo, Z.-S., & Fan, S.-C. (2021). An ultrahigh-sensitivity graphene resonant gyroscope. *Nanomaterials (Basel, Switzerland)*, *11*(8), 1890. DOI: 10.3390/nano11081890 PMID: 34443720

Macosko, C. (2010). Graphene/Polymer Nanocomposites. APS March Meeting Abstracts

Maji, A., Kuila, C., Murmu, N. C., & Kuila, T. (2024). Facile fabrication of covalently grafted functionalized graphene-PVDF composites as strain sensor with tunable sensitivity: Application to the structural health and human motion monitoring. *Sensors and Actuators. A, Physical*, *369*, 115081.

Mardanshahi, A., Sreekumar, A., Yang, X., Barman, S. K., & Chronopoulos, D. (2025). Sensing Techniques for Structural Health Monitoring: A State-of-the-Art Review on Performance Criteria and New-Generation Technologies. *Sensors (Basel)*, *25*(5), 1424. DOI: 10.3390/s25051424 PMID: 40096243

Mason, M. (2022). Recent Langley Aerothermodynamic Laboratory Testing (2019-2022). Thermal and Fluids Analysis Workshop 2022

Meng, Q., Zhao, J., Zhou, Z., Han, S., Feng, Y., Han, Q., & Liu, T. (2024). Damage monitoring of aircraft structural components based on large-area flexible graphene strain sensors. *Sensors and Actuators. A, Physical*, *369*, 115092.

Mirabedini, A., Ang, A., Nikzad, M., Fox, B., Lau, K. T., & Hameed, N. (2020). Evolving strategies for producing multiscale graphene-enhanced fiber-reinforced polymer composites for smart structural applications. *Advancement of Science*, *7*(11), 1903501. DOI: 10.1002/advs.201903501 PMID: 32537406

Mishra, R. K., Thomas, M. G., Abraham, J., Joseph, K., & Thomas, S. (2018). Electromagnetic interference shielding materials for aerospace applications: A state of the art. Advanced materials for electromagnetic shielding: fundamentals, properties, and applications, 327-365.

Mohapatra, D., Behera, D. R., & Mishra, D. K. (2023). Prospective of functionalized graphene as shape memory and self-healing polymer: A review. *Materials Today: Proceedings*.

Nagaraj, P. (2023). Study of Fatigue Damage Detection by Current Information of Measurement System Riga Technical University Rīga, LV, USA].

Nawaz, M., Islam, M., Nazir, M. A., Bano, I., Gul, I., & Ajmal, M. (2021). Transport properties in spinel ferrite/graphene oxide nanocomposites for electromagnetic shielding. *Ceramics International*, *47*(18), 25505–25513. DOI: 10.1016/j.ceramint.2021.05.274

Novoselov, K. S., Colombo, L., Gellert, P., Schwab, M., & Kim, K. (2012). A roadmap for graphene. nature, 490(7419), 192-200.

Novoselov, K. S., Geim, A. K., Morozov, S. V., Jiang, D.-e., Zhang, Y., Dubonos, S. V.,...Firsov, A. A. (2004). Electric field effect in atomically thin carbon films. science, 306(5696), 666-669.

Oliveira, F. M., Luxa, J., Bouša, D., Sofer, Z. K., & Gusmao, R. (2022). Electromagnetic interference shielding by reduced graphene oxide foils. *ACS Applied Nano Materials*, *5*(5), 6792–6800. DOI: 10.1021/acsanm.2c00785

Oluremi, D. (2025). Industrial Waste as a Source for Graphene Nanosheets.

Panghal, A., Kumar, Y., Mishra, P. K., Mathur, A., & Srivastava, A. K. (2025). Graphene-Based Polymer Composites for Aerospace, Electronic, Energy, and Biomedical Applications. *Multifunctional Materials*, •••, 81–104. https://doi.org/https://doi.org/1002/9781394234134.ch4

Patnaik, T. K. (2024). 13 Graphene-Based Materials. *Green Nanobiotechnology*, *9*, 177.

Pei, S., & Cheng, H.-M. (2012). The reduction of graphene oxide. *Carbon*, *50*(9), 3210–3228. DOI: 10.1016/j.carbon.2011.11.010

Pina, S., Oliveira, J. M., & Reis, R. L. (2015). Natural-based nanocomposites for bone tissue engineering and regenerative medicine: A review. *Advanced Materials*, *27*(7), 1143–1169. DOI: 10.1002/adma.201403354 PMID: 25580589

Prater, T., Tilson, W., Vickers, J., Singhal, S., Zakrajsek, J., & Ransom, J. (2020). Breakthrough Materials for Space Applications Workshop. Workshop at Jackson Center

Price, M. R. (2017). Quantifying the Protective Properties and Adhesion to the Substrate of Ultra-thin Multilayer Diamond-like Carbon Coatings Using Molecular Dynamics Simulation The University of Utah].

Qiu, Y., Wang, M., Zhang, W., Liu, Y., Li, Y. V., & Pan, K. (2018). An asymmetric graphene oxide film for developing moisture actuators. *Nanoscale*, *10*(29), 14060–14066. DOI: 10.1039/C8NR01785A PMID: 29999058

Ramesh, R. D., Santhosh, A., & Syamala, S. R. N. A. (2020). Implementation of Nanotechnology in the Aerospace and Aviation Industry. In *Smart Nanotechnology with Applications* (pp. 51–69). CRC Press. DOI: 10.1201/9781003097532-4

Reghat, M., Mirabedini, A., Tan, A. M., Weizman, Y., Middendorf, P., Bjekovic, R., & Fuss, F. K. (2021). Graphene as a piezo-resistive coating to enable strain monitoring in glass fiber composites. *Composites Science and Technology*, *211*, 108842. DOI: 10.1016/j.compscitech.2021.108842

Reina, A., Son, H., Jiao, L., Fan, B., Dresselhaus, M. S., Liu, Z., & Kong, J. (2008). Transferring and identification of single-and few-layer graphene on arbitrary substrates. *The Journal of Physical Chemistry. C, Nanomaterials and Interfaces*, *112*(46), 17741–17744. DOI: 10.1021/jp807380s

Riaz, A., Khan, S., & Arslan, T. (2023). Design and modelling of graphene-based flexible 5G antenna for next-generation wearable head imaging systems. *Micromachines*, *14*(3), 610. DOI: 10.3390/mi14030610 PMID: 36985015

Ryan, E. A., Seibers, Z. D., Reynolds, J. R., & Shofner, M. L. (2024). Electrically conducting polymers and composites for applications in space exploration. *Journal of Applied Polymer Science*, *141*(15), e55225. DOI: 10.1002/app.55225

Sabato, A., Dabetwar, S., Kulkarni, N. N., & Fortino, G. (2023). Noncontact sensing techniques for AI-aided structural health monitoring: A systematic review. *IEEE Sensors Journal*, *23*(5), 4672–4684. DOI: 10.1109/JSEN.2023.3240092

Sabet, M. (2024). Unveiling advanced self-healing mechanisms in graphene polymer composites for next-generation applications in aerospace, automotive, and electronics. *Polymer-Plastics Technology and Materials*, *63*(15), 2032–2059. DOI: 10.1080/25740881.2024.2365282

Sadeghi, A., Bellavista, P., Song, W., & Yazdani-Asrami, M. (2024). Digital twins for condition and fleet monitoring of aircraft: Towards more-intelligent electrified aviation systems. *IEEE Access : Practical Innovations, Open Solutions, 12*, 99806–99832. DOI: 10.1109/ACCESS.2024.3371902

Safian, M. T., Haron, U. S., & Ibrahim, M. M. (2020). A review on bio-based graphene derived from biomass wastes. *BioResources, 15*(4), 9756–9785. DOI: 10.15376/biores.15.4.Safian

Saleh, H. M., & Hassan, A. I. (2023). Synthesis and characterization of nanomaterials for application in cost-effective electrochemical devices. *Sustainability (Basel), 15*(14), 10891. DOI: 10.3390/su151410891

Santos-Ceballos, J. C., Salehnia, F., Romero, A., & Vilanova, X. (2024). Application of digital twins for simulation based tailoring of laser induced graphene. *Scientific Reports, 14*(1), 10363. DOI: 10.1038/s41598-024-61237-6 PMID: 38710895

Sarlioglu, B., & Morris, C. T. (2015). More electric aircraft: Review, challenges, and opportunities for commercial transport aircraft. *IEEE Transactions on Transportation Electrification, 1*(1), 54–64. DOI: 10.1109/TTE.2015.2426499

Scalia, T., Bonventre, L., & Terranova, M. L. (2023). From protosolar space to space exploration: The role of graphene in space technology and economy. *Nanomaterials (Basel, Switzerland), 13*(4), 680. DOI: 10.3390/nano13040680 PMID: 36839048

Schwierz, F. (2010). Graphene transistors. *Nature Nanotechnology, 5*(7), 487–496. DOI: 10.1038/nnano.2010.89 PMID: 20512128

Shalabi, O. R. Y. (2022). Manufacturing of smart composite aircraft skins using 2D material coated glass fiber reinforcements Khalifa. *Universitas Scientiarum.*

Sihver, L., Ploc, O., Puchalska, M., Ambrožová, I., Kubančák, J., Kyselová, D., & Shurshakov, V. (2015). Radiation environment at aviation altitudes and in space. *Radiation Protection Dosimetry, 164*(4), 477–483. DOI: 10.1093/rpd/ncv330 PMID: 25979747

Somanath, S., & Spaceport, K. Indian Space Research Organisation.

Song, P., Liu, B., Qiu, H., Shi, X., Cao, D., & Gu, J. (2021). MXenes for polymer matrix electromagnetic interference shielding composites: A review. *Composites Communications, 24*, 100653. DOI: 10.1016/j.coco.2021.100653

Souza, D. F. d. (2024). Design and development of an electric vertical take off and landing aircraft system control using sliding mode control system Universidade Tecnológica Federal do Paraná].

Stankovich, S., Dikin, D. A., Dommett, G. H., Kohlhaas, K. M., Zimney, E. J., Stach, E. A.,...Ruoff, R. S. (2006). Graphene-based composite materials. nature, 442(7100), 282-286.

Steinke, K., Groo, L., & Sodano, H. A. (2021). Laser induced graphene for in-situ ballistic impact damage and delamination detection in aramid fiber reinforced composites. *Composites Science and Technology*, *202*, 108551. DOI: 10.1016/j.compscitech.2020.108551

Su, J., He, K., Li, Y., Tu, J., & Chen, X. (2025). Soft Materials and Devices Enabling Sensorimotor Functions in Soft Robots. *Chemical Reviews*, *125*(12), 5848–5977. DOI: 10.1021/acs.chemrev.4c00906 PMID: 40163535

Sudan, K. (2024). Development and processing of shape memory polymer composites (SMPCS) for application in structural robotics and robotic sensors.

Sun, Y., & Zhang, J. (2021). Strategies for scalable gas-phase preparation of free-standing graphene. *CCS Chemistry*, *3*(4), 1058–1077. DOI: 10.31635/ccschem.020.202000289

TabkhPaz, M., Park, D.-Y., Lee, P. C., Hugo, R., & Park, S. S.TabkhPaz. (2018). Development of nanocomposite coatings with improved mechanical, thermal, and corrosion protection properties. *Journal of Composite Materials*, *52*(8), 1045–1060. DOI: 10.1177/0021998317720001

Taha, T. H., Elnouby, M. S., Abu-Saied, M., & Alamri, S. (2020). The green exfoliation of graphite waste and its suitability for biosensor applications. *RSC Advances*, *10*(16), 9347–9355. DOI: 10.1039/C9RA09602G PMID: 35497257

Tong, C. (2022). *Advanced Materials for Printed Flexible Electronics*. Springer. DOI: 10.1007/978-3-030-79804-8

Valorosi, F., De Meo, E., Blanco-Varela, T., Martorana, B., Veca, A., Pugno, N., & Bertocchi, F. (2020). Graphene and related materials in hierarchical fiber composites: Production techniques and key industrial benefits. *Composites Science and Technology*, *185*, 107848. DOI: 10.1016/j.compscitech.2019.107848

Voros, R. (2021). Generalizing Aspects of System Safety to Broaden Applicability. SAE International Journal of Advances and Current Practices in Mobility, 3(2021-01-0037), 1272-1284.

Wang, F., Ko, S. Y., Park, J. O., Park, S. H., & Kee, C. D. (2015). Electroactive polymer actuator based on PVDF and graphene through electrospinning. *Advanced Materials Research*, *1105*, 311–314. DOI: 10.4028/www.scientific.net/AMR.1105.311

Wang, J., Liu, Y., Fan, Z., Wang, W., Wang, B., & Guo, Z. (2019). Ink-based 3D printing technologies for graphene-based materials: A review. *Advanced Composites and Hybrid Materials*, *2*(1), 1–33. DOI: 10.1007/s42114-018-0067-9

Wang, J., Xing, Z., Chen, X., Cheng, Z., Li, X., & Liu, T. (2020). Recent progress in waveguide-integrated graphene photonic devices for sensing and communication applications. *Frontiers in Physics (Lausanne)*, *8*, 37. DOI: 10.3389/fphy.2020.00037

Wang, T., Huang, D., Yang, Z., Xu, S., He, G., Li, X., Hu, N., Yin, G., He, D., & Zhang, L. (2016). A review on graphene-based gas/vapor sensors with unique properties and potential applications. *Nano-Micro Letters*, *8*(2), 95–119. DOI: 10.1007/s40820-015-0073-1 PMID: 30460270

Wu, T., Shen, J., Li, Z., Xing, F., Xin, W., Wang, Z., Liu, G., Han, X., Man, Z., & Fu, S. (2021). Microfluidic-integrated graphene optical sensors for real-time and ultra-low flow velocity detection. *Applied Surface Science*, *539*, 148232. DOI: 10.1016/j.apsusc.2020.148232

Wu, Y., An, C., & Guo, Y. (2023). 3D printed graphene and graphene/polymer composites for multifunctional applications. *Materials (Basel)*, *16*(16), 5681. DOI: 10.3390/ma16165681 PMID: 37629973

Xie, X., Zhou, Y., & Huang, K. (2019). Advances in microwave-assisted production of reduced graphene oxide. *Frontiers in Chemistry*, *7*, 355. DOI: 10.3389/fchem.2019.00355 PMID: 31214562

Xu, J., Li, R., Ji, S., Zhao, B., Cui, T., Tan, X., & Qiao, Y. (2021). Multifunctional graphene microstructures inspired by honeycomb for ultrahigh performance electromagnetic interference shielding and wearable applications. *ACS Nano*, *15*(5), 8907–8918. DOI: 10.1021/acsnano.1c01552 PMID: 33881822

Xu, L., Lu, H., Zhou, Y., Chi, Z., Li, Z., Md, Z. I., Dong, Y., Fu, Y., Zhu, Y., & Ni, Q. (2021). Ultrathin, ultralight, and anisotropic ordered reduced graphene oxide fiber electromagnetic interference shielding membrane. *Advanced Materials Technologies*, *6*(12), 2100531. DOI: 10.1002/admt.202100531

Yang, J., Luo, S., Zhou, X., Li, J., Fu, J., Yang, W., & Wei, D. (2019). Flexible, tunable, and ultrasensitive capacitive pressure sensor with microconformal graphene electrodes. *ACS Applied Materials & Interfaces*, *11*(16), 14997–15006. DOI: 10.1021/acsami.9b02049 PMID: 30869860

Yavuz, M. T., & Ozkol, I. (2024). Thermal Structural Design Aspects of Military Aircraft. *Journal of Aeronautics and Space Technologies*, *17*(1), 124–158.

Zhang, J., Wang, X., Chen, X., Xia, X., & Weng, G. J. (2022). Piezoelectricity enhancement in graphene/polyvinylidene fluoride composites due to graphene-induced $\alpha \rightarrow \beta$ crystal phase transition. *Energy Conversion and Management*, *269*, 116121. DOI: 10.1016/j.enconman.2022.116121

Zhong, Y. L., Tian, Z., Simon, G. P., & Li, D. (2015). Scalable production of graphene via wet chemistry: Progress and challenges. *Materials Today*, *18*(2), 73–78. DOI: 10.1016/j.mattod.2014.08.019

Zitoun, A., Fakis, D., Jayasree, N., Omairey, S., Oikonomidis, F., Stoeva, Z., & Kazilas, M. (2022). Graphene-based strain sensing in composites for structural and health monitoring applications. *SN Applied Sciences*, *4*(2), 58. DOI: 10.1007/s42452-022-04940-1

Chapter 7
Graphene in Electromagnetic Interference Shielding:
From Fundamentals to Aerospace and 5G Applications

S. Khalid
https://orcid.org/0009-0005-6894-0087
Lincoln University College, Malaysia

Shashi Kant Gupta
https://orcid.org/0000-0001-6587-5607
Lincoln University College, Malaysia

Midhun Chakkaravarthy
https://orcid.org/0000-0002-0107-885X
Lincoln University College, Malaysia

Dharmendra Prakash
https://orcid.org/0009-0000-8039-3657
SRM University, Lucknow, India

Alkesh Agrawal
SRM University, Lucknow, India

Dinesh Kumar Nishad
https://orcid.org/0009-0003-5270-0922
Dr. Shakuntala Misra National Rehabilitation University, Lucknow, India

ABSTRACT

Electromagnetic interference (EMI) has become a critical challenge in modern electronics, telecommunications, aerospace, and defense systems due to the proliferation of wireless technologies and high-frequency devices. Traditional metallic shielding materials suffer from limitations including excessive weight, rigidity, and corrosion susceptibility. This chapter presents a comprehensive examination of graphene-based EMI shielding materials, which offer exceptional electrical conductivity, lightweight characteristics, mechanical flexibility, and tunable surface chemistry. The discussion encompasses EMI fundamentals, graphene's unique prop-

DOI: 10.4018/979-8-3373-3862-0.ch007

erties relevant to shielding applications, advanced fabrication strategies including CVD synthesis, composite integration, and 3D porous structures, and underlying shielding mechanisms involving reflection, absorption, and multiple scattering. Key applications across consumer electronics, automotive and aerospace systems (including UAVs), 5G telecommunications, and defense technologies are analyzed, demonstrating graphene's multifunctionality.

INTRODUCTION

Electromagnetic interference (EMI) has emerged as one of the most pressing challenges in modern electronics, telecommunications, and advanced technological systems. As the digital landscape continues to expand with the proliferation of wireless devices, high-frequency communication networks, and sophisticated electronic systems, the demand for effective EMI shielding solutions has reached unprecedented levels (Watanabe et al., 2024). The exponential growth in electronic device density, coupled with the miniaturization of components and the advent of next-generation technologies such as 5G networks, Internet of Things (IoT) devices, and autonomous systems, has created a complex electromagnetic environment where unwanted interference can significantly compromise system performance, reliability, and safety.

The fundamental nature of EMI stems from both intentional and unintentional electromagnetic emissions that can disrupt the normal operation of electronic devices and systems. These electromagnetic disturbances originate from diverse sources, including natural phenomena such as lightning and solar radiation, as well as man-made systems including power transmission lines, switching circuits, wireless communication devices, and motor-driven equipment (Hao et al., 2022). The challenge becomes particularly acute in densely populated electronic environments where multiple devices coexist, each potentially contributing to what researchers term "self-jamming" scenarios, where devices inadvertently interfere with their own or neighboring systems' operations.

In critical applications such as aviation and unmanned aerial vehicles (UAVs), EMI poses significant risks to operational safety and mission success. Modern aircraft and UAV systems rely heavily on sophisticated electronic systems for navigation, communication, flight control, and autonomous decision-making (Khalid et al., 2025). The complex interactions between various onboard subsystems, including motors, sensors, communication modules, and AI-powered navigation systems, create multiple potential sources of electromagnetic interference that can degrade performance or, in extreme cases, lead to system failures. This is particularly concerning given the increasing deployment of AI-driven UAV systems for applications ranging from fog dispersal at airports to military reconnaissance, where communication reliability and system integrity are paramount (Prakash et al., 2024).

Traditional approaches to EMI shielding have predominantly relied on metallic materials such as copper, aluminum, and steel, which provide effective electromagnetic wave attenuation through reflection mechanisms. However, these conventional solutions face significant limitations that increasingly constrain their applicability in modern electronic systems. Metallic shields typically exhibit substantial weight penalties, which is particularly problematic in aerospace applications where every gram of additional weight translates to increased fuel consumption and reduced payload capacity (Wang et al., 2021). Furthermore, traditional metallic shields suffer from rigidity issues that make them unsuitable for flexible and wearable electronics, susceptibility to corrosion in harsh environments, and processing challenges that limit their integration into complex geometries and miniaturized devices.

The emergence of 5G and next-generation wireless communication systems has further exacerbated the EMI challenge by introducing millimeter-wave frequencies and ultra-high data rates that demand more sophisticated shielding solutions. These advanced communication systems operate at frequency ranges where traditional shielding approaches may prove inadequate or impractical, necessitating the development of novel materials and shielding strategies that can effectively attenuate electromagnetic waves across broad frequency spectrums while maintaining the lightweight, flexible, and scalable characteristics required for modern applications (Ryu et al., 2022).

Graphene, a revolutionary two-dimensional material composed of sp^2-hybridized carbon atoms arranged in a hexagonal lattice structure, has emerged as a transformative solution to the EMI shielding challenge. Discovered through the groundbreaking work that earned Andre Geim and Konstantin Novoselov the Nobel Prize in Physics, graphene exhibits an extraordinary combination of properties that make it uniquely suited for electromagnetic interference shielding applications. Its exceptional electrical conductivity, surpassing that of copper and silver, enables highly effective reflection of electromagnetic waves. Simultaneously, its atomically thin structure and large surface area facilitate superior absorption mechanisms through defect engineering and hybridization strategies (Gutiérrez et al., 2024).

The remarkable versatility of graphene extends beyond its intrinsic electromagnetic properties to encompass mechanical flexibility, thermal stability, and chemical resistance that address many limitations of traditional shielding materials. Unlike rigid metallic shields, graphene-based materials can be engineered into flexible, transparent, and lightweight forms that are compatible with emerging applications in flexible electronics, wearable devices, and aerodynamically sensitive aerospace systems (Hong et al., 2012). Furthermore, graphene derivatives such as graphene oxide (GO) and reduced graphene oxide (rGO) offer tunable surface chemistry and processing characteristics that enable their integration into diverse composite matrices and manufacturing processes.

The potential applications of graphene-based EMI shielding span multiple critical sectors, each with unique requirements and challenges. In consumer electronics, graphene coatings and composites promise to enable the next generation of ultra-thin, flexible, and transparent devices while providing superior electromagnetic protection. The automotive industry, particularly in the context of electric vehicles and autonomous driving systems, stands to benefit significantly from lightweight graphene shields that protect sensitive battery management systems and sensor arrays without compromising vehicle efficiency or performance (Verma et al., 2015).

Perhaps most significantly, the aerospace and defense sectors represent frontier applications where graphene's unique combination of properties offers unprecedented opportunities. Advanced aircraft, satellites, and military systems require EMI shielding solutions that can operate reliably in extreme environments while minimizing weight and space penalties. The integration of graphene-based materials into these systems promises to enable more sophisticated electronic capabilities, improved stealth characteristics, and enhanced operational reliability in electromagnetically challenging environments.

The telecommunications sector, driven by the global deployment of 5G networks and the anticipation of 6G technologies, presents another critical application domain. Graphene's ability to provide effective shielding across broad frequency ranges, combined with its compatibility with miniaturized device architectures, positions it as an enabling technology for the next generation of high-frequency communication systems and IoT devices.

Figure 1 illustrates how electromagnetic interference (EMI) permeates multiple technological sectors. In consumer electronics, EMI disrupts smartphones, wearables, and IoT devices, leading to performance degradation and connectivity issues. In aviation and UAVs, EMI affects avionics, navigation reliability, and air traffic control systems, posing risks to flight safety. 5G and next-generation telecom networks are especially vulnerable at millimeter-wave frequencies, where EMI can compromise signal integrity and throughput. In defense and security, EMI interference can undermine radar stealth, secure communication, and battlefield electronics. Finally, healthcare technologies such as pacemakers, implants, and diagnostic tools face safety risks when exposed to uncontrolled EMI.

This flowchart underscores the cross-sectoral importance of EMI shielding and motivates the exploration of graphene as a disruptive material for lightweight, flexible, and high-performance EMI protection.

Figure 1. Impact of Electromagnetic Interference (EMI) Across Industries

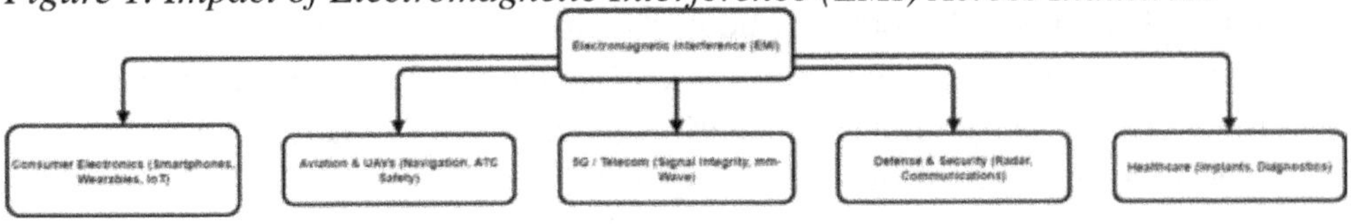

This chapter provides a comprehensive exploration of graphene-based EMI shielding, bridging fundamental scientific principles with practical engineering applications across diverse technological sectors. Beginning with an examination of electromagnetic interference fundamentals and the unique properties of graphene relevant to shielding applications, the discussion progresses through fabrication strategies, shielding mechanisms, and detailed analyses of sector-specific applications. The chapter critically evaluates current challenges and limitations while projecting future developments in this rapidly evolving field. Through this comprehensive treatment, the chapter aims to serve as both a scientific reference and a practical guide for researchers, engineers, and technology developers working to harness graphene's transformative potential in addressing the EMI challenges of tomorrow's electronic systems.

FUNDAMENTALS OF ELECTROMAGNETIC INTERFERENCE (EMI)

Electromagnetic interference (EMI) represents one of the most pervasive and challenging phenomena in modern electronic systems, fundamentally arising from the unintentional generation, propagation, and reception of electromagnetic energy that disrupts the normal operation of electronic devices and systems. As the electromagnetic environment becomes increasingly complex due to the proliferation of wireless devices, high-frequency communication systems, and sophisticated electronic equipment, understanding the fundamental principles governing EMI has become essential for engineers, researchers, and system designers across multiple technological domains.

Definition and Nature of EMI

Electromagnetic interference occurs when electromagnetic disturbances interrupt, obstruct, or otherwise degrade the performance of an electronic device or system. These disturbances can manifest as unwanted electromagnetic energy that couples into electronic circuits through various pathways, potentially causing malfunctions, data corruption, reduced performance, or complete system failure (Watanabe et al., 2024). The phenomenon becomes particularly critical in densely populated electronic environments where multiple devices coexist, each potentially contributing to what researchers term "self-jamming" scenarios, where devices inadvertently interfere with their own or neighboring systems' operations.

The complexity of EMI is amplified by the diverse nature of electromagnetic sources and the multiple coupling mechanisms through which interference can propagate. Unlike deliberate electromagnetic emissions designed for communication or sensing purposes, EMI typically represents parasitic electromagnetic energy generated as an unintended byproduct of electronic circuit operation. This parasitic energy can span a broad frequency spectrum, from low-frequency power line harmonics to high-frequency switching transients, making comprehensive EMI management a multifaceted challenge requiring sophisticated understanding and mitigation strategies.

Sources and Classification of EMI

EMI sources can be broadly categorized into natural and artificial origins, each presenting unique characteristics and challenges. Natural sources include atmospheric phenomena such as lightning strikes, solar radiation, and cosmic rays, which generate broadband electromagnetic disturbances across multiple frequency ranges. These natural phenomena can create transient electromagnetic pulses capable of disrupting electronic systems over vast geographical areas, particularly affecting sensitive communication and navigation systems.

Artificial EMI sources, however, represent the more prevalent and controllable category in modern technological environments. These sources include switching power supplies, digital clock circuits, motor drives, wireless communication devices, and computing systems. The rapid switching transitions inherent in digital circuits generate harmonically rich electromagnetic spectra that can extend well beyond the fundamental operating frequencies. For instance, high-speed digital processors operating at gigahertz frequencies can generate significant electromagnetic emissions at harmonics reaching into the microwave spectrum.

Hao et al. (2022) emphasize the importance of accurately identifying EMI characteristics associated with basic commutation units, elaborating on how specific waveforms, such as asymmetric trapezoidal waveforms, can be utilized to understand the behavior of EMI sources. Their research demonstrates that analyzing these sources using time-frequency characteristic parameters enhances the ability to predict interference effects, which is crucial for developing effective countermeasures in complex electronic systems.

EMI Coupling Mechanisms

Understanding how electromagnetic interference propagates from sources to susceptible devices requires comprehensive knowledge of coupling mechanisms. EMI coupling can occur through four primary pathways: conducted coupling, radiated coupling, inductive coupling, and capacitive coupling.

Conducted EMI propagates through physical connections such as power lines, signal cables, and ground planes. This type of interference travels along conductors and can affect devices connected to the same power distribution network or communication bus. Conducted emissions are typically more prominent at lower frequencies, where wavelengths are large compared to circuit dimensions, and direct conduction provides an efficient propagation path.

Radiated EMI occurs when electromagnetic energy propagates through free space as electromagnetic waves. This mechanism becomes increasingly important at higher frequencies where circuit dimensions become comparable to electromagnetic wavelengths. Radiated interference can affect devices without any direct physical connection to the source, making it particularly challenging to control in complex electromagnetic environments.

Inductive coupling results from magnetic field interactions, where time-varying currents in one circuit induce voltages in nearby circuits through mutual inductance. This coupling mechanism is particularly problematic in situations involving high-current switching circuits or transformers operating in close proximity to sensitive electronic equipment.

Capacitive coupling occurs through electric field interactions, where time-varying voltages create displacement currents that can affect nearby circuits through mutual capacitance. This mechanism is especially relevant in high-impedance circuits and situations involving rapid voltage transitions.

Frequency Characteristics and Spectral Analysis

The frequency characteristics of EMI sources play a crucial role in determining interference effects and mitigation strategies. Different electronic systems exhibit distinct spectral signatures based on their operational characteristics, switching frequencies, and circuit topologies. Understanding these spectral characteristics enables engineers to predict potential interference scenarios and design appropriate countermeasures.

Digital systems typically generate EMI spectra characterized by harmonics of fundamental clock frequencies, with spectral content extending well beyond the basic operational frequencies. The harmonic content depends on the rise and fall times of digital transitions, with faster switching speeds generally producing broader spectral distributions. Modern high-speed digital systems can generate significant electromagnetic emissions across frequency ranges spanning from kilohertz to gigahertz.

Power electronic systems, including switching power supplies and motor drives, generate EMI spectra dominated by switching frequency harmonics and associated sidebands. The spectral characteristics of these systems depend on switching topology, modulation schemes, and load characteristics. Pulse-width modulation (PWM) systems, commonly used in power electronics, generate particularly complex spectral distributions with energy distributed across multiple frequency bands.

EMI Effects on Electronic Systems

The effects of electromagnetic interference on electronic systems can range from subtle performance degradation to complete system failure, depending on the interference characteristics, system susceptibility, and operational environment. Understanding these effects requires consideration of both the interference mechanisms and the response characteristics of affected systems.

In communication systems, EMI can manifest as increased noise floors, reduced signal-to-noise ratios, bit error rate degradation, and complete communication link failures. Digital communication systems are particularly susceptible to EMI because interference can cause bit errors that propagate through digital processing chains, potentially causing system-wide malfunctions.

Control systems, including those used in automotive, aerospace, and industrial applications, can experience dangerous malfunctions due to EMI. Interference affecting sensor signals, control commands, or feedback loops can lead to unstable operation, reduced performance, or catastrophic failures. The safety-critical nature of many control applications makes EMI susceptibility a paramount concern in system design and validation.

For unmanned aerial vehicles (UAVs), EMI poses particularly significant challenges due to the complex interactions between various onboard subsystems, including motors, sensors, communication modules, and AI-powered navigation systems. As noted by Watanabe et al. (2024), the complex interactions of various internal subsystems, such as motors and navigational aids, can lead to performance degradation if not properly managed, where communication reliability is paramount for safe and effective operation.

Measurement and Quantification

Accurate measurement and quantification of EMI require sophisticated instrumentation and standardized procedures to ensure reproducible and meaningful results. EMI measurements typically involve specialized equipment including spectrum analyzers, EMI receivers, near-field probes, and anechoic chambers designed to provide controlled electromagnetic environments.

The measurement of EMI involves both emission testing, which quantifies the electromagnetic energy generated by a device under test, and susceptibility testing, which determines the device's response to applied electromagnetic disturbances. These measurements must be conducted according to standardized procedures defined by regulatory bodies and industry organizations to ensure consistency and comparability of results.

Quantification of EMI effects typically involves statistical analysis of interference events, measurement of performance degradation metrics, and assessment of system availability and reliability under various electromagnetic environments. These quantitative approaches enable engineers to establish design margins, validate mitigation strategies, and demonstrate compliance with regulatory requirements.

Mitigation Principles and Strategies

Effective EMI mitigation requires a systematic approach incorporating multiple strategies applied at various levels of system design and implementation. The fundamental principles of EMI mitigation include source suppression, transmission path interruption, and receptor protection.

Source suppression involves reducing electromagnetic emissions at their origin through techniques such as proper circuit design, component selection, and layout optimization. This approach represents the most effective EMI mitigation strategy because it addresses the root cause of interference problems. Techniques include the use of spread-spectrum clocking, soft switching in power electronics, and proper grounding and shielding practices.

Hariyawan et al. (2016) demonstrate that the implementation of spread-spectrum techniques represents an effective method for minimizing conducted EMI in systems such as LED luminance controls. These techniques work by spreading the signal over a wider bandwidth, thus reducing the potential for destructive interference and providing significant improvements in electromagnetic compatibility.

Transmission path interruption focuses on preventing the propagation of electromagnetic interference from sources to susceptible devices. This approach includes the use of EMI filters, shielding, isolation, and proper cable management. Verma et al. (2018) highlight that EMI filters, which are generally low-pass filters, serve as essential components in power converters, allowing desired power frequencies to pass while effectively attenuating higher-frequency noise that could lead to system malfunctions.

Receptor protection involves hardening susceptible devices against electromagnetic interference through techniques such as input filtering, circuit topology selection, and proper system architecture design. This approach focuses on reducing the coupling efficiency between interference sources and sensitive circuits, thereby maintaining system performance in challenging electromagnetic environments.

Regulatory Framework and Standards

The management of electromagnetic interference is governed by comprehensive regulatory frameworks and technical standards designed to ensure electromagnetic compatibility (EMC) in the increasingly complex electromagnetic environment. These regulations establish emission limits for electronic devices and define immunity requirements to ensure adequate performance in the presence of electromagnetic disturbances.

International standards organizations, including the International Electrotechnical Commission (IEC), the Federal Communications Commission (FCC), and the European Telecommunications Standards Institute (ETSI), have developed detailed standards covering EMI measurement procedures, emission limits, and immunity requirements for various device categories and applications. These standards provide the foundation for EMC compliance testing and certification processes that ensure electronic devices can coexist without mutual interference. Table 1 shows the classification of EMI by frequency range, type, sources, and impacts.

Table 1. Classification of electromagnetic interference (EMI) by frequency range, type, sources, and impacts.

Frequency Range	Type of EMI	Typical Sources	Impact on Devices	Example Systems Affected
0 – 150 kHz	Conducted (Low-Freq)	Power lines, ground loops, switching power supplies	Noise coupling into circuits, power instability	Household appliances, chargers
150 kHz – 30 MHz	Conducted EMI	Motors, converters, industrial drives	Signal distortion, heating in cables	EV battery systems, UAV motors
30 MHz – 1 GHz	Radiated EMI	Wireless transmitters, UAV avionics, digital circuits	Crosstalk, antenna interference	Radios, GPS, IoT devices
1 – 10 GHz	Radiated EMI	4G/5G base stations, radar, microwave ovens	Signal degradation, data loss	Telecom networks, UAV communication
>10 GHz (mm-wave)	Radiated EMI	5G/6G antennas, satellite comms, radar systems	Severe disruption of high-speed communication	5G small cells, aerospace radar, IoT

Both conducted and radiated EMI affect electronics across a wide spectrum, from household devices to aerospace and 5G communication systems.

The continuous evolution of these regulatory frameworks reflects the dynamic nature of the electromagnetic environment and the ongoing development of new technologies with unique EMI characteristics. As new applications such as 5G communication systems, autonomous vehicles, and Internet of Things (IoT) devices proliferate, regulatory bodies must adapt existing standards and develop new requirements to address emerging EMI challenges while facilitating technological innovation and deployment.

Figure 2 illustrates the diverse sources and propagation modes of electromagnetic interference (EMI). Natural sources such as lightning, solar flares, and cosmic rays inject strong bursts of electromagnetic energy that can disturb sensitive electronics. Man-made sources — including switching devices, power converters, motors, and wireless circuits (notably UAV avionics) — continuously generate EMI due to rapid switching frequencies and electromagnetic coupling.

Understanding these fundamental aspects of electromagnetic interference provides the essential foundation for developing effective EMI shielding solutions and implementing comprehensive electromagnetic compatibility strategies in modern electronic systems. The complexity and pervasive nature of EMI challenges require interdisciplinary approaches combining electromagnetic theory, materials science, and practical engineering solutions to ensure reliable operation in increasingly challenging electromagnetic environments.

Figure 2. Sources and Propagation Pathways of Electromagnetic Interference (EMI)

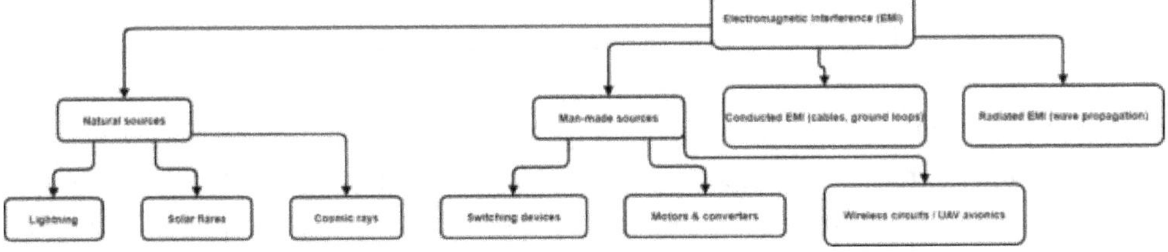

GRAPHENE AND ITS PROPERTIES RELEVANT TO EMI SHIELDING

Graphene, a single layer of carbon atoms arranged in a two-dimensional hexagonal lattice, has emerged as a revolutionary material for electromagnetic interference (EMI) shielding applications due to its extraordinary combination of electrical, mechanical, and thermal properties. As electronic systems become increasingly sophisticated and densely packed, particularly in aerospace applications such as unmanned aerial vehicles (UAVs), the demand for lightweight, flexible, and highly effective EMI shielding materials has reached unprecedented levels. Understanding the fundamental properties of graphene and how they translate into superior EMI shielding performance is crucial for developing next-generation electronic protection systems.

Table 2. Comparison of Conventional Metals, Polymers, and Graphene for EMI Shielding

Material	Density (g/cm^3)	Electrical Conductivity (S/m)	Flexibility	Typical SE (dB)	Remarks
Copper (Cu)	8.96	~5.8×10^7	Rigid	80–120	Excellent shielding but heavy; prone to corrosion.
Aluminum (Al)	2.70	~3.5×10^7	Rigid	60–100	Lightweight metal but still rigid; corrosion issues.
Polymers	0.9 – 1.4	~$10^{-12} - 10^{-8}$	Highly flexible	<20	Lightweight and flexible but poor shielding unless filled with conductive additives.
Graphene	~0.002 (per layer)	~$10^6 - 10^8$ (depending on quality)	Flexible	30–90 (films/ composites); >100 (foams/hybrids)	Combines conductivity with flexibility; scalable composites can outperform metals in SE/density ratio.

Table 2 presents the comparison of copper, aluminium, polymers, and graphene as EMI shielding materials. metals provide high shielding but suffer from weight and rigidity. polymers offer flexibility but lack conductivity. graphene uniquely combines lightweight, conductivity, and flexibility, enabling high shielding effectiveness per unit weight.

Electrical Conductivity and Charge Transport

The exceptional electrical conductivity of graphene represents its most significant advantage for EMI shielding applications. With theoretical conductivity values exceeding 10^6 S/m, graphene demonstrates conductivity levels comparable to or superior to many conventional metals while maintaining a thickness of just one atomic layer. This remarkable property arises from graphene's unique electronic structure, where electrons behave as massless Dirac fermions with extremely high mobility values reaching 200,000 $cm^2/V{\cdot}s$ at room temperature.

The high electrical conductivity enables graphene to effectively reflect electromagnetic waves through the formation of mobile charge carriers that respond rapidly to incident electromagnetic fields. When electromagnetic radiation encounters a graphene surface, the free electrons within the material oscillate in response to the electric field component of the wave, creating opposing electromagnetic fields that effectively reflect the incident radiation back toward its source. This reflection mechanism becomes particularly effective across a broad frequency spectrum, from radio frequencies through microwave and even into the terahertz range.

Hong et al. (2012) demonstrated that even a single monolayer of graphene can provide measurable EMI shielding effectiveness, with shielding values of approximately 2.3 dB in the gigahertz frequency range. This finding is particularly significant because it establishes that extremely thin graphene films can provide meaningful electromagnetic protection without the weight and thickness penalties associated with traditional metallic shields. The ability to achieve effective shielding with minimal material thickness makes graphene especially attractive for aerospace applications where weight reduction is paramount.

Mechanical Properties and Structural Integrity

Graphene exhibits extraordinary mechanical properties that complement its electrical characteristics for EMI shielding applications. With a Young's modulus of approximately 1 TPa and tensile strength reaching 130 GPa, graphene is among the strongest materials ever measured. These exceptional mechanical properties enable graphene-based EMI shields to maintain structural integrity under demanding operational conditions, including the vibrations, thermal cycling, and mechanical stresses encountered in aerospace environments.

The flexibility of graphene represents another crucial advantage for modern electronic systems. Unlike rigid metallic shields that can crack or fail under mechanical stress, graphene maintains its electrical and shielding properties even when subjected to significant bending, stretching, or twisting. This flexibility enables the development of conformal EMI shields that can be integrated into curved surfaces, flexible electronics, and complex three-dimensional structures without compromising performance.

For UAV applications, where aerodynamic considerations are critical, the ability to create lightweight, flexible EMI shields that conform to aircraft surfaces without disrupting airflow represents a significant technological advancement. Graphene-based shields can be engineered to follow the contours of UAV components, providing effective electromagnetic protection while maintaining optimal aerodynamic performance.

Thermal Stability and Environmental Resistance

The thermal properties of graphene contribute significantly to its suitability for EMI shielding in challenging environments. Graphene exhibits exceptional thermal stability, maintaining its structural and electrical properties at temperatures exceeding 2000°C in inert atmospheres. This thermal stability ensures that graphene-based EMI shields continue to function effectively across the wide temperature ranges encountered in aerospace applications, from the extreme cold of high-altitude flight to the elevated temperatures generated by electronic components and propulsion systems.

The high thermal conductivity of graphene, approaching 5000 W/m·K, provides an additional benefit by facilitating heat dissipation from electronic components. This thermal management capability helps prevent temperature-induced performance degradation in sensitive electronic systems while simultaneously providing electromagnetic protection. The combination of EMI shielding and thermal management in a single material represents a significant advantage over traditional approaches that require separate solutions for these challenges.

Chemical stability represents another crucial property for practical EMI shielding applications. Pristine graphene exhibits excellent resistance to chemical degradation, maintaining its properties in various environmental conditions. However, the chemical inertness of pristine graphene can present challenges for processing and integration with other materials, leading to the development of functionalized graphene derivatives.

Graphene Derivatives and Tunability

The development of graphene derivatives, particularly graphene oxide (GO) and reduced graphene oxide (rGO), has significantly expanded the practical applications of graphene in EMI shielding. Graphene oxide, produced through the oxidation of graphite, introduces oxygen-containing functional groups that enhance dispersibility in various solvents and polymer matrices. While GO exhibits lower electrical conductivity compared to pristine graphene due to the disruption of its conjugated π-electron system, it offers improved processability for large-scale manufacturing applications.

Reduced graphene oxide represents a compromise between the exceptional properties of pristine graphene and the processability of graphene oxide. Through controlled reduction processes, much of the electrical conductivity can be restored while maintaining sufficient functionalization for practical processing. Gutiérrez et al. (2024) demonstrated that printed graphene films based on rGO can achieve EMI shielding effectiveness comparable to traditional materials while offering significant advantages in terms of thickness and weight.

The tunability of graphene's properties through controlled functionalization enables the optimization of EMI shielding performance for specific applications. By adjusting the degree of reduction, the type and concentration of functional groups, and the processing conditions, researchers can tailor graphene materials to meet specific conductivity, mechanical, and processing requirements.

Surface Area and Absorption Mechanisms

The large specific surface area of graphene, theoretically reaching 2630 m^2/g for single-layer materials, provides numerous opportunities for electromagnetic wave interactions beyond simple reflection. This extensive surface area enables the development of absorption-based shielding mechanisms through the incorporation of defects, functional groups, and hybrid structures that can dissipate electromagnetic energy through dielectric losses.

Defect Engineering and Energy Dissipation

Defect engineering in graphene creates localized states that can interact with electromagnetic radiation, converting electromagnetic energy into thermal energy through various loss mechanisms. These defects can be introduced intentionally through controlled processing or chemical treatments, allowing for the optimization of absorption properties for specific frequency ranges.

The combination of reflection and absorption mechanisms in graphene-based materials enables the development of versatile EMI shields that can be optimized for different frequency ranges and applications. Wang et al. (2021) demonstrated that graphene-reinforced MgO composites achieve enhanced EMI shielding through multiple reflections and scattering facilitated by the lamellar structure of embedded graphene platelets.

Integration with Composite Systems

The versatility of graphene enables its integration into various composite systems to create multifunctional EMI shielding materials. Polymer-graphene composites combine the processability and mechanical properties of polymers with the electrical and thermal properties of graphene, resulting in materials that can be manufactured using conventional processing techniques while providing superior EMI shielding performance.

Polymer Composite Applications

Verma et al. (2015) demonstrated the development of graphene-polyurethane composites that provide both EMI shielding and electrostatic dissipation capabilities. These multifunctional materials address multiple electromagnetic compatibility challenges simultaneously, reducing system complexity and weight compared to solutions requiring separate materials for each function.

Ceramic Matrix Composites

The incorporation of graphene into ceramic matrices, as demonstrated by Wang et al. (2021), enables the development of high-temperature EMI shielding materials suitable for extreme environments. These ceramic-graphene composites maintain shielding effectiveness at temperatures where polymer-based systems would fail, expanding the range of applications for graphene-based EMI shields.

Scalability and Manufacturing Considerations

The practical implementation of graphene-based EMI shielding requires consideration of scalability and manufacturing processes. Various synthesis methods, including chemical vapor deposition (CVD), liquid-phase exfoliation, and chemical reduction of graphene oxide, offer different advantages and limitations for large-scale production. The selection of appropriate synthesis and processing methods significantly impacts the properties and cost-effectiveness of the final EMI shielding materials.

Processing and Production Methods

Ryu et al. (2022) highlighted the importance of graphene processability in developing millimeter-scale composites for 5G EMI shielding applications. Their work demonstrates that reduced graphene oxide can enhance the processability of graphene while maintaining sufficient conductivity for effective EMI shielding, making large-scale manufacturing more feasible.

The development of printing and coating technologies for graphene-based materials has opened new possibilities for cost-effective production of EMI shields. These processing approaches enable the deposition of graphene layers onto various substrates using techniques compatible with existing manufacturing infrastructure, facilitating the integration of graphene-based EMI shielding into commercial products.

Future Perspectives and Applications

The unique combination of electrical, mechanical, thermal, and chemical properties exhibited by graphene and its derivatives positions these materials as ideal candidates for next-generation EMI shielding applications. The ability to tune these properties through functionalization and composite formation enables the development of optimized solutions for specific applications, from lightweight aerospace systems to flexible consumer electronics. As manufacturing processes continue to improve and costs decrease, graphene-based EMI shielding materials are poised to revolutionize electromagnetic compatibility solutions across numerous industries.

The ongoing research into smart and adaptive EMI shielding systems incorporating graphene demonstrates the potential for materials that can respond dynamically to changing electromagnetic environments. These advanced systems could provide optimized protection while minimizing weight and power consumption, representing the next evolutionary step in electromagnetic interference management.

FABRICATION OF GRAPHENE-BASED EMI SHIELDING MATERIALS

The development of effective graphene-based electromagnetic interference (EMI) shielding materials requires sophisticated fabrication approaches that can harness graphene's exceptional properties while addressing the practical challenges of large-scale production, cost-effectiveness, and integration into existing manufacturing processes. The fabrication strategies employed must carefully balance multiple competing requirements including shielding performance, mechanical robustness, environmental stability, and economic viability. This comprehensive examination of fabrication methodologies demonstrates how various synthesis and processing techniques can be optimized to create next-generation EMI shielding solutions for diverse applications ranging from consumer electronics to aerospace systems.

Graphene Synthesis Methods

The foundation of any graphene-based EMI shielding material lies in the quality and characteristics of the graphene itself. Chemical vapor deposition (CVD) represents one of the most promising approaches for producing high-quality graphene films suitable for EMI shielding applications. CVD-grown graphene typically exhibits superior electrical conductivity and structural integrity compared to other synthesis methods, making it particularly valuable for applications requiring maximum shielding effectiveness. The process involves the decomposition of carbon-containing precursors on transition metal substrates at elevated temperatures, typically between 800-1000°C, under controlled atmospheric conditions.

Liquid-phase exfoliation offers an alternative approach that enables large-scale production of graphene nanosheets in solution. This method involves the mechanical or chemical separation of graphite layers in appropriate solvents, producing graphene flakes that can be readily incorporated into composite materials or processed into films. While the resulting graphene may have more defects compared to CVD-grown material, liquid-phase exfoliation provides excellent scalability and cost-effectiveness for industrial applications.

Chemical reduction of graphene oxide represents another widely adopted synthesis route, particularly for applications where processability is more critical than pristine electronic properties. This method begins with the oxidation of graphite to produce graphene oxide, which can then be reduced using chemical, thermal, or electrochemical methods to restore much of the original electrical conductivity. The resulting reduced graphene oxide (rGO) retains sufficient functional groups to facilitate dispersion in various matrices while providing adequate conductivity for EMI shielding applications.

Film and Coating Deposition Techniques

The transformation of synthesized graphene into functional EMI shielding films requires sophisticated deposition techniques that can maintain material integrity while achieving desired thickness and coverage. Spray coating has emerged as a particularly versatile approach, allowing for the deposition of graphene suspensions onto various substrates with excellent control over thickness and uniformity. This technique is especially valuable for coating complex geometries and large surface areas, making it suitable for applications in automotive and aerospace industries where conformal coatings are essential.

Spin coating provides exceptional uniformity for flat substrates and is particularly valuable in electronics manufacturing where precise thickness control is crucial. The centrifugal forces generated during spin coating ensure even distribution of graphene dispersions, resulting in films with consistent electrical and shielding properties across the entire surface. This technique has proven especially effective for creating transparent conducting films for display applications and flexible electronics.

Inkjet printing represents a revolutionary approach that enables precise patterning of graphene-based EMI shielding materials with minimal material waste. This additive manufacturing technique allows for the creation of complex patterns and selective shielding in specific areas, making it ideal for applications in printed electronics and smart textiles. The ability to deposit graphene inks with micron-scale precision opens new possibilities for creating multifunctional devices that integrate EMI shielding with other electronic functions.

Dip-coating offers simplicity and scalability for coating three-dimensional objects and continuous processing of flexible substrates. This technique involves immersing the substrate in a graphene dispersion and controlling the withdrawal speed to achieve desired film thickness. The process can be easily scaled for industrial production and is particularly suitable for creating protective coatings on metal substrates and fabric materials.

Polymer-Graphene Composite Fabrication

The integration of graphene into polymer matrices represents one of the most promising approaches for creating lightweight, flexible EMI shielding materials suitable for a wide range of applications. The fabrication of these composites requires careful attention to graphene dispersion, interfacial interactions, and processing conditions to achieve optimal performance. Solution mixing techniques involve dispersing graphene in solvents compatible with the target polymer, followed by polymer dissolution and subsequent solvent removal. This approach allows for excellent control over graphene distribution and can produce composites with tailored properties.

Melt blending offers an environmentally friendly alternative that avoids the use of organic solvents while enabling the processing of high-viscosity polymers. The mechanical shearing forces generated during melt processing help to disperse graphene throughout the polymer matrix, though care must be taken to prevent degradation of the graphene structure. This technique is particularly suitable for thermoplastic polymers and can be readily integrated into existing polymer processing equipment.

In-situ polymerization provides another approach where graphene is dispersed in monomer solutions before polymerization occurs. This technique can result in superior interfacial adhesion between the graphene and polymer phases, leading to improved mechanical properties and more efficient charge transport pathways. The approach is especially valuable for creating thermoset polymer composites with high graphene loadings.

Graphene-Metal Hybrid Systems

The incorporation of metallic nanoparticles into graphene-based systems represents an advanced fabrication approach that can significantly enhance EMI shielding effectiveness while maintaining the lightweight advantages of graphene. The decoration of graphene surfaces with silver, copper, or nickel nanoparticles creates synergistic effects that combine the high conductivity of metals with the flexibility and corrosion resistance of graphene.

Electrochemical deposition techniques enable precise control over metal particle size, distribution, and loading on graphene surfaces. This approach allows for the creation of hybrid materials with tailored properties optimized for specific frequency ranges and shielding requirements. The electrochemical process can be easily scaled for industrial production and enables the creation of gradient structures with varying metal content across the material thickness.

Chemical reduction methods offer an alternative approach for creating graphene-metal hybrids through the simultaneous reduction of metal salts and graphene oxide. This one-step process can produce intimate mixing between the metal and graphene phases, resulting in materials with enhanced electrical and thermal conductivity compared to simple physical mixtures.

Three-Dimensional Porous Structures

The fabrication of three-dimensional graphene structures, including foams and aerogels, represents a cutting-edge approach to EMI shielding that maximizes the surface area available for electromagnetic wave interaction while maintaining light-weight characteristics. These structures operate through multiple scattering mechanisms that effectively attenuate electromagnetic waves over broad frequency ranges.

Template-assisted methods involve using sacrificial templates such as polymer foams or metal frameworks to create three-dimensional graphene networks. After graphene deposition or growth on the template surface, the template is removed through chemical etching or thermal decomposition, leaving behind a free-standing graphene structure that replicates the original template geometry.

Freeze-drying techniques offer an alternative approach for creating porous graphene structures from aqueous dispersions. By controlling the freezing rate and ice crystal formation, it is possible to create materials with tailored pore structures and densities optimized for specific EMI shielding applications.

Quality Control and Optimization

The successful fabrication of graphene-based EMI shielding materials requires comprehensive quality control measures to ensure consistent performance and reliability. Characterization techniques including scanning electron microscopy, X-ray photoelectron spectroscopy, and electrical conductivity measurements are essential for monitoring material properties throughout the fabrication process. These analytical tools enable the optimization of processing parameters to achieve maximum shielding effectiveness while maintaining acceptable mechanical and environmental stability.

Each fabrication technique presents unique advantages and limitations that must be carefully considered in the context of specific application requirements. The ultimate goal is to balance shielding performance with scalability, transparency, and mechanical robustness to create commercially viable solutions that can compete with traditional metallic shields while offering significant advantages in weight, flexibility, and multifunctionality.

Figure 3. Flowchart linking graphene fabrication methods (CVD, exfoliation, GO reduction) to processing pathways (films/coatings, polymer composites, metal–graphene hybrids, 3D foams/aerogels) and downstream EMI shielding applications across consumer electronics, automotive/aerospace & UAVs, 5G/mm-wave, and defense/healthcare.

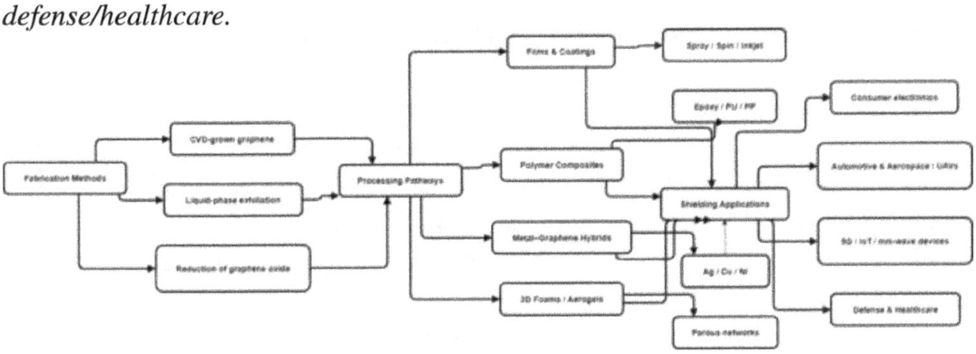

This flowchart shown in figure 3 tracks the manufacturing pipeline from graphene creation to deployable EMI shields. CVD graphene produces continuous, high-conductivity films ideal for ultrathin coatings and transparent shields. Liquid-phase exfoliation yields scalable flakes for inks, paints, and composite fillers, while GO reduction (rGO) balances processability and conductivity, enabling solution processing at low cost. These materials are then processed into (i) films/coatings (spray, spin, or inkjet printer), (ii) polymer composites (epoxy/PU/PP) for lightweight structural parts, (iii) metal–graphene hybrids (Ag/Cu/Ni) for boosted surface conductivity, and (iv) 3D foams/aerogels whose porous networks enhance multiple scattering and absorption. The resulting formats map to application needs: consumer devices (thin, flexible layers), automotive/aerospace & UAVs (weight-sensitive structural shields), 5G/mm-wave electronics (conformal, high-frequency compatible films), and defense/healthcare (robust, corrosion-resistant, or biocompatible solutions).

MECHANISMS OF EMI SHIELDING IN GRAPHENE-BASED SYSTEMS

The electromagnetic interference (EMI) shielding effectiveness of graphene-based materials stems from their unique ability to interact with electromagnetic radiation through multiple synergistic mechanisms. Unlike conventional metallic shields that primarily rely on reflection, graphene-based systems offer a sophisticated combination of reflection, absorption, and scattering phenomena that can be precisely engineered to achieve optimal performance across specific frequency ranges. Understanding these fundamental mechanisms is crucial for designing next-generation shielding materials that can meet the demanding requirements of modern electronic systems while maintaining the lightweight, flexible, and multifunctional characteristics that make graphene-based solutions so attractive for advanced applications.

Electromagnetic Wave Reflection Mechanisms

The reflection mechanism in graphene-based EMI shielding materials is fundamentally governed by the interaction between incident electromagnetic waves and the mobile charge carriers within the graphene structure. When electromagnetic waves encounter a graphene layer, the high electrical conductivity of pristine graphene, which can exceed 10^6 S/m, creates a significant impedance mismatch with free space, resulting in substantial reflection of the incident radiation. This reflection occurs at the interface between the electromagnetic wave propagation medium and the conductive graphene surface, where the electric field component of the wave induces currents in the graphene sheet.

The effectiveness of reflection-based shielding is directly related to the surface conductivity of the graphene material, which can be controlled through various parameters including the number of graphene layers, defect density, and chemical functionalization. Single-layer graphene exhibits remarkable reflection properties despite its atomically thin structure, with theoretical calculations suggesting reflection coefficients that can exceed 50% for high-quality samples. Multi-layer graphene structures demonstrate enhanced reflection capabilities, with each additional layer contributing to the overall shielding effectiveness through constructive interference effects.

Temperature-dependent conductivity variations in graphene also influence reflection performance, with higher temperatures generally leading to reduced conductivity due to increased phonon scattering. This temperature dependence must be carefully considered in applications where thermal stability is critical, particularly in aerospace and automotive environments where operating temperatures can vary significantly.

Absorption and Dielectric Loss Mechanisms

Absorption represents the second critical mechanism through which graphene-based materials attenuate electromagnetic waves. Unlike reflection, which redirects electromagnetic energy away from the protected region, absorption converts electromagnetic energy into thermal energy through various loss mechanisms. In graphene-based systems, absorption occurs primarily through dielectric losses, which are enhanced by the presence of defects, functional groups, and interfaces within the material structure.

Defect-induced absorption plays a particularly important role in practical graphene-based shielding materials. Structural defects such as vacancies, grain boundaries, and edge sites create localized states that can interact strongly with electromagnetic radiation, leading to enhanced absorption. These defects can be intentionally introduced through controlled oxidation, plasma treatment, or chemical functionalization processes to optimize absorption characteristics for specific frequency ranges.

The dielectric properties of graphene-based composites are significantly influenced by the interaction between graphene and the surrounding matrix material. In polymer-graphene composites, interfacial polarization effects contribute to dielectric losses, particularly at lower frequencies where dipolar relaxation processes become dominant. The orientation and dispersion of graphene sheets within the composite matrix also affect the overall dielectric response, with aligned graphene structures typically exhibiting anisotropic absorption characteristics.

Magnetic loss mechanisms can be introduced into graphene-based shielding materials through the incorporation of magnetic nanoparticles or magnetic substrates. These hybrid systems combine the excellent electrical properties of graphene with magnetic permeability enhancements, resulting in improved absorption performance across broader frequency ranges.

Multiple Scattering and Wave Interference

Three-dimensional graphene structures, including foams, aerogels, and hierarchical composites, exploit multiple scattering mechanisms to achieve superior EMI shielding performance. In these materials, electromagnetic waves undergo repeated scattering events as they propagate through the complex three-dimensional network, significantly extending the effective path length and increasing the probability of absorption or reflection at internal interfaces.

The porous structure of graphene foams creates numerous scattering centers where electromagnetic waves can undergo diffuse scattering, leading to randomization of wave propagation directions and enhanced attenuation. The size and distribution of pores within the foam structure can be tailored to optimize scattering efficiency for specific frequency ranges, with smaller pores generally more effective for higher frequency applications.

Wave interference effects also contribute to the overall shielding performance of multi-layer graphene structures. When electromagnetic waves encounter multiple graphene layers separated by specific distances, constructive and destructive interference patterns can be established that enhance or suppress transmission at particular frequencies. This phenomenon can be exploited to create frequency-selective shielding materials that provide enhanced protection at critical operating frequencies while maintaining transparency at others.

The interconnected nature of three-dimensional graphene networks ensures that scattered electromagnetic energy is efficiently captured and dissipated throughout the material volume, preventing the formation of standing wave patterns that could reduce shielding effectiveness.

Synergistic Effects in Hybrid Systems

Hybrid graphene-based shielding materials leverage synergistic effects between different components to achieve performance levels that exceed those of individual constituents. The combination of graphene with metallic nanoparticles, for example, creates materials that benefit from both the high conductivity of metals and the unique electromagnetic properties of graphene. These hybrid systems often exhibit non-linear improvements in shielding effectiveness due to enhanced charge transfer, modified electromagnetic field distributions, and improved impedance matching characteristics.

Polymer-graphene-metal hybrid composites represent a particularly promising class of materials where the polymer matrix provides mechanical flexibility and processability, graphene contributes lightweight conductivity and absorption, and metallic components enhance reflection and conductivity. The careful selection of component ratios and processing conditions allows for precise tuning of electromagnetic properties to meet specific application requirements.

Frequency-dependent synergistic effects are particularly important in these hybrid systems, as different mechanisms may dominate at different frequency ranges. At lower frequencies, reflection mechanisms typically predominate, while absorption becomes increasingly important at higher frequencies where skin depth effects become more pronounced.

Frequency-Dependent Performance Optimization

The electromagnetic shielding mechanisms in graphene-based materials exhibit strong frequency dependence, requiring careful consideration of target frequency ranges during material design. At microwave frequencies, the balance between reflection and absorption can be controlled through material thickness, conductivity, and dielectric properties. For millimeter-wave and terahertz applications, surface plasmon effects and quantum confinement phenomena become increasingly important, offering new opportunities for performance optimization.

The scalability of graphene-based shielding mechanisms from laboratory samples to industrial-scale materials remains a critical challenge. Maintaining uniform electromagnetic properties across large areas requires precise control of graphene distribution, orientation, and connectivity, particularly in composite materials where processing-induced variations can significantly impact performance.

Advanced characterization techniques, including vector network analysis, time-domain spectroscopy, and electromagnetic simulation, are essential for understanding and optimizing the complex interplay between different shielding mechanisms in graphene-based materials. These tools enable the development of predictive models that can guide material design and accelerate the development of next-generation EMI shielding solutions.

Figure 4. Shielding mechanisms in a graphene composite sheet.

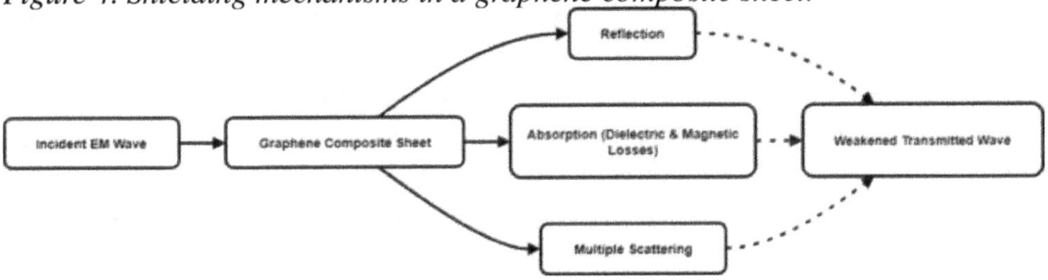

Figure 4 shows explains that the Incident electromagnetic (EM) waves interact with the graphene composite, where part of the energy is reflected due to high surface conductivity, a fraction is absorbed through dielectric and magnetic losses, and further multiple scattering occurs within the porous/defect-rich network. The residual energy that passes through emerges as a weakened transmitted wave, demonstrating the combined reflection–absorption–scattering shielding mechanism.

This synergistic combination of reflection, absorption, and scattering is what makes graphene composites superior to traditional metallic shields, especially in lightweight and flexible applications.

CONCLUSION

The focus of this research highlights, particularly in SDG 9 and SDG 13: Industry, Innovation and Infrastructure and Climate Action, respectively, that sustainable development is achieved through technological innovation and environmental stewardship in addressing climate-related challenges. The AI-powered UAV fog dispersal system represents a cutting-edge solution for aviation technology that promotes adaptive response to meteorological hazards in an environmentally conscious and technologically advanced manner. This is to say that future partnerships with aviation authorities, meteorological agencies, environmental organizations, and the technology innovators behind these systems become fundamental in scaling and disseminating this solution worldwide, contributing meaningfully towards the UN SDGs.

A comprehensive architecture of an AI-powered UAV fog clearing system—an innovative solution to fog's detrimental effects on the aviation industry. The integrated system demonstrates significant promise in enhancing visibility conditions required for safe flight operations, reducing flight delay costs within airport environments, and minimizing the environmental consequences of traditional fog management approaches. The overall architecture of the proposed solution combines multiple advanced simulation tools, including MATLAB/Simulink for UAV dynamics analysis, ANSYS Fluent/OpenFOAM for comprehensive fog behavior modeling, as well as AI-based algorithms for real-time decision-making, autonomous navigation optimization, and adaptive dispersal agent selection. The integration of UAV swarms capable of precise fog dispersion with superior adaptability and targeting accuracy significantly enhances the scalability and operational efficiency of the system compared to conventional approaches. Our proposed methodology fills critical gaps left by previous research and establishes the groundwork for future developments in AI-enabled environmental management and aviation safety enhancement.

However, challenges such as system scalability for large-scale fog events, regulatory compliance with aviation authorities, integration with existing air traffic management protocols, and the long-term environmental impact assessment of dispersal agents must be comprehensively addressed before widespread real-world implementation. Collaboration with international aviation authorities, development of environmentally sustainable and biodegradable seeding agents, and adherence to stringent air traffic management safety protocols are essential prerequisites for successful deployment. Furthermore, extensive physical prototyping, comprehensive field testing under various meteorological conditions, and rigorous validation against real-world fog scenarios are necessary to authenticate simulation results and refine the system architecture for large-scale operational deployment.

The proposed AI-driven UAV fog dispersal system represents a paradigmatic advancement in aviation technology, providing a scalable, cost-effective, and environmentally sustainable solution to fog-related operational disruptions. By systematically addressing the outlined technical challenges, regulatory requirements, and environmental considerations while continuously refining the system capabilities, this research possesses the transformative potential to revolutionize fog management strategies, contributing to safer, more efficient, and sustainable aviation operations on a global scale. The convergence of artificial intelligence, unmanned aerial systems, and advanced atmospheric modeling creates unprecedented opportunities for proactive weather management, ultimately enhancing the resilience and reliability of critical transportation infrastructure in an era of increasing climate variability and growing aviation demands.

REFERENCES

Gutiérrez, D., Doshi, P., Wong, H., Nordlund, D., & Gandhiraman, R. (2024). Printed graphene and its composite with copper for electromagnetic interference shielding applications. *Nanotechnology*, *35*(13), 135202. DOI: 10.1088/1361-6528/ad12e9 PMID: 38055996

Hao, B., Peng, C., Tang, X., & Zhao, Z. (2022). Analysis of time-frequency characteristic parameters of electromagnetic interference sources in the basic commutation unit. *IET Power Electronics*, *16*(4), 531–541. DOI: 10.1049/pel2.12403

Hariyawan, M. Y., Hidayat, R., & Firmansyah, E. (2016). The effects of spread-spectrum techniques in mitigating conducted EMI to LED luminance. [IJECE]. *International Journal of Electrical and Computer Engineering*, *6*(3), 1332–1343. DOI: 10.11591/ijece.v6i3.pp1332-1343

Hong, S., Kim, K., Kim, T., Kim, J., Park, S., Kim, S., & Ahn, J. H. (2012). Electromagnetic interference shielding effectiveness of monolayer graphene. *Nanotechnology*, *23*(45), 455704. DOI: 10.1088/0957-4484/23/45/455704 PMID: 23085718

Khalid, S., Chakkaravarthy, M., Gupta, S. K., Prakash, D., Agrawal, A., & Nishad, D. K. (2025). A conceptual framework for developing an AI-powered UAV system for effective fog management at airports. *SPAST Reports, 1*(1).

Prakash, D., Agarwal, A., Khalid, S., Nishad, D. K., & Singh, V. K. (2024). Innovative approaches to fog dispersal: The emerging role of unmanned aerial vehicles. In *2024 IEEE 11th Uttar Pradesh Section International Conference on Electrical, Electronics and Computer Engineering (UPCON).*

Ryu, S., Kim, H., Park, S., Kwon, S., Kim, S., Sundaray, B., & Park, C. (2022). Millimeter-scale percolated polyethylene/graphene composites for 5G electromagnetic shielding. *ACS Applied Nano Materials*, *5*(6), 8429–8439. DOI: 10.1021/acsanm.2c01544

Verma, M., Verma, P., Dhawan, S., & Choudhary, V. (2015). Tailored graphene based polyurethane composites for efficient electrostatic dissipation and electromagnetic interference shielding applications. *RSC Advances*, *5*(118), 97349–97358. DOI: 10.1039/C5RA17276D

Wang, W., Cui, R., Zhang, P., Wan, C., & Pan, W. (2021). High electromagnetic interference shielding effectiveness in MgO composites reinforced by aligned graphene platelets. *Journal of the American Ceramic Society*, *104*(6), 2868–2878. DOI: 10.1111/jace.17667

Watanabe, K., Sakai, R., Tanaka, S., Nagata, M., Osaka, H., Kuwahara, Y., & Fujii, T. (2024). Electromagnetic interference with the mobile communication devices in unmanned aerial vehicles and its countermeasures. *IEEE Access : Practical Innovations, Open Solutions*, *12*, 11642–11652. DOI: 10.1109/ACCESS.2024.3351216

Chapter 8
Graphene-Based Nanomaterials in Textiles for Advanced Aviation Applications

Sushila Sahani
https://orcid.org/0000-0002-5739-144X
Lovely Professional University, India

ABSTRACT

The aviation industry demands materials that offer superior performance in strength, thermal management, electrical conductivity, flame resistance, and sustainability. Graphene-based nanomaterials, present an unique opportunity to reinvent textile applications for aviation. This chapter presents a comprehensive exploration of the synthesis, functionalization, and integration of graphene nanomaterials into textile substrates. Detailed analyses reveal significant enhancements in tensile strength, thermal management, electromagnetic interference (EMI) shielding, flame retardancy, and sensor integration, all critical to the safety and efficiency of modern aviation systems. Environmental considerations, including life cycle assessment and recyclability, are also examined alongside the economic viability of large-scale production. Regulatory challenges are addressed, providing a holistic view of the hurdles and opportunities for commercialization. This chapter outlines the critical role of graphene nanotechnology in shaping the future of smart, adaptive textiles for the aviation sector.

DOI: 10.4018/979-8-3373-3862-0.ch008

INTRODUCTION

Graphene is a special planar (2D) Molecular structure based on the hexagonal planar structure of carbon atoms in graphite (carbon). Although most molecules either structure in order to make a linear or a three-dimensional (3D) network, the structure governed by the activity of the bonding agents, graphene structure generates a flat sheet of molecules. In textile applications, this structure may be its strength, and other times a problem. Ideally, it is expected that perfectly aligned graphene sheets can have outstanding axial strength and stiffness. However, their bad transverse properties would concentrate on disbonding just like graphite in a pencil or in a lubricant. Nevertheless, when used functionally, e.g., to produce graphene fibers as textiles, some disorder in the sheets is unavoidable (Hearle, 2001); sufficient disorder is then present to ensure that the fibers are cohesive enough to survive the spinning process, and are then ready to be used in clothing via solution-dropping methods (Holleck, 2008). Because of this, there can be found a balance between strength and structural stability. This enables the use of graphene in durable, flexible, and functional applications of textiles. The aeronautical industry needs high-performance textiles. Fabrics that will be able to provide excellent strength, thermal management, electrical conductivity, flame resistance, and even sustainability. Such attributes have to be in place to ensure safety, efficiency, and reduce ecological impact. Conventionally, these are used with aluminum, titanium, and carbon composites, but aluminum and titanium are limited in weight and expensive, whilst carbon composites can be unsustainable. Graphene-based nanomaterials, particularly graphene-enriched textiles, have emerged as one of such sustainable solutions. Worthy properties belong to graphene, which was discovered in 2004, i.e., a one-atom-thick carbon lattice (Katsnelson, 2007). As an example, it is around 200 times stronger than steel in terms of its weight (Zare et al, 2021). It has good thermal conductivity of up to ~5,000 W m -1 K -1, and has excellent electrical conductivity, approaching 200,000 cm 2 V -1 s -1 in electron mobility, yet it is highly lightweight and chemically inert (Balandin et al, 2008 & Bolotin et al, 2008). Graphene or/its deriving products (eg, graphene oxide) when embedded into fabrics could have a huge mechanic, thermal, and electrical impact. They have the capacity to be used as advanced thermal management systems within high-temperature areas. They are able to provide fire-retardant performance of cabin materials, e.g., seats and wall panels, through char formation layers. It is also capable of enabling the production of smart clothes that can be fitted with sensors to provide real-time structural health monitoring or electromagnetic protection. Besides, the wearable technology in pilots and crews with graphene textiles can introduce new horizons in the aviation industry. It could find use in high-load, flexible items such as cargo nets or parachutes. The textile materials are also recyclable, which is also in line with the increased use of sustainable business in the aviation industry. Graphene-enhanced textiles are, in general, a highly multidisciplinary innovation that is able to provide the strict performance requirements of the current aviation industry, and they can also sustain long-term performance, safety, and environmental responsibility requirements.

NEED FOR PERFORMANCE ENHANCEMENTS IN AVIATION TEXTILES

Flame Retardancy

Conventional textile fabrics will not be adequate in protecting the environment where temperatures might reach more than enough 1000°c in less than 20 seconds. This point in time is where graphene takes center stage as a material of interest in terms of outstanding thermal stability, flame resistance, and tensile behaviors (Revaiah et al., 2020). The risk associated with military aircraft fires is very severe as it entails excessive quantities of highly inflammable JP-4 fuel. The small escape openings put the crew at a disadvantage to survive, and this frequently gives rise to burns, impact injuries, asphyxiation, or a mixture of all three. Fires of post-crash JP-4 fuel are characterized by the maximum intensity in 20 seconds, and the thermal environment in the area of the fire, with the temperatures of 927 1260 °C and heat flux 2.79 7.45 cor/cm2/sec. Extraction experiments provided by the US Army Aeromedical Research Laboratory (USAARL) underline how different the severity of the fire may be, depending on several factors, such as wind, terrain, and vaporization of the fuel. Under the maximally possible scenario, the temperature may increase up to 1149°C and the heat flux may total 5.5 cal/cm 2/sec. This heat is obtained by radiant energy approximately 90 percent (Knox et al., 1979). It is essential that there be escape time, and it has been documented in the literature that an unprotected pilot needs to escape within 1-5 seconds (Stoll et al., 1979), but protective clothing can buy a couple of seconds of lifesaving. Nevertheless, being fortunate to evade the fuel fire that can be 50-60 feet with a temperature of 1200 °C in a short span of time may be a dream considering the intense heat flux (Crown & Capjack, 2005). The verge of death response (e.g., ignition, charring) can have a lot of restrictions on the survivability. Exclusive protective measures demand special consideration during such dangerous situations. Nonetheless, the protective clothing should also strike a balance between thermal protection, comfort, and affordability.

Sensor Integration

Sensor-enabled intelligent textiles are also revolutionizing smart textiles in the aviation sector, both in the structural evaluating of the plane and safety monitoring of the crew and passengers. The new smart fabrics would have sensors for stress, temperature, and damage embedded into furniture/cockpit fabric, cabin wall cladding, and even on the uniforms of pilots. In the field of science, there have been a lot of studies that show new ways of incorporating the use of grapheme-based sensors in textiles. In another of the studies, strain, pressure, and impact-sensitive yarns are enabled by a polydopamine (PDA)-aided dipping-coating technique of graphene which is pristine graphene. These yarns perform consistently in incorporated parts of these flexible epoxy panels and rigid PLA composition panels. Such yarns, in the aviation textile-application context, may be integrated into aircraft interior composite or structural textile-strengthening elements to enable real-time sensing. Cotton-based yarn embedded with PTFE and glass yarns covered with CMC-modified graphene exhibited a gauge factor of 4.2-9.8 and pressure sensitivity of 4 Omega (g x cm -2)-1, respectively, at a range of 0-12 percent tensile strain. They were able to identify as little as 0.5 J of impact energy, and this reveals their potential in aircraft skin or cabin panel surveillance. Sensors could be trusted with getting elastic, yield, fall, and fracture in tensile, compression, 3-point bending, cyclic bending, and impact studies. They were also able to sustain their work during 30,000 seconds of repetitive loading (Zhu et al., 2025). Such outcomes highlight the capacity of one economical manufacturing process in developing powerful multimodal sensing fabrics suited in terms of structural health monitoring that are aviation-related.

Electromagnetic Interference (EMI) Shielding

The aviation sector uses EM Interference (EMI) shielding fabrics in their airplane interiors because they are needed to avoid EMI interfering with cockpit instruments, airplane communications, and even airline passengers' electronic items. According to the literature by Liu et al. (2019 and Zhang et al. (2015, the use of graphene-based materials instead of the costly alternatives has a high potential in the EMI shielding aviation textiles development area. Its use was stressed specifically on applications in which electromagnetic interference is hazardous to sensitive avionics and communication systems. Liu et al. (2019 investigated how the combination of graphene and graphite, and waterborne polyurethane can be used to make a functional EMI shielding coating. Although developed on coated fabric and achieved the lower end result of 22 dB at 0-40 MHz, the coating of the fabric, making graphene uniformly dispersed, is difficult, and the cost of the material inhibits its commercial aviation use. Conversely, Zhang et al. (2015 improved the shielding effectiveness to 25 dB in X-band (8.2-12.4 GHz). In-situ polymerization of polypyrene and graphene on fabric was used for radar, satellite communication, and related frequencies. As aviation textile innovation, issues determined by a low rate of utilization of graphene and poor bonding of fabric/graphene combinations have to be overcome to achieve the aerospace requirements of durability, low weight, and a high safety standard.

Thermal Management

Thermal regulation and thermal protection of textiles are crucial to the health and survival of humans, especially when exposed to extreme conditions or environments, like in the aviation industry. Although conventional textiles can contribute to our physical security, they tend to fail to deliver high attributes to offer individual thermo-physiological comfort (Ge et al, 2022). Thermal regulation is the process by which a garment creates an ideal microclimate between a body and its surroundings and is responsive to the change of temperature between the body and the surrounding environment (Havenith, 2002). This is vital, particularly in individuals exposed to fluctuating temperatures or in any intensive physical exercise, as the temperature management capacity of the body has a direct bearing on comfort, functioning, and health. In order to remedy these inadequacies, superior textile materials are being produced, including smart fabrics, phase change materials (PCMs), and graphene-enhanced textiles, which have two benefits, i.e., active thermal control and increased flame resistance. The graphene-enhanced textiles are also capable of reflecting heat dissipated by smoke and insulating against heat, but without any weight or stiffening effect, and this is very critical in terms of protective clothing, like in the case of flight suits, firefighting suits, or military uniforms.

GRAPHENE NANOMATERIALS' PROPERTIES AND SYNTHESIS

Graphene consists of a single layer of carbon atoms, which has a flat honeycombed structure. It has attracted too much attention in science and technology due to its outstanding characteristics (Anjali & Kizhakayil, 2024). With time, the scientific investigation of graphene continues; it is vital to comprehend its features and the ways of its creation. In this section, this is examined why graphene is special. It also talks about the principal ways of its production. These fundamentals can also make us determine how applicable graphene can be in future advance engineering practices.

Fundamental Properties of Graphene

Whether or not to include out-of-plane deformations, graphene is considered to be a two-dimensional (2D) or quasi-2D material. It belongs to the carbon material that consists of the 3D forms such as diamond and graphite, 0D fullerenes, 1D carbyne, and quasi-1D carbon nanotubes (Boehm, 2010). The smaller dimensionality of graphene has a different morphology and reaction as seen in its bulk counterparts. Edges, surfaces, and interfaces are the most important material lines in low-dimensional systems (0D to 2D). New designs of material are made possible by surface and interface engineering, including the atom-by-atom doping via chemical reaction, irradiation, etc (Novoselov et al., 2004; Weng et al., 2008).

Mechanical strength

Graphene can be defined as a two-dimensional and a single layer of sp2-hybridized carbon, in a hexagonal lattice. The in-plane carbon bonds are sp2 hybridized, which results in strong sp2 in-plane σ bonds, making graphene have great mechanical strength (Geim & Novoselov, 2007). Because of the 2D nature of graphene, its mechanical behavior is determined by its physical dimensions and elastic moduli, which are issues of concern to its full extent. The thickness of grapes is not easy to define because of the quantum mechanical property of the grapes and the effect of the orbitals of 6. The layers of graphene in graphite are symmetrically layered with an interlayer spacing that is known to be 3.35 A. But in other mediums, this thickness can be substituted with interaction with the neighboring media (Short & Walker, 1963). It is difficult to measure the mechanical properties (at a nanoscale thickness) of graphene. The high-pressure mechanical behavior of graphite varies with its environment (e.g., suspended in fluid PTMs as pressure-transmitting media (PTM) or deposited on supports). Graphene can be converted into defective graphene or novel materials such as graphene oxide through functionalization (e.g., oxygen, hydrogen, or fluorine). These changes change mechanical properties, especially in the case when covalent bonds are created massively (Sun et al, 2021). The elastic modulus of graphene of about 1 TPa (terapascal) makes it one of the most powerful as well as stiffest known substances, greatly surpassing others (Sakhaee-Pour, 2009). This unique stiffness and its low density make graphene very exotic in the field of lightweight materials of high strength (aerospace and aviation materials, to mention just a few).

Thermal Conductivity

Thermal conductivity may be considerably lowered in bad conductors through atomic vibration and rotation. It happens because of the phonon scattering whereby phonon (lattice vibrations) engages with other components (other phonons, defects in materials, etc) and leads to a change in the direction of energy (Balandin et al, 2008). In the graphene-polymers composites, there are many interfaces because of the large values of their specific surface area of the graphene. The consequence is phonon scattering and a poor interfacial thermal resistance that does not promote effective heat transfer (Xu et al., 2012; Luo et al., 2012). Graphene TCE per wt% is about 50% which shows that low quantities of graphene can considerably enhance thermal conductivity. Nevertheless, the pure form of graphene has difficulties surpassing 100% TCE per wt%. This may be attained through surface modification either by covalent or non-covalent bonds, as seen by adding a titanite coupling agent that can take TCE due to increased interfacial forces, and a decrease in thermo-resistance to close to 100% of TCE per wt%, in fact (Han & Fina, 2011). Certain orientations, like aligned graphene, segregated structures, or 3D structures, contribute effectively to the thermal conductivity more than random orientations. These structures establish an effective thermal reach in the polymer matrix, leading to a systematic heating transfer (Lim et al., 2013; Yao et al, 2016). The research was conducted to examine the thermal conductivity of graphene sheets-epoxy composites with no percolation threshold increasing with an increase in graphene loading (Moore & Shi, 2014). These results were similar to graphene/octadecanol composites (Chen et al, 2016), and it was also seen that below a percolation threshold of 0.7 wt%, the polymer molecules leave an open space that restricts the thermal conductivity. Beyond this energy point, they develop a topological contact between the graphene sheets, which results in a sharp increase in the thermo-conductivity [Balandin et al, 2008]. Thermal energy storage is enhanced by using Graphene derivatives to improve the thermal conductivity of phase change materials (PCMs). In one such study, it was found that by embedding ultrathin graphite foams inside PCMs, it is capable of enhancing thermal conductivity was enhanced 18-fold (Alkan et al, 2011). The potential application of the superlative thermal conductivity of graphene in the polymer composite is high. Surface modification, preplanning arrangement, as well as optimal loading can lower surface resistance and larger the transfer of heat. These thermal conductivity upgrades of the graphene-polymer composite have the mind-blowing potential in upgraded aviation textiles.

Electrical Conductivity

Organic conducting polymers - Organic conducting polymers are a type of material that can conduct electricity, with implementation in computer displays, fuel cells, and microsurgical instruments (Awuzie, 2017). They have the benefit that they can be processed by means of dispersion. They cannot, however, generally be readily thermo-formed and have other disadvantages, including high manufacturing costs, toxicity, poor solubility, material inconsistencies, and poor mechanical properties relative to other commercially available polymers. Talented polymers Thermal and electrical characteristics (Peng et al, 2007; Clingerman, 2001) could be enhanced by combining conductive polymers with carbon-based imbibitions such as carbon nanotubes (CNTs), graphite, and graphene (Kaur et al, 2015). The lightweight property, the strength to create conductive networks, and its oxidation resistance make graphene particularly popular. Graphene exhibits excellent thermal conductivity and simulation estimates of 6000 W m-1K-1 and the experimental value of 2000- 5000 W m-1K-1 at room temperature (Alemour et al, 2018). Conductive polymers comprised of graphene are able to provide a high-performance, yet lightweight answer to thermal regulation of aviation fabrics, surmounting significant weaknesses of conventional conductive polymers.

Synthesis Methods for Graphene-Based Nanomaterials

Chemical Vapor Deposition (CVD)

Chemical Vapor Deposition (CVD) has emerged as one of the most practically productive processes in creating high-quality graphene. Some of the several CVD techniques include Low–Pressure CVD (LPCVD) and Atmospheric-Pressure CVD (APCVD). Both of them have different advantages depending on the process conditions and on the end-user requests.

Low-Pressure Chemical Vapor Deposition (LPCVD)

Low-Pressure Chemical Vapor Deposition (LPCVD) is a process that utilizes low pressure. This helps in making the single-layer graphene, as the supply of carbon precursor was constrained, and the surface diffusion was promoted. Nonetheless, it needs elaborate vacuum systems, which increases complexity in operation, besides limiting scalability. Use of accurate pressure control is necessary to get repeatable results. LPCVD is especially well-positioned in situations when single-layer graphene with high quality needs to be used (Lee et al, 2020).

Atmospheric-Pressure CVD (APCVD)

APCVD has an ambient pressure. This gives more carbon precursor concentration in the reaction atmosphere. Such availability of carbon precursors helps in growing multiple-layer graphene. Nonetheless, in APCVD, higher flow rates of methane are possible, resulting in more particulates formed through gas-phase reactions. This can possibly affect the quality of the end product (graphene). That notwithstanding, the APCVD is easier to set up and does not require the use of vacuum systems, and may be more practical with respect to large-scale manufacturing. Applications that are generally associated with it are conductive films and composites made of graphene (Pham, 2018). APCVD-grown graphene can be applied to textiles, where graphene can be deposited as a conductive layer into fabric to make conductive fabric and wearable electronic devices, where multiple layers of graphene can enhance electrical conductivity, durability, and electromagnetic properties.

Exfoliation Techniques

• Mechanical Exfoliation

Graphene is taken out of bulk graphite via mechanical exfoliation of the layers and fracture of the Van der Waals force between them. This may come in two principal ways, i.e., the use of normal force to elevate the layers vertically or shear force to move them out laterally. Although this technique produces high-quality, large-area graphene, it is time-consuming, labor-intensive, and not scalable at the industry level (Yi & Shen, 2015). To ensure simplification of the manual work, a group of researchers created a lathe-like experimental system of cleaving highly oriented pyrolytic graphite (HOPG) to produce graphene flakes. It is a setup that can be scaled by the utilization of standard lathe methods. The resulting flakes, however, have flakes of thickness in the tens of nanometer range, and to yield high-quality graphene, fine control of the diamond wedge, requiring the process, is still needed (Jayasena & Subbiah, 2011). There is another micromechanical technique that was invented, which was based on the conventional Scotch tape technique. It involves the application of a three-roll mill machine and a polymer adhesive. In this technique, there is exfoliation and dispersion in the adhesive medium. Despite the prevalence of the three-roll mill as an industrial device used mainly in the rubber industry and the relative ease of using it, it is difficult to remove all of the residual polyvinyl chloride (PVC) and dioctyl phthalate (DOP) by the use of the mill. Such further processing makes the creation of high-purity grapheme more intricate (Chen et al., 2012). When discussing the use of aviation textiles, these micromechanical techniques can provide a basis for infusing graphene into fibers and coatings, which may be used to increase conductivity, strength, and their place in thermo-regulation.

These methods have, however, to increase the scalability, purity, and uniformity of their products to achieve the standards of processing and operational performance of the textile industry.

• Liquid-Phase Exfoliation

Liquid Phase Exfoliation (LPE) of graphene, which is the ultrasonication of graphite in organic solvents or aqueous medium in the presence or the absence of surfactants. The technique creates a scalable path towards the production of graphene flakes (Ciesielski & Samor, 2014). This may be appropriate in textile functionalization. The aqueous dispersions can be especially useful in coating or printing operations as part of constructing smart textiles, where it is important that the graphene coating is relatively uniform on fiber surfaces to provide electrical conductivity, thermal management, or even sensing potential (Afroj et al., 2019).

• Electrochemical exfoliation

Electrochemical exfoliation can be used to make graphene by putting an electric potential in graphite electrodes immersed in an electrolyte that results in the intercalation of the ions and subsequent detachment of the layers of graphene (Liu et al., 2019). The use of the method is beneficial due to its flexibility, affordability, and the possibility to provide high-quality graphene with a controlled set of properties (Prekodravac et al, 2021). Those properties render the electrochemically exfoliated graphene particularly suitable in textile applications, where the affordability of mass quantities of graphene is crucial. The developed units of classification that come about during exfoliation enhance dispersion in polymer matrices. It is also easily adsorbed on textile fibers, thus it is suitable for the production of conductive yarns and wearable sensors. The approach is, therefore, viable towards the creation of low-cost and high-performance smart textiles in the aviation field.

• Reduction of Graphene Oxide

Graphene oxide (GO) is a nonconductive material containing oxygen-based functional groups that are structurally different from pristine graphene. Reduction aims to eliminate these oxygen-bearing functionalities, forming a material commonly referred to as reduced graphene oxide (rGO) that replicates the outstanding electrical and mechanical characteristics of graphene (Pei & Cheng, 2012). Chemical reduction through hydrazine or derivatives (ex, hydrazine hydrate) is one of them, which enhances conductivity. This can be achieved by the addition of polymers or ammonia to keep the material stable in a liquid state. This feature can be used in the design of porous constructions, such as batteries or detectors. The other method is thermal reduction by heating GO to remove oxygen groups. As a result of one of the studies, calculations demonstrate that epoxy groups are less difficult to eliminate compared to the hydroxyl ones, though it is difficult to achieve a high C/O ratio (between 75% and 6.25% (C/O ratio 16:1) (Park et al, 2011). These enhancements indicate that rGO can be used in electric gadgets. All methods of reduction have tradeoffs, eg, chemical methods are low cost, but have a low risk of leaving some oxygen behind, thermal methods require high temperatures, and advanced techniques like CVD are highly effective but expensive and complicated to perform. Accordingly, selection will be based on application. Reduced graphene oxide (rGO) improves the conductivity and mechanical properties of smart fabrics, wearable electronics, and sensors in textiles. As an illustration, when used as aviation uniforms, reduced graphene oxide (rGO) coatings can be a source of electrostatic discharge protection, thermal control, and/or EMI protection that improves safety and comfort.

INTEGRATION OF GRAPHENE INTO TEXTILE SUBSTRATES

Functional textiles will fulfill the end-specific needs of users, like wearable electronics, sensors, and smart clothes. Graphene and its derivatives in textile substrates have created new avenues for creating next-generation functional fabrics. These positive developments, notwithstanding, are subject to practical application, with a thought that must be given to the way the use of graphene relates to the textile substrates. This segment will look at the simple approaches being applied, which include coating processes, spinning of fibers, and the formation of nanocomposites. It also determines such long-standing issues as successful uniform dispersion and good adhesion, and the assessment of the scalability of the methods to industrial textile manufacturing.

Coating Techniques

Graphene-type materials make it possible to manufacture conducting elastic materials, which are used in many technical realms as well as items of clothing. Its imperially barrier layer prevents penetration of gases such as helium, but it allows permeation and propagation of water vapor. The Graphene Oxide (GO) contains oxygen-containing functional groups that disperse carbon materials in a surfactant way and has the capacity to undergo chemical functionalization (Kuila et al, 2012). GO can also be used as a catalyst in oxidation and coupling reactions (carbon catalysis) that provide non-metallic catalytic potential to industries and the surroundings (Ahmad & Nishina, 2020). The addition of GO and rGO to the textile treatments brings different capabilities. GO and rGO coating has a significant effect in enhancing the thermo-conductivity of fabrics. As an example, the thermal conductivity of GO-coated polyethylene glycol (PEG)-grafted cotton fabric was 10 times higher than bare cotton. Likewise, it was seen that by use of a knife-over-roll technique, rGO-coated polyester fabric enhanced thermal conductivity compared to 0.10 W/(m K), which resulted in 0.42 W/(m K). Nevertheless, not many challenges emerged due to these methods of application. It led to the lifeless brownish appearance of GO/rGO-coated fabrics that inhibit dyeing operations and consumer attraction (Liu et al., 2023). GO-coated fabrics of cotton and linen contain anti-bacterial characteristics because they interfere with bacterial metabolism (Liu et al., 2011). These facilities are supplemented by a reduction of GO to rGO and a deposition of silver nanoparticles (Ag NPs) that further advance UV protection and photocatalytic degradation of stains (Farouk et al., 2020; Zuo et al., 2021). The reversibility of wettability is performed by UV light irradiation in graphene/titania surfaces or sonication of graphene-covered fabrics (Yang et al., 2021). Some methods of preparation. Conductive textiles are typically made in one of two ways. The first approach is the weaving or knitting of fabrics with conductive yarns, and the second approach involves impregnation of conductive substance onto an already woven or knitted structure by an appropriate coating technique. In contrast to the coating technique, the former is more universally applicable as it is cheaper and simple to conduct. The following are the coating techniques that may be applied to embed grapheme in the textile:

Dip Coating

The fabric in dip-coating is dipped in the coating solution in such a way that the mixture of graphene gets dispersed uniformly on it. Nevertheless, the fabric only absorbs a restricted quantity of coating (Abdi et al., 2024). Dip-coating (or pad-coating) is a cheap, simple, and easily up-scaled process to functionalize (or add conductive nanomaterials to) fibers, yarns, or fabric. Its primary strength points to the lack of special or costly equipment (Zhu et al., 2021; Tang & Yan, 2017). This can either be effectuated in one continuous or divided. It most commonly requires immersing the cloth in a coating solution (usually aqueous), briefly. To create conductive fabrics, this solution may include such materials as conductive polymers, carbon derivatives, metal nanostructures, or a combination of them (Tseghai et al., 2020). That results in a smooth layer of the conductive material on both sides of the fabric (Gualandi et al., 2021). Altogether, the process of dip-coating presents a convenient and non-problematic method of making textile conductive without many preparations and expenses.

Spray Coating

Graphene coating on irregular surfaces can be an interesting solution because spray coating of graphene on textile fabrics may provide an opportunity. It is a mature process of thin film formation, and it can be applied even to bigger surfaces. Hand-held sprayers, air brushes, and pneumatic nozzles can be applied as different types of spray tools that work with the help of air pressure. They are now under trials to use graphene and its composites on fabrics. The air-based techniques are cheap and they have the potential to cover vast areas within a short period; however, these techniques do not provide much control in regards to the uniformity of instances upon the coating (Pham, 2010). In lieu of it, ultrasonic spray coating applies high-frequency sound waves in order to atomize the liquid. This process gives greater control over droplet size, and a smoother and more uniform layer can be made. It can generate ultra-small droplets of the same or even smaller size in comparison to the ones created via inkjet printing. Sadanandan et al. (2020) were able to show a simple, low-cost, and scalable ultrasonic spray coating technique in order to produce conductive textile electrodes by depositing a water-based suspension of graphene nanoplatelets. It was used on meta-aramid, polyester, and nylon fabrics, with sheet resistances down to 4.5 k on the non-additive nylon and polyester (4.5 k and 5.5 k on doped and undoped nylon, respectively). To further enhance the adhesion of the coating, an extra treatment of UV-ozone was applied. These graphene textiles could have high potential for making wearable sensors, nanogenerators, supercapacitors, and thermoelectric materials.

Layer-By-Layer Assembly

Layer-by-layer (LbL) approach consists of coating graphane-oxide (GO) followed by a layer of adhesive. The most widespread application of graphene-coated fibers is conductivity enhancement. The adhesive layer is, however, insulating, thus the overall conductivity may be lowered by the layer. These fibers are normally manufactured through the 2-step process of conducting graphene-coated fibers. The fiber is first dip-coated in a GO solution, and then the GO is reduced with varying strategies to produce reduced graphene oxide (rGO), which regains its conductivity. Several cycles of dip-coating may be done to improve the already high conductivity by depositing additional graphene on the fibers. However, because of electrostatic repulsion against other GO layers, the amount of deposited GO is likely to stagnate early. To tackle this, a repetition of the coating and reduction step was established to enlarge the GO deposition on the fiber. The result of this experiment is are very conductive RGO-coated silk fiber. Nevertheless, such an approach is more time-consuming (ke et al., 2020). The creation of flexible conductive devices has been successfully achieved by anchoring layers of graphene oxide on the viscose fiber using the layer-by-layer (LbL) technique, and then reducing the graphene oxide into the conductive graphene (Zhang et al, 2016). The other experiment came up with a different approach of adhering graphene to cotton through a layer-by-layer (LbL) coating process. The anionic polymer in this procedure was the polyvinyl phosphonic acid. The results demonstrated that the degree of hydrophobicity was increased with the coating cycle, as well as functionalized graphene nanosheets (FGN) content in the dispersion. Moreover, the treated cotton was more thermally stable, electrically conductive, UV protective, and EMI shielding compared with the untreated cotton. In general, considering the common electricity conductivity, this method of inorganic/organic nanocomposite coating seems to be a quite good method to provide various functions to cellulosic materials (Nooralian et al, 2016). In this way, the layer-by-layer (LbL) approach provides a rather universal and successful tactic to increase the multi-functionality of graphene-coated fibers and textiles.

Fiber Spinning

Its physical qualities are astonishing, which is why graphene can be viewed as a promising material, although its application in real life is constrained by the lack of means and methods to combine graphene sheets into coherent strong materials, such as fibers and yarns. Solution/dope spinning Also called: wet spinning, dry-jet spinning, and dry spinning can be used to build Graphene fibers. Nonetheless, due to the fact that the melting point of graphene is extremely high (about 4600 o C), it is hard to obtain pure graphene fibers by means of melt spinning (Al Faruque et al., 2021). The current techniques generally demand high concentrations of Graphene oxide (GO) dispersions and elaborate post processing procedures. This do not support scalability and affordability. Jalili et al, 2013 conducted a research leading to the discovery of a scalable single-step wet-spinning process based on liquid crystalline graphene oxide dispersions. The research demonstrated a new method on how to form strong, conductive graphene yarns and fibres with a very low quantity (0.075 wt%) which is considerably lower in comparison with the earlier requirements. The low concentration means that it is easy to blend graphene with other substances like carbon nanotube, metals and polymers, something which could not be done with thicker mixtures. They can also produce high lengths of yarn attesting that the process is practical to use on large scale. The study enhances our knowledge on how graphene behaves to become a dispersant and a spinning fiber and finds new ways of producing advanced fiber materials. Since the process is easy, harmless, and cost-efficient, it may extensively be applied in smart textile in terms of energy, environmental and medical technologies.

Nanocomposite Formation for Enhanced Textile Matrices

Graphene-based nanocomposites Textile applications require integrating graphene or one of its derivatives with polymers or fibers in order to make graphene-based nanocomposites. It employs the methods that offer uniform mixing and compatibility of materials used in textiles. The methods that are most suitable in the application of textiles are as follows:

• Solution Mixing

The solvent is added with graphene or graphene oxide (GO) to create a composite using a polymer solution. This blend is further used in treating textiles in order to form a resistive or a shielding coating. This approach leads to consistent mixing and good graphene-polymer interactions, which further promote the electrical conductivity of fabrics (previously, graphene/polyaniline composites). The possibility of using toxic or too costly solvents and the challenges of removing surfactants are also quite problematic (Araby et al., 2014; Raza et al.,2020).

• Electrospinning

Nanofibers are formed by ejecting a graphene-polymer solution using high high-voltage electric field. A mat of these nano-fibers is then deposited on a collector. The process enables direct uptake of flexible, conductive nanofiber layers into textiles as wearable electronics, sensors, and biomedical applications. On the one hand, the electrospinning technique provides excellent control and feasibility, but on the other hand, it needs the adjustment of specific parameters to avoid flaws within the formed fibers (Navarro-Pardo et al., 2016; Khorshidi et al., 2020; Kanjwal & Ghaferi, 2022).

• Melt Blending

In the melt blending method, Graphene is blended with molten thermoplastic polymers like PA6 without solvent and formed into fibers or films (Madhad et al., 2021; Trivedi & Rachchh, 2022). Studies pertaining to the manufacturing of fibers and the use of melt-synthesized textile fibers of polyamide 6 (PA6)/graphene composite materials are extremely limited, but based on available literature, it can be seen that the accomplishment of a high dispersion condition of graphene becomes a significant challenge. Investigations were carried out to assess the potential of using the commercially available multilayered graphene nanoplatelets (GnPs) to prepare PA6/graphene composite textile fibers through in situ dispersing conductions, GnPs dispersed in the melt of 1 and 2 wt% 1 in molten epsilon-caprolactam, in situ ring-opening polymerization, and in a pilot-scale melt turning process. This method is very applicable to aviation textiles where lightweight, high-performance, and fire-retardant materials are more important to the interior of the aircraft, seat upholstery, cabin walls, and the insulation layers. Introduction of GnPs had no marked influence on the melting point of PA6 but had a small influence on its crystallization point and the crystal ability properties, useful in support of a perpetuation of dimensional preservation and mechanical integrity at reducing cabin temperatures and cabin pressure. Nevertheless, the restricted dissociation and disaggregation of GnPs caused the inclusion of fine particles. This led to a clogged spinning pack filter and caused the fibers to lose their tensile strength. This implies that there is a need to have better dispersion processes to manufacture aviation-quality textile fibers (Vasiljevic et al., 2020).

• Nanostitching

In the recent past, a nano stitching strategy was devised to utilize nanomaterials in the process of fabricating preforms and extruding composites. In this method, multi-walled carbon nanotubes (MWCNT) are put in epoxy. This is achieved through a mixture of a magnetic stirrer and an ultrasonic mixer (200-Watt, 40 kHz) so that there is uniformity of dispersion. The resultant MWCNT/epoxy blend was there upon degassed in vacuum to remove air bubbles. Nano stitching yarns, carbon or aramid nanopreg fabrics were produced with the blending of carbon nanotube with graphene nanoplatelets through the hand lay-up method. These nanopregs were pre-heated in an air oven to retain their shape and stability. The treated layers were subsequently deposited and layered in the out-of-plane direction in an attempt to produce a nano-reinforced structure. The blend was put in a Teflon film-coated metallic form and pressed with the help of a hot press. After the cycle of curing is over, the last multi-nanoceramic composite laminate is demolded to be further investigated and used in practice (Bilisik & Sapanci, 2018; Bilisik et al., 2018). Although it has advantages, nano stitching is very complicated and requires very strict handling of the processing parameters. However, with nano stitching added to graphene, new perspectives of smart and structural textiles in aerospace engineering are created. However, grapheme incorporation into textiles has great prospective in aviation textiles, but some important issues that require proper solution must be tackled, including uniform dispersion of graphene in textile matrices and sufficient adhesion strength, and the feasibility of a scalable processing system should also be processed on an industrial level.

ENVIRONMENTAL AND ECONOMIC CONSIDERATIONS

Life cycle assessment (LCA has been extensively seen over the years as a means to furnish a reliable and systematic scientific foundation to the environmental performance of the design methods of systems. It has proved to be a resourceful system in the evaluation of the environment (Wernet et al, 2010). New electronics and sensor technology also seek to minimize a product on the environment and increase performance. Research was conducted and showed how wearable electronic textiles can be made by inkjet printing of PEDOT: PSS and graphene to monitor ECG signals and skin temperature with the accuracy of the high-performance materials. Biodegradability exhibited by the graphene-based textile electrodes was outstanding, and the characteristic tensile strength decreased to 98%, and overall degradation yielded a being 48 percent in four months. In addition, Tencel fabric and graphene have also been identified to enhance the development of soil microbes (bacteria and fungi). Compared to PEDOT: PSS and the standard reference substance applied in textile electrodes, comparative life cycle analysis revealed that the global heating potential of graphene was 70.8% and 97.5% less, respectively (Dulal et al., 2025). These findings prove the prospect of graphene with regard to sustainable aviation textiles. Graphene can also be used in textiles by being added to their products, which can make them last longer by making them stronger and wear-resistant. It will minimize the frequency of returns and accompanying wastages. The recycled fibers, which include recycled polyester, can also be reinforced with graphite, which makes them reinforced and lasting. That enables more recycled materials in textiles and less use of virgin material, and a smaller carbon footprint. Graphene-enhanced fabrics might be costly in the early phase due to the cost of materials as well as the production process. The cost of ownership, however, can be minimized due to their longer life and reduced requirements to be replaced. However, improvements in the fabrication of scalable, cost-effective, and sustainable graphene are required to lower the cost of processing and allow commercial applications to be made. To incorporate graphene in the industry, there should be a repeatable and affordable supply chain able to deliver a master batch to the customer. The problem in the current industry is the lack of a standard definition to describe graphene. This implies that the R&D and general purchasing departments may have no idea of what they are purchasing, as well as may not be able to replicate test results using the next batch of material (Gibbs, 2018). There is uncertainty that the industry requires standardization of material.

REGULATORY CHALLENGES AND CERTIFICATION

The possibilities advanced fabrics are providing in the sphere of aviation textiles involve an integration of a graphene-related material (GRM). But this has also raised major concerns of safety and regulation. Since nanomaterials such as graphene are complex, there are no particular regulations. In the European Union (EU), there are regulations, amongst others, such as REACH (Regulation (EC) No 1907/2006, 2006). Registration, Evaluation, Authorization, and Restriction of Chemicals (REACH) and CLP (Classification, Labeling, and Packaging) (Regulation (EC) No 1272/2008, 2008) were used to avoid the harm to people and the environment caused by chemicals, including nanomaterials such as graphene. However, there is a huge disparity in the global regulation of graphene-based materials (GRMs). Nanomaterials are not regulated in specific ways in most nations. In the majority of countries, the risk assessment generally starts at the point of exposure. In the United States, the Toxic Substances Control Act utilizes new Special Implementation Guidelines to provide a direction (EPA, 2019). More specifically, REACH is regarded as one of the most demanding chemical safeguards. It involves a thorough risk assessment at the beginning, which is not the case in many other countries [Allan et al, 2021]. Due to the increased number of graphene-based products being introduced to the market, in addition to several patents relating to the material (more than 26,000 applications have been filed globally, Khan, 2022), a coherent regulation is greatly needed. In the absence of synchronized standards, guidelines are understandably confusing to the manufacturers, hampering the rates of innovation and market penetration (Bleeker et al., 2023). Guidance on how to test graphene is also being updated by organizations like ISO (e.g., ISO229) (ISO, n.d.), the American Society of Testing and Materials (ASTM) (ASTM International, n.d.), and by the OECD Working Party on Manufactured Nanomaterials (WPMN). These progress updates are, however, slow, attributed to little funding by OECD member states. Regulators and industry worked out the Malta Priority List as a means of giving priority to this work. It formulates the most relevant guidelines that must first be transformed. Among the greatest concerns is the fact that the existing measures of safety testing are sometimes ineffective when applied to graphene materials. It is highly toxic and varies depending on certain physical and chemical properties, which could be very different relative to other materials, including other nanomaterials. The majority of the available safety tests have been conducted on soluble chemicals whose toxicity is calculated according to the quantity of the substance. Others are on metal nanoparticles, e.g., Test no. 318 of the OECD. Yet even such tests are unsuitable in the case of graphene (Heunisch et al., 2024). Due to its peculiar characteristics (high surface area, layer structure, small particle diameter, and the tendency to clump together), graphene demands special test procedures (Connolly et al., 2023). Exposure-based risk assessments also have problems. They are necessary according to EU regulations on food material (FCM) and medical devices (MDR). The gauge of graphene in the intricate media, including body fluids, water, or soil, is challenging because we are not yet in a dependable, consolidated model of discovering and gauging the grains at a minuscule breadth. Development of other methods is still in progress. There are a number of things that must occur in order to promote the safe use and sale of graphene-enriched material. To begin with, it should be in line with the regulatory norms nationally and even internationally. This entails coming up with a collective understanding of what graphene entails and making sure that all the safety measures face the same demands. This requires good coordination among such international organizations as OECD and ISO/CEN.

FUTURE DIRECTIONS AND OPPORTUNITIES

Future studies ought to focus on the advancement of economical and environment-friendly manufacturing methods. Such methods can be graphene oxide reduction methods (e.g., reduction using plant extract or reducing agents) and energy-efficient reduction approaches. Also, one should study graphene hybrid composites, providing the simultaneous effect of the combination of graphene and other nanomaterials (e.g., carbon nanotubes, MXene). As another example, graphene might be used together with phase-change materials (PCMs) to enhance heat management in pilot uniforms. Joining graphene with piezoelectric materials might allow energy-harvesting clothes that may even drive in-flight wearable sensors. The multifunction systems might lower the number of layers of specific fabrics that have to be used. This would, in turn, result in weight loss and efficiency of aeronautical applications. What is more, the progress of additive manufacturing (e.g., 3D printing of fibers with a high concentration of graphene) could allow creating aerospace textiles with precision and on-demand production. This will further minimize wastage and allow customization to the specific aircraft parts or to the specific practice of the crew. Since the aerospace industry still concerns itself with sustainability, the graphene textile has to adhere to the principles of the circular economy. Life cycle assessments (LCAs) to quantify the environmental effects of processes in the production and integration of graphene should be a target in our future activities. Reduction in pollution and production of endorsed net-zero emission industries can be achieved through innovation in recyclable graphene-polymer composites and biodegradable graphene coating (e.g., using Tencel or any other bio-based substrate). Academia-industry partnerships should also inquire into closed-loop recycling systems of graphene-incorporated fabrics. It should be aimed that the products at the end of their life span can be sold to other people and do not interfere with their performance.

CONCLUSION

Nanomaterials made of graphene prove to be a milestone in the emergence of advanced fabrics in the field of aviation. They create unmatched possibilities of better performance, safety, and durability. The unique characteristics of graphene enable one to make a fabric that helps satisfy the pressing demands of the aviation industry. The graphene textile may provide solutions as simple as flame-resistant interiors capable of surviving post-impact temperatures or as advanced as smart fabric that may include sensors to monitor the structure's integrity in real-time. The solutions that graphene textiles can provide might be, in terms of simplicity, chemical stability, and stability over time, comparable to the traditional materials, i.e., aluminum or carbon composite, and significantly lighter, cheaper, and less environmental-friendly to the industry. Nevertheless, even after such achievements, there exist major obstacles to the common use of graphene fabrics. The process of scaling up synthetic techniques becomes constrained in terms of cost and energy-intensive procedures. Simultaneously, even dispersion of the graphene and its adhesion to textile matrices should be enhanced. Environmental issues, such as the energy intensity of graphene manufacturing and, absence of a common recycling method, should also be implemented in accordance with a vision of sustainability. Certification questions and processes, which are hampered by regulatory issues, especially government non-harmonization of standards of graphene nanomaterials, must also be resolved to transform market entry needs. Venture research and development processes that incorporate interactions between scientists, industry, and regulators to promote the future of translational research and come up with standardized testing processes are also significant. Through public-private partnership, the development of prototypes of graphene fabrics can be accelerated and lead to the real-world flight test and protective clothing. They can also sustain the circular economy and closed-loop recycling systems since graphene fabrics can help the airline industry with its net-zero target. As to the future, the graphene-based in aviation textiles vision is an intelligent, stretchy, and sustainable technology. Flame-resistant, EMI-resistant, and sensor-integrated, multi-functional clothing will streamline aircraft design, weight, and efficiency. Through concentrated research, cooperation, and collaboration, graphene nanotechnology presents the future of aircraft clothing by addressing the challenges of the day to bring safer, smart, and more sustainable solutions that aircraft clothing will offer tomorrow.

REFERENCES

Abdi, B., Tarhini, A., Baniasadi, H., & Tehrani-Bagha, A. R. (2024). Developing Graphene-based Conductive Textiles Using Different Coating Methods. *Advanced Materials Technologies*, *9*(4), 2301492. DOI: 10.1002/admt.202301492

Afroj, S., Karim, N., Wang, Z., Tan, S., He, P., Holwill, M., Ghazaryan, D., Fernando, A., & Novoselov, K. S. (2019). Engineering graphene flakes for wearable textile sensors via highly scalable and ultrafast yarn dyeing technique. *ACS Nano*, *13*(4), 3847–3857. DOI: 10.1021/acsnano.9b00319 PMID: 30816692

Ahmad, M. S., & Nishina, Y. (2020). Graphene-based carbocatalysts for carbon–carbon bond formation. *Nanoscale*, *12*(23), 12210–12227. DOI: 10.1039/D0NR02984J PMID: 32510079

Al Faruque, M. A., Syduzzaman, M., Sarkar, J., Bilisik, K., & Naebe, M. (2021). A review on the production methods and applications of graphene-based materials. *Nanomaterials (Basel, Switzerland)*, *11*(9), 2414. DOI: 10.3390/nano11092414 PMID: 34578730

Alemour, B., Yaacob, M. H., Lim, H. N., & Hassan, M. R. (2018). Review of Electrical Properties of Graphene Conductive Composites. *International Journal of Nanoelectronics & Materials*, *11*(4).

Alkan, C., Sarı, A., & Karaipekli, A. (2011). Preparation, thermal properties and thermal reliability of microencapsulated n-eicosane as novel phase change material for thermal energy storage. *Energy Conversion and Management*, *52*(1), 687–692. DOI: 10.1016/j.enconman.2010.07.047

Allan, J., Belz, S., Hoeveler, A., Hugas, M., Okuda, H., Patri, A., Rauscher, H., Silva, P., Slikker, W., & Tong, W. (2021). Regulatory landscape of nanotechnology and nanoplastics from a global perspective. *Regulatory Toxicology and Pharmacology*, *122*, 104885. DOI: 10.1016/j.yrtph.2021.104885 PMID: 33617940

Anjali, C., & Kizhakayil, R. N. (2024). Graphene-The Material of the Era. In *Graphene-Based Photocatalysts for Hydrogen Production and Environmental Remediation* (pp. 3–47). Springer Nature Switzerland. DOI: 10.1007/978-3-031-68464-7_1

Araby, S., Meng, Q., Zhang, L., Kang, H., Majewski, P., Tang, Y., & Ma, J. (2014). Electrically and thermally conductive elastomer/graphene nanocomposites by solution mixing. *Polymer*, *55*(1), 201–210. DOI: 10.1016/j.polymer.2013.11.032

ASTM International. (n.d.). Standards & publications. https://www.astm.org/standards-and-solutions/standards-publications#standard-products

Awuzie, C. I. (2017). Conducting polymers. *Materials Today: Proceedings*, *4*(4), 5721–5726. DOI: 10.1016/j.matpr.2017.06.036

Balandin, A. A., Ghosh, S., Bao, W., Calizo, I., Teweldebrhan, D., Miao, F., & Lau, C. N. (2008). Superior thermal conductivity of single-layer graphene. *Nano Letters*, *8*(3), 902–907. DOI: 10.1021/nl0731872 PMID: 18284217

Bilisik, K., Erdogan, G., & Sapancı, E. (2018). Flexural behavior of 3D para-aramid/phenolic/nano (MWCNT) composites. *RSC Advances*, *8*(13), 7213–7224. DOI: 10.1039/C7RA13437A PMID: 35540322

Bilisik, K., & Sapanci, E. (2018). Experimental determination of fracture toughness properties of nanostitched and nanoprepreg carbon/epoxy composites. *Engineering Fracture Mechanics*, *189*, 293–306. DOI: 10.1016/j.engfracmech.2017.11.033

Bleeker, E. A., Swart, E., Braakhuis, H., Cruz, M. L. F., Friedrichs, S., Gosens, I., Herzberg, F., Jensen, K. A., von der Kammer, F., Kettelarij, J. A., & Navas, J. M. (2023). Towards harmonisation of testing of nanomaterials for EU regulatory requirements on chemical safety–A proposal for further actions. *Regulatory Toxicology and Pharmacology*, *139*, 105360. DOI: 10.1016/j.yrtph.2023.105360 PMID: 36804527

Boehm, H. P. (2010). Graphene—How a laboratory curiosity suddenly became extremely interesting. *Angewandte Chemie International Edition*, *49*(49), 9332–9335. DOI: 10.1002/anie.201004096 PMID: 20865715

Bolotin, K. I., Sikes, K. J., Jiang, Z., Klima, M., Fudenberg, G., Hone, J., Kim, P., & Stormer, H. L. (2008). Ultrahigh electron mobility in suspended graphene. *Solid State Communications*, *146*(9-10), 351–355. DOI: 10.1016/j.ssc.2008.02.024

Chen, H., Ginzburg, V. V., Yang, J., Yang, Y., Liu, W., Huang, Y., Du, L., & Chen, B. (2016). Thermal conductivity of polymer-based composites: Fundamentals and applications. *Progress in Polymer Science*, *59*, 41–85. DOI: 10.1016/j.progpolymsci.2016.03.001

Chen, J., Duan, M., & Chen, G. (2012). Continuous mechanical exfoliation of graphene sheets via three-roll mill. *Journal of Materials Chemistry*, *22*(37), 19625–19628. DOI: 10.1039/c2jm33740a

Ciesielski, A., & Samorì, P. (2014). Graphene via sonication assisted liquid-phase exfoliation. *Chemical Society Reviews*, *43*(1), 381–398. DOI: 10.1039/C3CS60217F PMID: 24002478

Clingerman, M. L. (2001). *Development and modelling of electrically conductive composite materials*. Michigan Technological University.

Connolly, M., Molés, G., Carniel, F. C., Tretiach, M., Caorsi, G., Flahaut, E., Soula, B., Pinelli, E., Gauthier, L., Mouchet, F., & Navas, J. M. (2023). Applicability of OECD TG 201, 202, 203 for the aquatic toxicity testing and assessment of 2D Graphene material nanoforms to meet regulatory needs. *NanoImpact*, *29*, 100447. DOI: 10.1016/j.impact.2022.100447 PMID: 36563784

Crown, E. M., & Capjack, L. (2005). Flight suits for military aviators. Textiles for protection, 678-698.

Dulal, M., Modha, H. R. M., Liu, J., Islam, M. R., Carr, C., Hasan, T., & Karim, N. (2025). Sustainable, Wearable, and Eco-Friendly Electronic Textiles. *Energy & Environmental Materials*, *8*(3), e12854. DOI: 10.1002/eem2.12854

Farouk, A., Saeed, S. E. S., Sharaf, S., & Abd El-Hady, M. M. (2020). Photocatalytic activity and antibacterial properties of linen fabric using reduced graphene oxide/silver nanocomposite. *RSC Advances*, *10*(68), 41600–41611. DOI: 10.1039/D0RA07544B PMID: 35516560

Ge, Q., Chu, J., Cao, W., Yi, F., Ran, Z., Jin, Z., Mao, B., Li, Z., & Novoselov, K. S. (2022). Graphene-based textiles for thermal management and flame retardancy. *Advanced Functional Materials*, *32*(42), 2205934. DOI: 10.1002/adfm.202205934

Geim, A. N. D. R. E., & Novoselov, K. (2007). Graphene calling. *Nature Materials*, *6*(169), 10–1038. PMID: 17330074

Gibbs, R. (2018). Near term commercial applications for graphene, composites and conductive pastes. *Journal of Material Science & Engineering*, *7*, 7. DOI: 10.4172/2169-0022-C4-101

Gualandi, I., Tessarolo, M., Mariani, F., Possanzini, L., Scavetta, E., & Fraboni, B. (2021). Textile chemical sensors based on conductive polymers for the analysis of sweat. *Polymers*, *13*(6), 894. DOI: 10.3390/polym13060894 PMID: 33799437

Han, Z., & Fina, A. (2011). Thermal conductivity of carbon nanotubes and their polymer nanocomposites: A review. *Progress in Polymer Science*, *36*(7), 914–944. DOI: 10.1016/j.progpolymsci.2010.11.004

Havenith, G. (2002). Interaction of clothing and thermoregulation. *Exogenous Dermatology*, *1*(5), 221–230. DOI: 10.1159/000068802

Hearle, J. W. (Ed.). (2001). *High-performance fibres* (pp. 7–8). Elsevier.

Heunisch, E., Bleeker, E., Kelly, S., Matzke, M., Pohl, A., Tommaso, S., & Suarez-Merino, B. (2024). Priority list for making OECD test guidelines and guidance documents applicable for nanomaterials and advanced materials. The Malta Initiative. https://malta-initiative.org/MaltaInitiative_UPLOADS/20240301_The_Malta_Initiative_Priority_List.pdf

ISO ISO/TC 229 Nanotechnologies available at: www.iso.org/committee/381983.html

Jalili, R., Aboutalebi, S. H., Esrafilzadeh, D., Shepherd, R. L., Chen, J., Aminorroaya-Yamini, S., Konstantinov, K., Minett, A. I., Razal, J. M., & Wallace, G. G. (2013). Scalable one-step wet-spinning of graphene fibers and yarns from liquid crystalline dispersions of graphene oxide: Towards multifunctional textiles. *Advanced Functional Materials*, *23*(43), 5345–5354. DOI: 10.1002/adfm.201300765

Jayasena, B., & Subbiah, S. (2011). A novel mechanical cleavage method for synthesizing few-layer graphenes. *Nanoscale Research Letters*, *6*(1), 1–7. DOI: 10.1186/1556-276X-6-95 PMID: 21711598

Kanjwal, M. A., & Ghaferi, A. A. (2022). Graphene incorporated electrospun nanofiber for electrochemical sensing and biomedical applications: A critical review. *Sensors (Basel)*, *22*(22), 8661. DOI: 10.3390/s22228661 PMID: 36433257

Katsnelson, M. I. (2007). Graphene: Carbon in two dimensions. *Materials Today*, *10*(1-2), 20–27. DOI: 10.1016/S1369-7021(06)71788-6

Kaur, G., Adhikari, R., Cass, P., Bown, M., & Gunatillake, P. (2015). Electrically conductive polymers and composites for biomedical applications. *RSC Advances*, *5*(47), 37553–37567. DOI: 10.1039/C5RA01851J

Ke, F., Song, F., Zhang, H., Xu, J., Wang, H., & Chen, Y. (2020). Layer-by-layer assembly for all-graphene coated conductive fibers toward superior temperature sensitivity and humidity independence. *Composites. Part B, Engineering*, *200*, 108253. DOI: 10.1016/j.compositesb.2020.108253

Khan, M. (2022, July 12). The environmental impact of graphene nanomaterials [Editorial feature]. AZoNano. Retrieved from https://www.azonano.com/article.aspx?ArticleID=6223

Khorshidi, S., Mohebbali, M., Imani, R., Mahmoodi, M., & Solouk, A. (2020). Electrospun fibroin/graphene oxide nanocomposite mats: An optimization for potential wound dressing applications. *Fibers and Polymers*, *21*(3), 480–488. DOI: 10.1007/s12221-020-9465-z

Knox, F. S., Sauermilch, P. W., Wachtel, T. L., Mccahan Jr, G. R., & Trevethan, W. P. (1979). A Fire Simulator/Shutter System for Testing Protective Fabrics and Calibrating Thermal Sensors (No. USAARL794).

Kuila, T., Bose, S., Mishra, A. K., Khanra, P., Kim, N. H., & Lee, J. H. (2012). Chemical functionalization of graphene and its applications. *Progress in Materials Science*, *57*(7), 1061–1105. DOI: 10.1016/j.pmatsci.2012.03.002

Lee, B., Chu, W., & Li, W. (2020). Effects of process parameters on graphene growth via low-pressure chemical vapor deposition. *Journal of Micro-and Nano-Manufacturing*, *8*(3), 031005. DOI: 10.1115/1.4048494

Lim, H. S., Oh, J. W., Kim, S. Y., Yoo, M. J., Park, S. D., & Lee, W. S. (2013). Anisotropically alignable magnetic boron nitride platelets decorated with iron oxide nanoparticles. *Chemistry of Materials*, *25*(16), 3315–3319. DOI: 10.1021/cm401488a

Liu, F., Wang, C., Sui, X., Riaz, M. A., Xu, M., Wei, L., & Chen, Y. (2019). Synthesis of graphene materials by electrochemical exfoliation: Recent progress and future potential. *Carbon Energy*, *1*(2), 173–199. DOI: 10.1002/cey2.14

Liu, S., Zeng, T. H., Hofmann, M., Burcombe, E., Wei, J., Jiang, R., Kong, J., & Chen, Y. (2011). Antibacterial activity of graphite, graphite oxide, graphene oxide, and reduced graphene oxide: Membrane and oxidative stress. *ACS Nano*, *5*(9), 6971–6980. DOI: 10.1021/nn202451x PMID: 21851105

Liu, Y., Fei, B., & Xin, J. H. (2023). Functionalization of fabrics with Graphene-based coatings: Mechanisms, Approaches, and Functions. *Coatings*, *13*(9), 1580. DOI: 10.3390/coatings13091580

Liu, Y. J., Liu, X. L., & Zhang, Y. Q. (2019). Study on the electromagnetic properties and mechanical properties of graphene/graphite single-layer coating fabric. *Journal of Textile Science & Engineering*, *36*(2), 1–5.

Luo, T., & Lloyd, J. R. (2012). Enhancement of thermal energy transport across graphene/graphite and polymer interfaces: A molecular dynamics study. *Advanced Functional Materials*, *22*(12), 2495–2502. DOI: 10.1002/adfm.201103048

Madhad, H. V., Mishra, N. S., Patel, S. B., Panchal, S. S., Gandhi, R. A., & Vasava, D. V. (2021). Graphene/graphene nanoplatelets reinforced polyamide nanocomposites: A review. *High Performance Polymers*, *33*(9), 981–997. DOI: 10.1177/09540083211011216

Moore, A. L., & Shi, L. (2014). Emerging challenges and materials for thermal management of electronics. *Materials Today*, *17*(4), 163–174. DOI: 10.1016/j.mattod.2014.04.003

Navarro-Pardo, F., Martinez-Hernandez, A. L., & Velasco-Santos, C. (2016). Carbon nanotube and graphene based polyamide electrospun nanocomposites: A review. *Journal of Nanomaterials*, *2016*(1), 3182761. DOI: 10.1155/2016/3182761

Nooralian, Z., Gashti, M. P., & Ebrahimi, I. (2016). Fabrication of a multifunctional graphene/polyvinylphosphonic acid/cotton nanocomposite via facile spray layer-by-layer assembly. *RSC Advances*, *6*(28), 23288–23299. DOI: 10.1039/C6RA00296J

Novoselov, K. S., Geim, A. K., Morozov, S. V., Jiang, D. E., Zhang, Y., Dubonos, S. V., ... & Firsov, A. A. (2004). Electric field effect in atomically thin carbon films. science, 306(5696), 666-669.

Park, S., An, J., Potts, J. R., Velamakanni, A., Murali, S., & Ruoff, R. S. (2011). Hydrazine-reduction of graphite-and graphene oxide. carbon, 49(9), 3019-3023.

Pei, S., & Cheng, H. M. (2012). The reduction of graphene oxide. *Carbon*, *50*(9), 3210–3228. DOI: 10.1016/j.carbon.2011.11.010

Pham, P. V. (2018). Atmospheric pressure chemical vapor deposition of graphene. *Chem. Vap. Depos. Nanotechnol*, *6*, 115–134.

Pham, V. H., Cuong, T. V., Hur, S. H., Shin, E. W., Kim, J. S., Chung, J. S., & Kim, E. J. (2010). Fast and simple fabrication of a large transparent chemically-converted graphene film by spray-coating. *Carbon*, *48*(7), 1945–1951. DOI: 10.1016/j.carbon.2010.01.062

Prekodravac, J. R., Kepić, D. P., Colmenares, J. C., Giannakoudakis, D. A., & Jovanović, S. P. (2021). A comprehensive review on selected graphene synthesis methods: From electrochemical exfoliation through rapid thermal annealing towards biomass pyrolysis. *Journal of Materials Chemistry. C, Materials for Optical and Electronic Devices*, *9*(21), 6722–6748. DOI: 10.1039/D1TC01316E

Raza, M. A., Maqsood, M. F., Rehman, Z. U., Westwood, A., Inam, A., Sattar, M. M. S., Ghauri, F. A., & Ilyas, M. T. (2020). Thermally reduced graphene oxide-reinforced acrylonitrile butadiene styrene composites developed by combined solution and melt mixing method. *Arabian Journal for Science and Engineering*, *45*(11), 9559–9568. DOI: 10.1007/s13369-020-04845-4

Regulation (EC) No 1907/2006 of the European Parliament and of the Council of 18 December 2006 concerning the registration, evaluation, authorisation and restriction of chemicals (REACH) Official Journal of the European Union L 396/1

Regulation (EC) No 1272/2008 of the European Parliament and of the Council of 16 December 2008 on classification, labelling and packaging of substances and mixtures Official Journal of the European Union L 353/1

Revaiah, R. G., Kotresh, T. M., & Kandasubramanian, B. (2020). Technical textiles for military applications. *Journal of the Textile Institute, 111*(2), 273–308. DOI: 10.1080/00405000.2019.1627987

Sadanandan, K. S., Bacon, A., Shin, D. W., Alkhalifa, S. F., Russo, S., Craciun, M. F., & Neves, A. I. (2020). Graphene coated fabrics by ultrasonic spray coating for wearable electronics and smart textiles. *JPhys Materials, 4*(1), 014004. DOI: 10.1088/2515-7639/abc632

Sakhaee-Pour, A. (2009). Elastic properties of single-layered graphene sheet. *Solid State Communications, 149*(1-2), 91–95. DOI: 10.1016/j.ssc.2008.09.050

Short, M. A., & Walker, P. L.Jr. (1963). Measurement of interlayer spacings and crystal sizes in turbostratic carbons. *Carbon, 1*(1), 3–9. DOI: 10.1016/0008-6223(63)90003-4

Stoll, A. M., CmANTA, M. A., & PIEROALLINI, J. (1979). Thermal r Human Skin. *Aviation, Space, and Environmental Medicine,* •••, 779.

Sun, Y. W., Papageorgiou, D. G., Humphreys, C. J., Dunstan, D. J., Puech, P., Proctor, J. E., Bousige, C., Machon, D., & San-Miguel, A. (2021). Mechanical properties of graphene. *Applied Physics Reviews, 8*(2), 021310. DOI: 10.1063/5.0040578

Tang, X., & Yan, X. (2017). Dip-coating for fibrous materials: Mechanism, methods and applications. *Journal of Sol-Gel Science and Technology, 81*(2), 378–404. DOI: 10.1007/s10971-016-4197-7

Trivedi, D. N., & Rachchh, N. V. (2022). Graphene and its application in thermoplastic polymers as nano-filler-A review. *Polymer, 240*, 124486. DOI: 10.1016/j.polymer.2021.124486

Tseghai, G. B., Malengier, B., Fante, K. A., Nigusse, A. B., & Van Langenhove, L. (2020). Development of a flex and stretchy conductive cotton fabric via flat screen printing of PEDOT: PSS/PDMS conductive polymer composite. *Sensors (Basel), 20*(6), 1742. DOI: 10.3390/s20061742 PMID: 32245034

U.S. Environmental Protection Agency (EPA). (2019). Significant new use rules on certain chemical substances (FRL No. 9991–19; Docket ID No. EPA-HQ-OPPT-2017-0575). https://www.epa.gov/reviewing-new-chemicals-under-toxic-substances-control-act-tsca/significant-new-use-rules-certain

Vasiljević, J., Demšar, A., Leskovšek, M., Simončič, B., Čelan Korošin, N., Jerman, I., Šobak, M., Žitko, G., Van de Velde, N., & Čolović, M. (2020). Characterization of polyamide 6/multilayer graphene nanoplatelet composite textile filaments obtained via in situ polymerization and melt spinning. *Polymers, 12*(8), 1787. DOI: 10.3390/polym12081787 PMID: 32785048

Weng, L., Zhang, L., Chen, Y. P., & Rokhinson, L. P. (2008). Atomic force microscope local oxidation nanolithography of graphene. *Applied Physics Letters, 93*(9), 093107. DOI: 10.1063/1.2976429

Wernet, G., Conradt, S., Isenring, H. P., Jiménez-González, C., & Hungerbühler, K. (2010). Life cycle assessment of fine chemical production: A case study of pharmaceutical synthesis. *The International Journal of Life Cycle Assessment, 15*(3), 294–303. DOI: 10.1007/s11367-010-0151-z

Xu, P., Loomis, J., Bradshaw, R. D., & Panchapakesan, B. (2012). Load transfer and mechanical properties of chemically reduced graphene reinforcements in polymer composites. *Nanotechnology, 23*(50), 505713. DOI: 10.1088/0957-4484/23/50/505713 PMID: 23196792

Yang, S., Sha, S., Lu, H., Wu, J., Ma, J., Wang, D., Hou, C., & Sheng, Z. (2021). Graphene oxide and reduced graphene oxide coated cotton fabrics with opposite wettability for continuous oil/water separation. *Separation and Purification Technology, 259*, 118095. DOI: 10.1016/j.seppur.2020.118095

Yao, Y., Zeng, X., Wang, F., Sun, R., Xu, J., & Wong, C. P. (2016). Significant Enhancement of Thermal Conductivity in Bioinspired Freestanding Boron Nitride Papers Filled with Graphene Oxide. *Chemistry of Materials, 80*(4), 1357–1359. DOI: 10.1021/acs.chemmater.5b04187

Yi, M., & Shen, Z. (2015). A review on mechanical exfoliation for the scalable production of graphene. *Journal of Materials Chemistry. A, Materials for Energy and Sustainability, 3*(22), 11700–11715. DOI: 10.1039/C5TA00252D

Zare, P., Aleemardani, M., Seifalian, A., Bagher, Z., & Seifalian, A. M. (2021). Graphene oxide: Opportunities and challenges in biomedicine. *Nanomaterials (Basel, Switzerland), 11*(5), 1083. DOI: 10.3390/nano11051083 PMID: 33922153

Zhang, H., Cao, J., Wu, W., Cao, Z., & Ma, H. (2016). Layer-by-layer assembly of graphene oxide on viscose fibers for the fabrication of flexible conductive devices. *Cellulose (London, England), 23*(6), 3761–3770. DOI: 10.1007/s10570-016-1088-6

Zhang, S. L. (2015). Polypyrrole/graphene oxide films based microwave absorbing fabric through layer-by-layer assembly.

Zhu, C., Zhang, Y., & He, G. (2025). Graphene-Based Textile Sensors for Intelligent Structural Health Monitoring. *Polymers, 17*(11), 1484. DOI: 10.3390/polym17111484 PMID: 40508727

Zhu, S., Wang, M., Qiang, Z., Song, J., Wang, Y., Fan, Y., & Ye, C. (2021). Multifunctional and highly conductive textiles with ultra-high durability through 'green' fabrication process. *Chemical Engineering Journal*, *406*, 127140. DOI: 10.1016/j.cej.2020.127140

Zuo, D., Li, G., Ling, Y., Cheng, S., Xu, J., & Zhang, H. (2021). Durable UV-blocking property of cotton fabrics with nanocomposite coating based on graphene oxide/ZnO quantum dot via water-based self-assembly. *Fibers and Polymers*, *22*(7), 1837–1843. DOI: 10.1007/s12221-021-0750-2

Chapter 9
Graphene Integration in Avionics:
Barriers, Safety Protocols, and Environmental Considerations

Sitesh Kumar Singh
https://orcid.org/0000-0002-7108-0808
College of Engineering, National University of Science and Technology, Oman

Shyam Agasthya Hande Ramachandra
https://orcid.org/0009-0001-9382-9971
College of Engineering, National University of Science and Technology, Oman

Leevesh Kumar
https://orcid.org/0000-0001-9476-0841
World University of Bangladesh, Bangladesh

M. K. Harikeerthan
https://orcid.org/0000-0001-9738-2495
Dayanand Sagar Academy of Technology and Management, India

ABSTRACT

The extraordinary electrical, thermal and mechanical characteristics of graphene offer a revolutionary prospect in the next-generation avionics systems. The current chapter presents the diverse possibilities of graphene in aerospace electronics showing all the functional capabilities of EMI shielding, heat management and structural sensing. It also takes a critical look on major integration issues such as variability of synthesis, compatibility of manufacture and constraints of regulations. These detailed discussions on occupational safety, risk assessment patterns and environmental sustainability which are backed up by the lifecycle analysis give a

DOI: 10.4018/979-8-3373-3862-0.ch009

complete picture of the future of graphene in the aviation industry. The chapter ends with strategic suggestions, which are safe, efficient and responsible integration of graphene in avionics, according to the industry standards and sustainability.

INTRODUCTION ABOUT GRAPHENE IN AVIONICS

Fundamental Properties

Graphene is a carbon atom monolayer form of material that is hexagonally loaded and has raised so many concerns in the field of material science because of its excellent electrical, mechanical, and thermal properties. As a zero-gap semiconductor, and thus made of the massless Dirac fermions, graphene is estimated to have a charge carrier mobility of more than 15,000 cm^2/V (s), more than the traditional semiconductors used in avionics systems. This allows faster movement of electrons and reduction of power wastage in critical flight systems.

Graphene is a very good heat spreader thermally, with conductivities of more than 5000 W/m K, and one of the few known high-conductivity materials. This property aids in facilitating passive heat removal systems and sealing the operation temperatures of microelectronic systems of avionics systems, where the thermal density of components is increasingly becoming better (Shahil & Balandin, 2012). Graphene is a better reinforcement agent in composite structures in vibrational and pressurized flight regimes due to its Young modulus, which is slightly under 1 TPa, and strength with an upper limit of 130 GPa.

The surface of the graphene (approximately 2630 m^2) is specific and can be easily modified to fit in sensors, energy storage, and electronic interfaces, through oxidation and reduction or by attaching functional groups to the surface. It is also because it is optically transparent (~97.7%), atomic depth that can be incorporated into the transparent display and embedded cockpit instrumentation without any loss of visibility or ergonomics (Tang et al., 2021). Such multi-functional properties make it a good platform in the development of next-generation avionics systems that require the combination of lightweight technology and high performance with multi-functionalities.

Despite graphene being outstanding in terms of its inherent properties, e.g., carrier mobility of over 15,000 cm2/V+s and thermal conductivity of over 5000 W/mK, are generally obtained only under ideal conditions of using a defect-free sample in the laboratory. When graphene is added as a component in the form of a composite or a thin film into avionics, performance can be dramatically lowered by defect density, grain boundaries, surface functionalization, thermal resistance at an interface, and so forth. To give a few examples, an order of magnitude of charge carrier mobility in device-grade graphene is often reduced when transferred to substrates, whereas thermal conductivity can often be limited in polymer or ceramic systems by long-range interactions between matrices. It is necessary to identify such discrepancies to match the expectations of research with the realistic performance of the engineering in the aerospace conditions.

Figure 1. Multifunctional properties of graphene relevant to avionics applications.

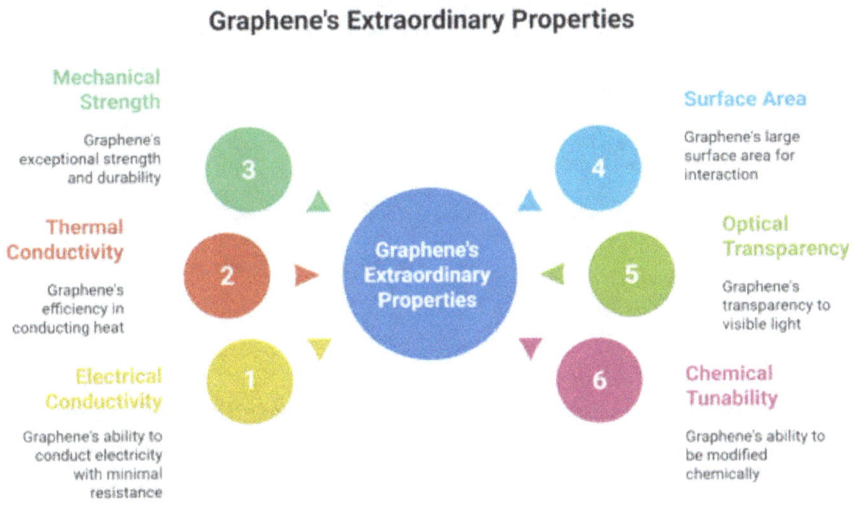

These combined properties, as shown in **Figure 1**, make graphene a promising material for multifunctional integration in advanced aerospace systems.

Applications of Graphene in Avionics Systems

The applicability of graphene in avionics has led the way to improvements in the performance of various sub-systems.

Figure 2. Overview of primary application areas for graphene in avionics systems.

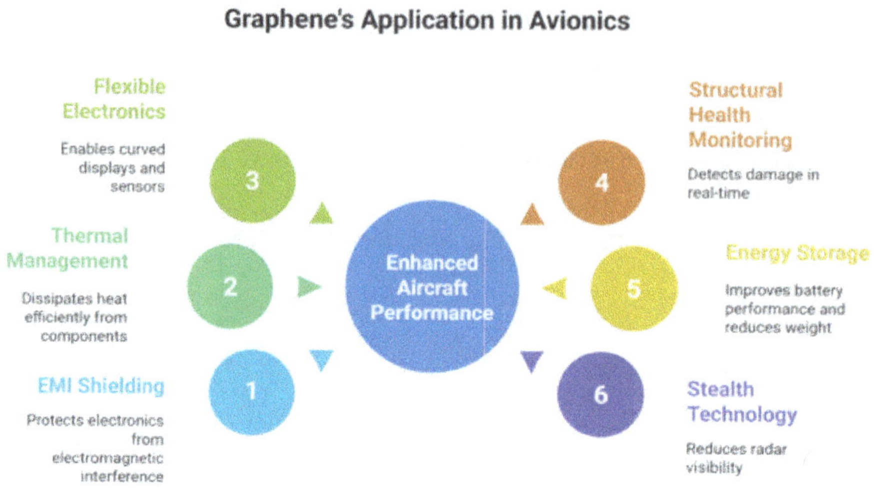

Optimally, the various points of integration evident in Figure 2 demonstrate the role of graphene in enhancing the performance and functionality of contemporary avionics. One of the potential uses that can be cited is in electromagnetic interference (EMI) shielding. In other aerospace uses, radar, telemetry, and high-frequency digital make it significant to have large EMI protection. Graphene is highly conductive and has tenable dielectric properties and can be an excellent shielding agent when incorporated into polymer matrices. It was discovered that graphene-enhanced composites can have shielding effectiveness over 60 dB and can also reduce the weight of the whole system compared to their metallic counterparts in metal enclosures (Kausar et al., 2023). Whereas it is true that graphene-based composites can offer lighter substitutes to conventional metallic enclosures, to give the same shielding effectiveness, more material thickness is often required. Indicatively, thin metallic foil can offer sufficient EMI attenuation at only a fraction of the thickness of a graphene-polymer composite. Such thickness may reduce or even cancel the anticipated weight reduction, especially in high-frequency damage, as the shielding effectiveness of the composite architecture is heavily dependent on the surface conductivity (Jovanovic et al., 2023). Attestations of weight decrease should be handled as part of the entire shielding setup and not on a material replacement level.

It is worth noting that the values of graphene-enhanced composites reported as being shielding effective are frequency-dependent and differ over the MHz to GHz spectrum of avionics. Tests in the laboratory, more than 60 dB, usually must be optimized in terms of thickness, controlled flake orientation, and good processing conditions. Graphene composites, in most instances, must be significantly thicker than metallic foils to attain the same attenuation, which can be compensated for by the predicted weight reduction. Besides, interfacial contact resistance, dispersing of fillers, and environmental stability add some variation that should be taken into consideration when seeking to be certified for aerospace.

The second useful use is thermal management. The avionics modules are particularly used in the flight controls and data processing, which generate significant heat during operation. The thermal conductivity of graphene enables the dispersion of heat in composite structures or heat sink layers. Graphene is also shown to assist in reducing junction temperatures in microelectronic components used in high-altitude, low-pressure applications where convective cooling can be limited, and it is possible to use hybrid composites (Qin et al., 2019). Even though early experiments indicate that graphene-based composites could help distribute localized heating loads, assertions of reduced junction temperatures in high-pressure, low-altitude avionics systems remain mostly theoretical. Real-world thermal management of avionics involves complex design considerations, including limited airflow, multilayer packaging, and high reliability. Available experimental evidence on graphene composites in such cases is limited, and most studies are based on laboratory-scale testbeds rather than flight-qualified systems. Thus, the possible application of graphene to reduce junction temperatures could be considered an emergent research pathway, but not an approved engineering methodology.

The elasticity and mechanical properties of graphene have given the world impetus to move towards the use of invariable electronics, which are more feasible in avionics systems that can be modified with wearable flight gear, responsive surfaces, and miniaturized counter-conformal electronics. Its combination with roll-to-roll printed circuits and flexible interconnects helps in the creation of reconfigurable cockpit controls and body-worn navigation devices. In addition, the properties of graphene as a conductor and a sensor can be utilized to imagine single-material systems that are simultaneously transmitters and sensors that read environmental/structural information.

Structural health monitoring (SHM) has become another significant field of application. Graphene-based sensors, especially those with graphene oxide (GO) or reduced graphene oxide (rGO), are very strain, temperature, and pressure sensitive. These may be fitted to fuselage composites or the control surfaces to ensure that there is constant and continuous monitoring of the fatigue, formation of microcracks, etc, and the next predictive life allotment. Of particular importance is this in military and commercial aircraft, which undergo heavy-duty in high-cycle loading and complex maneuvers, in which the loss of material that occurs unexpectedly can negatively affect safety. Although strain and temperature sensitivity have been demonstrated with graphene-based sensors, signal-to-noise ratio, drift, and long-term stability with cyclic loading and flight conditions have not been well studied. Quantitative norms of the detection thresholds (e.g., micro-crack growth rates or fatigue cycles) remain inaccessible compared to the existing aerospace SHM systems. To ensure the reliability of calibration and environmental stability, the implementation of the graphene sensors in the laboratory needs to be met with applications before they become viable in the field.

Beyond the conventional systems, supercapacitor-based energy storage options using graphene are being explored in avionics modules with high power transient requirements and rapid rechargeability. The devices could do a backup of flight data recorders or emergency avionics equipment. This type of application is still in its infancy, but it is an indication that there is a trend towards multifunctional and high-reliability energy platforms in aircraft design.

The article that provides the description of the work of Bhanuprakash, Varghese, and Singh (2022) on the glass fiber reinforced epoxy composites modified by the incorporation of graphene nanofillers shows that the mechanical and electrical optimization could be conducted with the help of the incorporation of graphene. They have shown in their studies that they possess an augmented dielectric behavior and structural modulus that is of foremost importance in avionics enclosures, where shielding and mechanical strength are needed. This empirical experiment makes the graphene-based reinforced composite even more helpful in the aerospace sphere, where more emphasis is laid on the presence of multifunctional properties and the resilience of the composite.

Moreover, stealth technology is gradually being applied to using graphene in radar-absorbing material (RAM) since it has better concealment properties (about electromagnetic attenuation). As the layered graphene structures are introduced in the subsystems of avionics, such as signal detection and noise reduction, both band-specific absorption at frequencies and signal reflections can be selective extended to extend mission survivability and clarity of signal received.

Emerging Trends and Industrial Interest in Graphene Integration

The aerospace sector is already conducting research in graphene with an understanding of its disruptive power in designing, manufacturing, as well as in optimization performance. It is also worth noting that Airbus has partnered with the National Graphene Institute and has been able to come up with prototypes in which graphene has been applied in thermal de-icing of wing surfaces. The Joule heating effect of graphene is used in these systems to avoid the formation of ice, and this has provided a lightweight and energy-saving alternative to conventional de-icing systems.

Lightweight radar and communication modules made with graphene foams and coatings are sources of electromagnetic shielding but also act as structural fillers of the module, hence lowering the total weight of the aircraft. The co-location of weight loss and improvements in functionality is consistent with modern priorities in civil aerospace as well as defense aerospace, particularly in the more electric aircraft and reduced-emission aviation requirements.

One of them is the growing interest in using simulation and artificial intelligence (AI) when designing and validating graphene-based systems. Digital twins and machine learning applications are under implementation to forecast the behavior of graphene composites under different stresses, such as temperature gradients, pressure variations, and mechanical fatigue. These resources can also make graphene integration more of a reality within the customary aerospace development cycles due to the increase in the speed of material certification and cutting down on the overhead of experimentation.

Notwithstanding such developments, there are still challenges. The absence of standardization in the production and differentiation of graphene can be considered one of the major barriers. Flake size, defect density, and surface chemistry variability between production batches have an impact on the reproducibility of composite properties. This inconsistency presents challenges in avionics products, which require high tolerances and high reliability. The long-term performance of the graphene-enhanced materials needs to be confirmed under the conditions of aerospace loading, as Sharma et al. (2022) stress that they should be tested in conditions of exposure to cosmic radiation and thermal cycles.

There also exists the challenge of integration into the current avionics manufacturing processes. The supply chains and automated assembly infrastructure support the use of traditional materials: copper, aluminum, and FR4 laminates. Introduction of the elements of graphene can require alteration of the process parameters, inspection procedure, and bonding technique, which, consequently, influence cost and scalability.

However, the trend line is evident: the graphene is moving out of the laboratory level research towards pilot-level integration into aerospace applications. This trend is enhanced by the interplay between material science and nanotechnology, and avionics engineering, with policy incentives in terms of sustainable and high-performance aircraft materials. With the onset of the global aviation standards to integrate the safety measures of nanomaterials and lifecycle evaluation, the role of graphene can probably gain even further in military and civilian aircraft systems.

Introducing the additive manufacturing technologies, especially the direct ink writing and laser sintering, has led to an increased variety of options in graphene integration. These technologies allow a demand-based manufacturing of complicated avionics elements, including integrated graphene circuits or sensors, local repair, design/customization, and speedy prototyping. It has found significant use in defense logistics, remote aerospace operations, and other areas where the conventional supply chains might not be sufficient.

To conclude, the use of graphene in avionics systems is not a hypothetical concept anymore but a reality on the path of development. Although industrial-scale usage is conditioned by the need to overcome technical and regulatory barriers, the multifunctionality of the material, its lightweight, and the ability to be used in conjunction with the technologies of the latest generation of aerospace make it a central power, facilitating the next generation of aerospace innovation. Graphene can transform the material standards and system properties in the avionics platforms as the sector becomes electrified, automated, and resilient.

TECHNICAL CHALLENGES AND BARRIERS TO GRAPHENE INTEGRATION

Variability in Graphene Production and Quality Control

The structural quality and consistency of graphene have a very complex relationship with the performance of the avionics systems in which it is being employed. However, a large discrepancy that exists in the process of graphene manufacturing is one of the greatest barriers. All the methods involve chemical vapor deposition (CVD), chemical exfoliation, mechanical exfoliation, and the reduction of graphene oxide needs another method to achieve the material with individual physical properties a layer number, a density of defects, a lateral size, and oxygen content, which directly determine electrical, thermal, and mechanical properties (Panghal et al., 2025). Table1compares the suitability of the various methods and the challenges of synthesizing graphene by synthesizing key methods of synthesizing graphene, their characteristics, advantages, limitations, and their relevance to avionics. As pointed out in Figure 3, both synthesis methods have specific advantages and drawbacks that affect the integration of avionics systems.

It is not homogeneous, thus complicating the standardization and certification of graphene for aerospace uses. In avionics, the performance of components is strictly regulated, and therefore, the batch-to-batch variation poses a threat to the reproducibility and trust of systems. After advancements in metrology, still, no standard has been established that specifies the low-quality mark of aerospace-grade graphene has been established. There have been no such standard descriptors in the aerospace industry to quantify the anisotropy and defect tolerance, and polymer or metal interface properties of graphene, thus posing a challenge in vindicating the suppliers of such commodities (Kim et al., 2010).

Table 1. Comparative Overview of Graphene Synthesis Methods for Avionics Applications

Synthesis Method	Key Features	Benefits for Avionics	Limitations	References
Chemical Vapor Deposition (CVD)	High-quality, large-area monolayers	Excellent electrical and thermal properties	High cost; complex transfer processes	Panghal et al., 2025; Kim et al., 2010
Mechanical Exfoliation	Simple lab-scale method	High-purity samples for basic research	Poor scalability; batch variability	Panghal et al., 2025
Chemical Exfoliation	Bulk production via oxidation-reduction processes	Mass production potential; cost-effective	Introduces defects; lower conductivity	Lopez et al., 2022
Electrochemical Exfoliation	Greener process using aqueous systems	Environmentally friendly; scalable	Still under refinement; variable quality	Shaheen et al., 2022
Liquid Phase Exfoliation	Uses solvents to delaminate graphite	Simple and scalable	Limited flake size control; solvent toxicity	Arvidsson, 2017

Although Table 1 points out the important synthesis methods, quality-cost-scalability trade-offs are more complicated. Indicatively, CVD prices differ significantly by substrates and throughput, and current developments on growth without transfers are making the processing needs less difficult. These inventions especially apply to avionics, where the control of contamination and reproducibility is essential. Therefore, assessments in the future must take into consideration both the existing constraints and the new methods that can increase economic and practical viability.

Figure 3. Comparison of graphene synthesis methods for avionics applications.

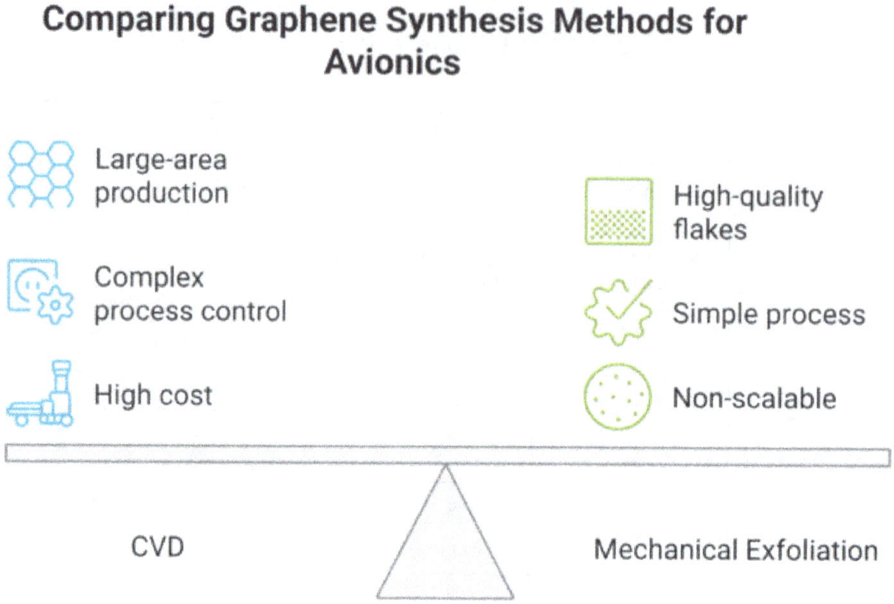

Compatibility with Aerospace Manufacturing and Design

The integration of graphene into the existing avionics production is very logistically and technologically demanding. The traditional materials that include copper, aluminum, and aramid-fiber composites are naturally accommodative to the industrial procedures, which include autoclave treatment, CNC machining, and solder reflow. Conversely, the graphene-reinforced films or composites are generally hard to fit, the new conditions, like vacuum deposition, high temperature sintering, or surface functionalization, cannot be incorporated into the existing avionics manufacturing infrastructure (Panghal et al., 2025).

Lack of knowledge is also present regarding the mechanical response of the graphene composites to the aerospace-assembly type of stresses; the stresses under thermal cycling, press fit assembly, and the stresses under vibrational bonding. In hybrid laminate structures or flexible PCBs with graphene, delamination, warping due to a mismatch of thermal expansion coefficient, and debonding of the interface can be a problem. All these demands require the re-engineering of the production and increase the cost of qualifying aerospace suppliers.

Modeling Challenges and Simulation Limitations

Avionics engineering is an important area that requires the use of computational mathematical modeling to ascertain the response of the material to be used under operating conditions. Whereas simulation parameters in finite element simulation (FEA) have always been a long-established property of traditional materials, the peculiarities in scale of graphene as a material make it impossible not to use multiscale conclusive methodologies. Molecular dynamics (MD) simulations can be atomistical simulated to provide a molecular level, but this is computationally expensive as well as technically immature to integrate them with mesoscale and macroscale models (Singh, Jain, Palaniappan, Parmar, et al., 2023).

The directionality of conductivity, damage, and contact to the matrices in the graphene is an obstacle to consistent establishment of simulations of the elements of the avionics systems under load or thermal-equilibrium circuitry. Furthermore, it also has a limitation in the checking of the simulated data at the real flight conditions, altitude, pressure, UV radiation, and rapid heating of the thermal flux. Validation of the digital twins of the graphene-based systems could be a valuable contribution because, otherwise, risk modeling is one of the main challenges.

Reliability in Harsh Aerospace Environments

Unreasonable environmental conditions, to which the avionics systems are subjected, are extreme temperature variations, high frequency, and vibrations, as well as long-term radiation exposure. There is no intensive study on such stresses with respect to the long-term behavior of graphene. It is postulated (on early observations) that fatigue and premature failure can be introduced by such factors as the growth of defects, oxidation, concentration of stress at the graphene-matrix interface, and by structural sensors or EMI shielding foams (Vinci et al., 2023).

As an illustration, exposures to the aggressive reactive oxygen species, as well as exposures to the UV radiations, airplanes, and stratosphere travelers can experience this at some distance, query the vulnerability of the graphene to the oxidation process. Moreover, thermal cycling can result in the development of interfacial strain, as well as mechanical and electrical integrity. These unanswered reliability issues still pest the way of having graphene officially accepted to be used in primary avionics systems, wherein a long-life span and a non-failure design are a mandatory necessity.

Electromagnetic Compatibility and Signal Interference Risks

Although the graphene composites can have potential electromagnetic interference (EMI) shielding, the need not to add the graphene composites to the avionics electronics is due to the possible unwanted electromagnetic effect. Graphene has a high conductivity rating that is high, which is beneficial as a shielding material but may interfere with the signaling between neighboring circuits, especially when using multilayer printed circuit boards (PCBs) or radiofrequency (RF) subsystems. Due to the use of high-speed data and data at many communication frequencies, as may be experienced in the avionics systems, system integrity can be destroyed by a small impedance error or a capacitance that is not entirely intended (Kim et al., 2010).

Extensive electromagnetic compatibility (EMC) tests are one of the ways of averting such risks. However, the EMC testing standards under consideration are not optimally friendly to test 2D constructions made of nanomaterials. Furthermore, the behavior of graphene due to time-dependent fields and time-dependent sources of EMI is underdescribed. Without the information, the flight-critical systems cannot be certified by the FAA or EASA.

Supply Chain and Standardization Limitations

The global supply chain of graphene is not well established, which is one of the barriers to its application in avionics, which is hardly felt by the world. Even though graphene of high quality is produced in large quantities in mostly research-scale facilities, it is incapable of reaching the volume and quality bureaucracy checks and traceability procedures to be aerospace certified (Vinci et al., 2023). Absence of a standardized set of criteria of how to define graphene, such as those in the CFRPs industry or thermoplastics, only further restricts its adoption.

This has been somewhat bridged by standards organizations, including the ISO and the ASTM, but most of the currently relevant frameworks (e.g., the ISO/TS 80004) have yet to address aerospace factors (e.g., flammability, off-gassing, and long-term outgassing under vacuum). In addition to this, owing to a deficiency of the potent supplier qualification strategy, at least, aerospace OEMs are not keen to incorporate graphene in the lengthy design cycles, especially those systems that are mission-critical.

Safe-by-Design Strategies and Predictive Risk Modeling

Mechanisms like safe-by-design are continually being adopted to remedy these problems. Its strategies involve using ideas of safety, toxicity, life cycle assessment, and regulatory compliance in the front end of material development. In this case, machine learning technology and tools are also finding a convenient role in the prediction of the behavior of graphene composites in diverse conditions of flight. According to Singh et al., data-driven models can augment the designing, verification, and optimization of the materials, and they can reduce the reliance on expensive physical prototyping.

Rao, Singh, and Kumar (2021) also place much stress on the benefits of the case-based reasoning (CBR) methodology as a way of finding patterns in the materials breakdown history and transferring them to new nanomaterial systems. Incorporating the safety heuristics into the design pipeline, developers will be able to predict the cycles of failure and counter them proactively by redesigning the structure or improving the process.

SAFETY PROTOCOLS, RISK ASSESSMENTS, AND REGULATORY FRAMEWORKS

Occupational and Operational Safety Hazards in Graphene-Enabled Avionics

The multifunctional nature of graphene has rendered it very interesting to be deployed in aerospace systems, particularly in avionics. However, the excitement must be kept within a proper limit, and the way to go about it, how to process and utilize it, all must be carefully investigated. Like most engineered nanomaterials, at the nanoscale, graphene can act physiochemically in a manner that is not necessarily predictable when scaled up to materials in the macro scale. The tendency of the graphene sheets to aggregate or aerosolize, as well as high aspect ratio and surface reactivity, are the sources of unusual occupational health and operational safety concerns in the aviation environment that are unusual.

Graphene might be manipulated in a fabrication setup via substances release or emission during manufacturing, dispersion, blending, or treatment procedures. In several studies, it has been demonstrated that aerosols of graphene can cause pulmonary inflammation and oxidative stress and even fibrosis, especially when a few-layer graphene or graphene oxide derivative is used (Batool et al., 2025). The nature of graphene as fibrous is also an issue in the evaluation of asbestos in the respiratory tissues. It might sound harmful, but the chronic symptoms of short-term exposures have not been sufficiently researched, and it is therefore worthwhile to exercise caution and consider engineering protection measures and environmental issues.

The mitigation measures that should be taken are the skin contact and accidental ingestion. Due to its lipophilicity and permeability of cell membranes, it is hypothetically hazardous that graphene can lead to cytotoxicity through the transdermal patch or absorption (Fadeel et al., 2018). There is, therefore, the requirement of personal protective gear (PPE), including face shields, face masks, and localized exhaust ventilation, during the use of powdered or suspended graphene in production laboratories and integration laboratories.

At the working level, the use of graphene in aircraft systems and particularly in printed circuits, composite panels, and thermal interface materials creates more issues of concern. In case their structures are damaged in some way, or a fire takes place because of an accident, the graphene-enhanced substances could collapse and pour tiny particles of the substance into the Boeing cabin or the avionics space. In graphene, C_2H_2O functional groups (like in graphene oxide) or residual catalyst (like metals) can be used to increase reactivity, even though the material is primarily carbon-based, when using CVD. Fire behavior has shown that some graphene composites, particularly those based on thermoplastics, can have ignition thresholds or smoldering behavior that is dissimilar to the standard substances (Sethuraman & Alagar, 2014). This creates an additional risk of emissions of smoke, toxic products of combustion, which remain in the fireplace following the completion of the combustion process, as well as corrosion of surrounding systems.

One grave technical blank that still exists is to establish what can fail in the extreme conditions, such as lightning hits, electromagnetic bursts, or sudden depressed air, where graphene-enhanced avionics performs. Even though graphene can enhance the reliability of the system in normal operation, the behavior of graphene in the event of a catastrophic event has not been properly evaluated and thus fails to meet any aviation certification standards. This demand is a swift focus on the scenario-equivalent risk simulations, as well as experiments during which flight-representative assessments are performed.

Risk Assessment Frameworks and Toxicological Evaluation

In fact, to coordinate the floating set of safety concerns in an organized fashion, a methodological process of risk assessment must be applied in all stages of the cycle of graphene and dispersion of composites integration, further all the way to the level of employment and final disposal. Table 2 briefly outlines the safety and risk issues in the various areas and the proposed mitigation measures. Currently, little is done with traditional risk assessment models on nanomaterials, often because they are generally insufficient, given the size-specific behaviors that nanomaterials display, and the limited data available on the subject, usually on toxicology. This is the reason why the specially designed frameworks, the Control Banding Nanotool, NanoRisk Framework, are also translated into industrial applications in Europe or North America (Visser et al., 2022). Figure 4 represents a detailed risk assessment framework that should be used to address these new safety issues and apply to graphene in avionics applications.

Table 2. Risk Domains and Mitigation Strategies for Graphene-Enhanced Avionics Systems

Risk Domain	Key Challenge	Potential Impact	Mitigation Strategy	References
Occupational Safety	Aerosolization and dermal exposure during handling	Respiratory and skin toxicity risks	PPE, ventilation, closed system processing	Batool et al., 2025; Fadeel et al., 2018
Operational Safety	Fire behavior and particle release during failure	In-flight hazards and toxic combustion byproducts	Flame-retardant composites, predictive testing	Sethuraman & Alagar, 2014
Environmental Impact	Disposal and end-of-life scenarios	Soil/water contamination, air quality deterioration	Lifecycle analysis (LCA), green synthesis	Arvidsson, 2017; Keller et al., 2013
Regulatory Compliance	Lack of harmonized graphene-specific standards	Certification delays; legal uncertainty	ISO/ASTM engagement, early regulatory dialogue	Visser et al., 2022; Giese et al., 2018
Electromagnetic Compatibility	Signal distortion in high-frequency avionics	Performance degradation in PCBs and sensors	Extensive EMC testing, shielding design optimizations	Kim et al., 2010

Figure 4. Graphene-specific risk domains and mitigation strategy framework across the avionics lifecycle.

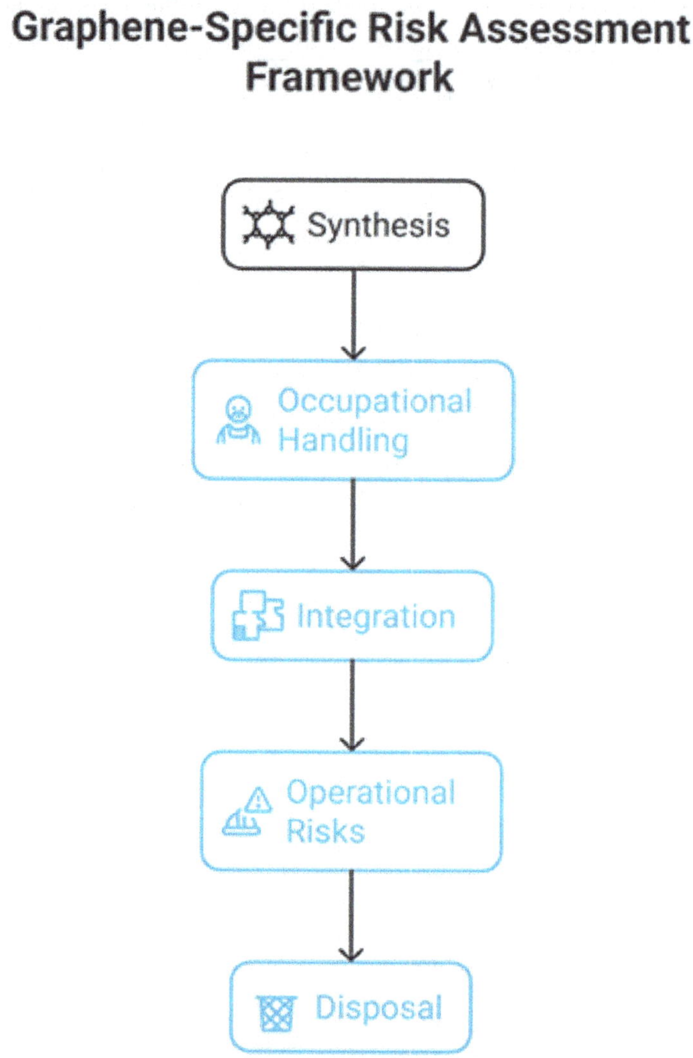

The most significant deficiency of the current risk assessment practice is the absence of harmonized graphene-specific toxicological data. Synthetic pathways cause materials to have vastly different surface areas, oxidation states, and impurity compositions, with all of them potentially influencing bio-reactivity and toxicity. A good example of this is the modified graphene oxide, reduced graphene oxide (rGO) indicated to induce further damage to the mitochondria than pristine graphene due to high concentration of surface defects and oxygen concentration (Lopez et al., 2022). Such variations thus interfere, complicating the development of dose-response models or occupational exposure limits (OELs), which are the fundamental aspects of quantitative risk assessment in aerospace manufacturing.

Efforts to close this data gap are being supported by international efforts at promoting systematically planned in vitro and in vivo testing, e.g., models of lung deposition, skin penetration tests, and immunological profiling, including the Graphene Flagship (EU) and the National Nanotechnology Initiative (US). Notwithstanding this enhancement, a recognized OEL on graphene on the global level is not available. This kind of regulatory gap will force the industries to work with conservative amounts or with generic levels of exposures to the particles, which would not reflect the actual risk profile of the graphene.

Another significant issue of risk management that is related to lifecycle assessment is that LCA calculates all the human and environmental risks that a certain material is likely to be exposed to in its life cycle. The LCA of the systems containing graphene is supposed to target waste generated in the synthesis of this material (e.g., the solvent disposal, the use of energy) as well as the dangers of the processing (e.g., dust, thermal degradation) and the risks at the end-of-life (e.g., recycle, incineration, the resultants). A full LCA can reveal certain unseen liabilities, such as the use of unclean reducing agents in wet-chemical procedures or embedded graphene being non-recoverable because of the composite grid, and this may influence regulatory likenesses and green PR (Keller et al., 2013).

Safe-by-Design is a relatively new phenomenon as a proactive approach to nanomaterials development. This ideology is to think of safety concerns during the early phase of research studies, where benign precursors are chosen, surface chemistry is modified to reduce reactive capacity, or the ability to entrap or wrap graphene materials in a protective scaffold to avoid exposure. This paradigm also allows meeting the evolving regulatory requirements and not only reduces downstream liabilities. An illustration of that would be to make the graphene epoxy composites through covalent bonding to reduce the shedding of the particles during the mechanical stress to enhance safety to the occupational labor force, besides the functional life.

The recent examination of Archana et al. (2023) is relevant to the situation at that. They found that the application of algorithms developed by the Yolov3 framework could identify and predict the degradation characteristics of materials in real-time, exploiting the concrete uses of a smart city. These models may be used to check and oversee the size of safety limits, premature failures, and restricted human exposure during inspection and upkeep when introduced into the graphene-based aerospace systems. Their complexity and the need to encourage safe use of nanomaterials in other fields, including aviation, therefore necessitate the AI tools to be innovated into risk management as a dire need.

Regulatory Landscape, Certification Challenges, and International Standards

The regulatory requirements for the use of graphene in avionics are not standardized and are at an advanced level. Unlike the conventional aerospace materials, having tens of years of performance history and established qualification protocol, the novelty of graphene is a cause of a regulatory dilemma that can retard or even prohibit its integration into safety-critical systems. Guidelines on the treatment of the peculiarities of nanomaterials are still being developed by national and international agencies, most of which are within the framework of the chemical composition and a large-scale exposure in the workplace, rather than level-interconnected application in the electronics or aircraft composition sector.

There are twenty-five international standards currently under operation or being advanced, with the current baseline measure of international standards in action being ISO 80004 (nanotechnologies vocabulary), ISO/TR 16197 (nanomaterials risk assessment), and REACH regulations in the European Union. However, none of these has a direct solution to the requirements of the avionics systems, and they do not take into consideration flight operating conditions. In addition, the US Occupational Safety and Health Administration (OSHA) and the National Institute of Occupational Safety and Health (NIOSH) have already issued precautionary guidelines on carbon-based nanomaterials, but no limits or validation strategies have been formulated to apply to graphene (Visser et al., 2022).

One of the challenges unique to the aerospace industry is the demand for a multi-agency certification. Thought to illustrate, fitting a graphene-enhanced EMI shielding panel into a merchant airliner would require passing through the material safety regulations, but also the material certification arms, which include the electronic subsystem certification, the electromagnetic compatibility test facility, and the flight regulatory authority, such as the FAA and EASA. The absence of a harmonized certification system leads to bottlenecks in the regulatory system and different application of safety requirements across jurisdictions. Another level of compliance would be military aviation, in the form of defense procurement practices and classified standards, which increase the burden of adopting graphene in next-generation air systems.

Despite these difficult factors, some steps are being undertaken in the field of collaboration where the regulatory gaps may be bridged. Government-industry partnerships and consortia, such as ASTM International Committee on Nanotechnology and the European Chemicals Agency (ECHA), are now applied in the development of testing protocols and standard operating practices in graphene. A good example is that ASTM E3125 provides specifications on the determination of the number of graphene layers and lateral dimensions that are crucial in the allocation of performance and possible risk.

The recent coping mechanism to these regulatory wrinkles is the digital certification. In digital twins relying on AI, developers would be able to simulate the behavior in as many loading, environmental, and failure conditions as possible. If the validation of these simulations is done through some physical testing conducted on the same, performance equivalence or safety compliance can be demonstrated using these simulations. The undertaking that the authors have discussed can help save time and assist traffic and transportation systems in their working, and the same can be applied to aircraft certification frameworks, as time can be saved, and aviation innovations can be implemented at relatively lower prices due to the absence of conducting full size prototyping and making their component undergo the destructive testing in a graphene aircraft.

The third strategic measure entails bringing in questions of regulatory bodies at the R&D level early. Adaptive testing protocols, phased approvals, and real-time reaction to the requirements on safety data can be facilitated by the adoption of co-development models, where the developers and certifiers work in parallel. This will not only increase the process of certification but also the establishment of institutional knowledge in agencies, thereby making it easier to expose advanced nanomaterials in the future.

The introduction of graphene into the avionics is not a mere modification of materials but more of a paradigm shift in how safety, functionality, and innovations are united in the aerospace engineering industry. Until the regulatory science is no longer playing the same tune as the material science, it is the prerogative of the researchers, as well as the industry stakeholders, coupled with the policymakers, to unite to create such a structure where, on one hand, the technological advancements are being made, and on the other hand, the safety is not compromised.

ENVIRONMENTAL CONSIDERATIONS AND LIFECYCLE ASSESSMENT

Environmental Impact of Graphene Production and Synthesis Pathways

Regardless of these predicaments, there is an effort being put in collaboration where the regulatory gap would be bridged. The government-industry partnerships and consortia, such as ASTM International Committee on Nanotechnology and

European Chemicals Agency (ECHA), are already utilized in the creation of certain testing procedures and specific operating practices within the sphere of graphene. To list but a few, ASTM E3125 provides guidelines to define the number of graphene layers and subsequent lateral dimensions, which are crucial to performance and possible risk determination.

The most recent adaptation for dealing with such regulatory kinks is through digital certification. In the case of AI-driven digital twins, the developers will be able to model the behavior in as many loading and environmental, and failure conditions as possible. They can be used to demonstrate performance equivalence or performance safety when such simulations are proven by some physical testing of them. In their undertaking, the authors have addressed how digital modeling and simulation systems can support the operation of traffic and transportation systems, and the same can be applied in aircraft certification systems, whereby time is saved and aviation innovations are attained at relatively lower prices since they need not undertake full-size prototyping and run destructive testing of components of a graphene aircraft.

The third strategic measure is the involvement of regulatory bodies at the early stage of R&D. Safety data requirements, adaptive testing procedures, and staged approvals and real-time responses to needs can be enabled by the adoption of co-development frameworks in which developers and certifiers work concurrently. This not only guarantees that the certification process is expedited, but it also leads to the institutionalization of knowledge in agencies, hence making it easier to expose advanced nanomaterials in the future.

Implementing graphene in avionics is a paradigm change rather than a material change in terms of how safety and functionality, and innovations are united within the aerospace engineering field. Until the regulatory science is not singing the same song as material science, both the researchers and the industry stakeholders, along with the policymakers, should join to come up with this type of framework where technological developments are enhanced on the one hand, and safety is not sacrificed on the other hand.

Recent procedures have considered greener synthesis procedures, such as electrochemical exfoliation and graphene precursors/biomass, in addition to water-based dispersion mechanisms. Though it might be successful, such possibilities still do not result in high-quality or stable graphene at this point in the development and thus may be insufficient for avionics applications requiring a high level of quality. Such a balance is to be discovered between the environmental compatibility and demands of the functional performance, and within the rigid lifecycle thinking and multi-criteria decision-making (Shaheen et al., 2022).

One example of a good precedent, in this regard, is the study of Verma et al. (2024) in silica fume-based geopolymer concrete. Therefore, their studies provide a model that can be generalized to aerospace composites by comparing the environmental and mechanical performances of new materials using graphene. The environmental metrics should be co-optimized to meet thermal, electrical, and mechanical demands to promote the responsible use of systems using graphene in an avionics design.

Lifecycle Assessment (LCA), End-of-Life Scenarios, and Circular Economy Considerations

The means of environmental assessment applied in LCA are imperative means of evaluating the environmental impact of graphene materials during the life of the materials (in production, usage, and at the end of it). Since, e.g., aluminum or copper graphene is not a traditional avionic material, the life cycle profile of the material has not yet been established, with sparse information on industrial scale available. However, the first study shows that the process of the manufacture of graphene composites may become rather heavy on the environment, which can be compensated for during the utilization stage or not, based on the application and the efficiency benefits across the entire system (Arvidsson, 2017).

The full LCA of an avionics component that has been integrated with a graphene element must consider upstream effects (extraction of raw materials (graphite mining or biomass processing), emissions while manufacturing, water and power usage, and chemicals). The transportation and formulation of composites and integration in aircraft systems are midstream processes. The examples of downstream issues are the problems with operational performance (e.g., weight savings, energy efficiency improvement), maintenance and disposal, or recycling.

The major issues of the evaluation of downstream are the low feasibility of disposing of or milling down graphene-additional fabrications. Graphene, in turn, can never be economical recovered in most cases in aerospace, where it is employed in polymer matrices or laminated constructions. The incineration has the possibility to produce fine particulates of graphene or incomplete combustion byproducts as well, which are health and environmental hazards. Leaching of additives or unreactive precursors into soil and water systems is possible during a slow process of gradual degradation of such materials when landed (National Academies Press (US), 2000).

In addition, a trend towards a paradigm shift in the design of a circular economy that encourages the practice of reuse, recycling, and minimal production of waste has increased. These models can make a significant positive contribution to the aerospace industry since the industry is vastly esteemed, and the life of the parts utilized is lengthy. However, the current state of graphene-enhanced products is not that circular, as there is no efficient delamination, separation, or purification.

To address these fears, studies are being done to come up with recyclable graphene composites through reversible covalent bonds, dynamic cross-linking networks, and thermo-reprocessable matrices. Additionally, attempts have been made to work on the development of a so-called sacrificial layer, because of which protective films can be removed and/or adhesives to the building material can be removed so that the material can be recovered. These gains, though promising, are still mainly in the research front, and they need to be demonstrated in aerospace-related conditions.

The other potential outcome of increasing the environmental performance is the adoption of modular design strategies, whereby the graphene-based subsystems would be adopted as replacements or modules, instead of permanent ones, which are fixed. In this, specific enhancements can be made, wastage of material can be reduced, and maintenance can be eased. Adding graphene electronics with sensors, such designs can support predictive maintenance and health monitoring of the material, which would indirectly boost the component life and decrease material consumption. Figure 5 shows that the sustainable material choice and the design circularity in avionics innovation should be directed by lifecycle thinking.

Figure 5. Lifecycle assessment of graphene-enhanced materials in avionics and sustainability pathways.

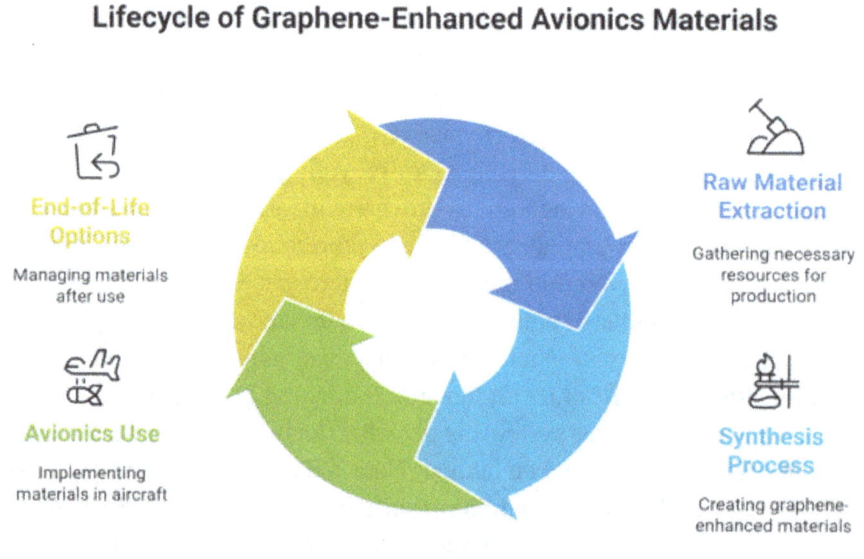

It is worth mentioning that sustainability assessment tools, such as the GREEN-SCOPE and TRACI (Tool for the Reduction and Assessment of Chemical and Other Environmental Impact) frameworks, have been adapted to the more sophisticated materials to offer more sophisticated insights into the environmental impact of graphene (Inayathullah & Buddala, 2024). These tools could help aerospace engineers thrive in the initial-stage design to make decisions on the selection of material, content refinement practice, and end-of-life incorporation.

Regulatory, Policy, and Sustainability Governance in Graphene Applications

The environmental regulation of graphene-enhanced avionics components occurs in the material science, aerospace engineering, and regulatory policy. The regulations that are in place to address nanomaterials are typically too general to consider the lifecycle and environmental dynamics through which graphene undergoes. Popular regulatory systems (like the European Union regulation, REACH (Registration, Evaluation, Authorisation and Restriction of Chemicals) and that of the US Environmental Protection Agency, TSCA (Toxic Substances Control Act) have offered a floor issue of the requirements of the chemical, but are insufficient in long-term weather conditions, bioaccumulation potential of the chemical, and intricate composite impact in end-use application (Giese et al., 2018).

To provide an example, graphene is entirely composed of carbon (it consists of carbon, carbon, carbon), but this nanoscale character, functionalization, and surface chemistry bring about certain interactions with the environment. These include altered mobility of the systems in the aquatic systems, potential biomagnification in the food chains, and impacts on the upside of the soil microbiota. In addition, the ecological trend of graphene derivatives such as GO and rGO cannot be easily controlled due to their common usage in the composite formulations.

Voluntary standards and industry consortia are taking the gap. The Organization for Economic Co-operation and Development (OECD) has voiced the principles of nanomaterial testing, and this has encompassed the stability of dispersions, measurement of the size, and degradation behavior of particles. The aim of ISO 14040 and 14044 standards is to direct the collection of LCA data, and such standards do not demand any indicators of graphene. In this vacuum, aerospace organizations will be left with no option but to embark on active environmental custodianship to manage the uncertainty of the laws and the demands of the stakeholders.

One way of such approaches involves putting environmental product declarations (EPDs) on graphene-based products. EPD provides a third party with a guaranteed overview of the environmental performance of a product in the entire lifecycle of the product, in this regard, to procurement transparency and adherence to green procurement approaches. EPDs may also prove particularly handy in a competition of defense or civil aviation contracts where environmental sustainability is currently included in the scoring system.

The use of artificial intelligence (AI) in helping to govern the environment is also increasing. AI tools can model the material degradation, and the environmental exposure situations can be anticipated, and optimization of the manufacturing parameters can be performed to minimize their effect on the environment. As an example, the synthesis parameters may be associated with the energy and water usage via the use of machine learning algorithms and, therefore, guide the optimization of the sustainable processes. In a similar study, Singh et al. (2022) reviewed the potential of machine learning to predict the trends of air quality in urban India. In production situations or lifecycle emissions, the approach can deliver real-time sustainability feedback and can also be used to document sustainability compliance.

In addition, the green concerns are also entering the scene, along with the ethical and social sides of aerospace innovation. The implementation of the systems with the graphene enriched in the civil aircraft or military aircraft must be conducted with the full multifunctionality of sustainability in mind, equal access to green technologies, the harmless conditions of working on the production process, and the minimal impact on the environment. As a high-profile industry, the aerospace business is being pushed towards integration in accordance with the universal environmental goals, such as the UN Sustainable Development Goals (SDGs) and net-zero emissions.

Along with such refinement, the life cycle thinking must be institutionalized in internal sustainability protocols, supplier scorecards, and continuous improvement frameworks. The evaluation of environmental impacts is already being part of the research and development phase during the early stages of some aerospace companies, which also require part and full disclosure of materials (FMD) to suppliers of any advanced material (including graphene) used in the development of such material.

One of the strategies that aerospace designers are recommended to employ to protect the environment in the future is the process of interaction with environmental scientists, material engineers, policy experts, and AI professionals, through cross-sectoral cooperation. This kind of cooperation can aid in a complete conception and feasible control of the environmental impact of the graphene application in the avionics.

CONCLUSION AND RECOMMENDATIONS

Introduction of graphene to avionics systems is one of the steps towards the development of the future of aerospace materials, where the requirements of the increased amount of information carriage, structural sparing, thermal, electrical, and mechanical strength, constantly adjust the requirements of the weight, strength, and thermal data and electrical signal provision. Scientific basis, technical limits, safety conditions, and environmental parameters have been mentioned in the chapter as far as the adoption of the use of graphene in the avionics sector is concerned. These dimensions, when introduced as a set, form a complex narrative and not only reveal the potential of graphene but also show how complex problems will be accompanied by the application of the material in an ultra-regulated sector where safety is the key.

Graphene has a high tensile strength, thermal conductivity, light weight, and extraordinary conductivity, which makes it a material that can be utilized at the material level, a multi-purpose unit in sophisticated aerospace systems. Its advanced conductive shield, thermal spreader, and mechanical reinforcement, and sensing substrate offer the opportunity to combine system levels and performance benefits. In avionics applications Avionics In avionics, where it is needed to be highly integrated with embedded electronics, structural reliability, and fidelity to real-time behavior, such properties may be manifested as reduced payload weight, increased durability, and increased energy efficiency. These benefits are aligned with the systematic aerospace goals of fuel-saving, electrification, and mission flexibility.

Nonetheless, a flow of ideas (i.e., material novelty) to implementation on the system level is a complex path. As mentioned in this chapter, synthesis protocols of graphene, including chemical vapor deposition, chemical exfoliation, and liquid-phase methods, are not uniform in production. The variation in the size of the flakes, the oxygen concentration, defect concentration, and the number of white layers develops variations in the performance and compatibility that are not desirable in the aerospace industry, where long-term reliability and repeatability are not a matter of concern. Also, current industrial manufacturing systems as well as avionics design pipelines are adjusted to metals and old composites, hence incorporation of graphene includes not only replacement of material, but also redesigning of processes, requalification of test, and rearranging the supply chain.

The fact that there are no comprehensive simulation models and multiscale predictive tools is among the most urgent, since they, in effect, could have been used to provide a foreboding of the behavior of graphene under operational conditions. Even though the studies of molecular dynamics have made it possible to characterize the atomistic interactions, they may not be easily coupled with the finite elements, which are large-scale scale because of assumptions that cannot be experimentally verified in the conditions of real aircraft and are computationally costly. Without the digital twins or AI-based performance predictive tools, the developers and regulators cannot be assured of predicting material performance, at flight loads, in the presence of electromagnetic fields, and at environmental extremes, e.g., cosmic radiation or rapid decompression.

The other area of concern is safety operation and occupational safety. Graphene or solvents, Graphene and highly reactive solvents, can be found in the atmosphere whenever it is manufactured; strong acids or carbonyl fluoride can be used as solvents when making it, and hydrazine can be used. The aspect ratio fibrous structure of graphene is high enough to raise concern about respiratory and dermal toxicity, but there are no epidemiologic investigations of long-term impacts to guide exposure limits. The graphene-enhanced materials will also require a special type of analysis of failure during aircraft operation, perhaps because the behavior of the materials under thermal runaway, or lightning strike of the aircraft accident, is quite different from the behavior of the conventional materials. These new failure modes are to be anticipated, tried, and opened to safety standards, which are, in current times, particular with nanomaterials.

And just as significant is the environmental and lifecycle impact of graphene deployment. Though theoretically it can be deployed to reduce the weight of planes and make them economical in terms of fuel usage (indirectly contributing to the reduction of emissions), some processes of graphene production are not only energy consuming but also use potentially dangerous substances. End of life is also not developed yet, particularly when embedded in the composite-based mat in graphene, where it is currently not separable and recyclable with ease. The sustainability of the materials employed in the sector, like graphene, will be put in more scrutiny in the sector that is facing intense criticism over its sustainability promises.

A fragmented regulatory environment is mixed up with such complexity. Some of the international bodies that have done a commendable job in coming up with terminologies and test methods to deal with nanomaterials include the ISO and OECD, but there seems to be a loophole regarding information concerning aviation. Official routes to certification have not been defined by the FAA, EASA, and other regulatory bodies or national defense to support the use of graphene-based systems. Lack of standardized standards brings doubt to the manufacturers and slugs the innovation mechanisms.

Nevertheless, there are also good developments that are discussed in this chapter. The efforts on safe-by-design procedures, risk modeling on AI, and other hybrid test procedures are being made to put infrastructure in place to present a safe and sustainable uptake of graphene. The industrial groups, inter-industrial alliances, and forums held by the policymakers are being carried out to fill the material innovation and application gap. Once these attempts are comprehensive and strategic, then the process of incorporating graphene in avionics can proceed to create a compromise between efficiency, security, and sustainability.

In conclusion, the application of graphene in avionics will not be a direct substitute for one substance with another one, and the change of paradigm in the creation of aerospace systems, their validation, and control. Its multi-purpose characteristic can reduce the number of components, enhance redundancy, and flexibility of system design. However, this can only be done if case scientific, engineering, regulatory, and ethical spheres can be synchronized with one another based on integrated systems. As the aerospace sector enters the paradigm of electrified, autonomous, and sustainable, graphene is a power enhancing enabler, at least in the case when the integration has been realised, with extraordinary rigour, foresight, and sensitive sense, bearing in mind the mission criticality of the aviation industry.

Recommendations

An integrated and multi-stakeholder strategy is necessary to achieve safe, efficient, and future-proof usage of graphene in avionics systems. The latter suggestions are communicated to the scientific community, industrialists, legal institutions, and policymakers to identify solutions to the hindrances and capitalize on the potential of graphene.

Standardization and Material Quality Control

Internationally accepted graphene standards of characterization are required in the aerospace industry. The variation will be controlled by setting the parameters such as the flake size distribution, the number of layers, purity, and the functionalization of the surface, which will result in cross-platform compatibility. Its cooperation with other organizations, such as ISO, ASTM, and SAE, needs to be enhanced to develop aviation-specific graphene materials standards.

Safe-by-Design and Risk-Informed Innovation

Safe-by-design strategy is important to implement graphene. That will entail the identification of the threats in the first phases of synthesis, processing, and roll-out, and it will be assisted by the predictive models and simulations that will involve AI. It is necessary to think of less exposure of the material to the environment and the workers by means of non-toxic synthetic labeling, sustainable functionalization, and operations under closed systems when designing the material.

Regulatory Alignment and Adaptive Certification

Governments such as the FAA, EASA, and ICAO will accelerate the process and the development of transparent requirements for certification of graphene-based components. Adaptive certification models can also be used to reduce time-to-market, such as non-alarming applications that can be tentatively approved and await full safety scrutiny. The task forces on industry-academia-regulatory must be co-established to coordinate the development of avionics-specific, avionics use-case-specific testing protocols.

Environmental and Lifecycle Considerations

The usage of graphene should also be guided by the concept of environmental sustainability. In reference to the synthesis process done by respective developers' lifecycle analyses (LCA) to end-of-life must be carried out through the application of various tools such as GREENSCOPE and TRACI. Recyclable systems of graphene compounds and modular avionics components could be invested in to improve carelines. The outcome of pursuing Environmental Product Declarations (EPDs) should be taken into consideration with regard to green procurement and compliance reporting.

Industry Readiness and Workforce Development

The implementation of graphene should be gradual, and the aerospace companies must implement it step by step in non-sensitive systems (e.g., enclosures, thermal interfaces) first before doing it in critical systems, depending on their performance. The supply systems should have traceability systems, quality, and labeling systems for the nanomaterial. Concurrently, nanomaterial safety education, such as hazardous substances in digital models, regulation compliance, should be niche educated either injected into the engineering curriculum and professional development patterns.

REFERENCES

Archana, U., Sharma, S., & Singh, S. K., R., S., P., S., & Babu, T. H. (2023). Analysis of concrete cracks and fatigue in smart cities using YOLOv3. *2023 IEEE International Conference on Research Methodologies in Knowledge Management, Artificial Intelligence and Telecommunication Engineering (RMKMATE)*, 1–6. https://doi.org/DOI: 10.1109/RMKMATE59243.2023.10368774

Arvidsson, R. (2017). Review of environmental life cycle assessment studies of graphene production. *Advanced Materials Letters*, *8*(3), 187–195. DOI: 10.5185/amlett.2017.1413

Batool, F., Muhammad, S., Muazzam, R., Waqas, M., Ullah, Z., Roy, S., Wang, K., & Guo, B. (2025). Advancements in Graphene-Based Composites: A review of the emerging applications in healthcare. *Smart Materials in Medicine*, *6*(1), 120–138. Advance online publication. DOI: 10.1016/j.smaim.2025.01.001

Bhanuprakash, L., Varghese, S., & Singh, S. K. (2022). Glass Fibre Reinforced Epoxy Composites Modified with Graphene Nanofillers: Electrical Characterization. *Journal of Nanomaterials*, *2022*(1), 4611251. Advance online publication. DOI: 10.1155/2022/4611251

Fadeel, B., Bussy, C., Merino, S., Vázquez, E., Flahaut, E., Mouchet, F., Evariste, L., Gauthier, L., Koivisto, A. J., Vogel, U., Martín, C., Delogu, L. G., Buerki-Thurnherr, T., Wick, P., Beloin-Saint-Pierre, D., Hischier, R., Pelin, M., Carniel, F. C., Tretiach, M., & Bianco, A. (2018). Safety Assessment of Graphene-Based Materials: Focus on human health and the environment. *ACS Nano*, *12*(11), 10582–10620. DOI: 10.1021/acsnano.8b04758 PMID: 30387986

Giese, B., Klaessig, F., Park, B., Kaegi, R., Steinfeldt, M., Wigger, H., Von Gleich, A., & Gottschalk, F. (2018). Risks, release and concentrations of engineered nanomaterial in the environment. *Scientific Reports*, *8*(1), 1565. Advance online publication. DOI: 10.1038/s41598-018-19275-4 PMID: 29371617

Inayathullah, S., & Buddala, R. (2024). Review of Machine Learning applications in Additive Manufacturing. *Results in Engineering*, *25*, 103676. DOI: 10.1016/j.rineng.2024.103676

Jovanović, S., Huskić, M., Kepić, D., Yasir, M., & Haddadi, K. (2023). A review on graphene and graphene composites for application in electromagnetic shielding. *Deleted Journal*, *8*(3–4), 59–80. DOI: 10.1007/s41127-023-00065-3

Kausar, A., Ahmad, I., Zhao, T., Aldaghri, O., Ibnaouf, K. H., Eisa, M. H., & Lam, T. D. (2023). Graphene Nanocomposites for Electromagnetic Interference Shielding—Trends and Advancements. *Journal of Composites Science*, *7*(9), 384. DOI: 10.3390/jcs7090384

Keller, A. A., Lazareva, A., Suh, S., & Davis, J. M. (2013). Global life cycle releases of engineered nanomaterials. *Journal of Nanoparticle Research*, *15*(6), 1–17. DOI: 10.1007/s11051-013-1692-4

Kim, H., Abdala, A. A., & Macosko, C. W. (2010). Graphene/polymer nanocomposites. *Macromolecules*, *43*(16), 6515–6530. DOI: 10.1021/ma100572e

Lopez, R. M., White, J. R., Truong, L., & Tanguay, R. L. (2022). Size- and Oxidation-Dependent toxicity of graphene oxide nanomaterials in embryonic zebrafish. *Nanomaterials (Basel, Switzerland)*, *12*(7), 1050. DOI: 10.3390/nano12071050 PMID: 35407167

National Academies Press (US). (2000). *Incineration processes and environmental releases*. Waste Incineration & Public Health - NCBI Bookshelf. https://www.ncbi.nlm.nih.gov/books/NBK233627/

Panghal, A., Kumar, Y., Mishra, P. K., Mathur, A., & Srivastava, A. K. (2025). Graphene-Based polymer composites for aerospace, electronic, energy, and biomedical applications. *Multifunctional Materials*, •••, 81–104. DOI: 10.1002/9781394234134.ch4

Qin, H., Pei, Q., Liu, Y., & Zhang, Y. (2019). Thermal transport in graphene-based layered materials: An analytical model validated with extensive molecular dynamics simulations. *Carbon*, *155*, 114–121. DOI: 10.1016/j.carbon.2019.08.062

Rao, P., Singh, S. K., & Kumar, V. S. S. (2021). Safety approach in construction industry using case based reasoning. *Journal of Real Estate. Construction Manager*, *36*(1), 47–54. DOI: 10.1177/2977657020210105

Sethuraman, K., & Alagar, M. (2014). Thermo-mechanical and dielectric properties of graphene reinforced caprolactam cardanol based benzoxazine–epoxy nanocomposites. *RSC Advances*, *5*(13), 9607–9617. DOI: 10.1039/C4RA14383C

Shaheen, S., Saeed, Z., Ahmad, A., Pervaiz, M., Younas, U., Khan, R. R. M., Luque, R., & Rajendran, S. (2022). Green synthesis of graphene-based metal nanocomposite for electro and photocatalytic activity; recent advancement and future perspective. *Chemosphere*, *311*, 136982. DOI: 10.1016/j.chemosphere.2022.136982 PMID: 36309056

Shahil, K. M., & Balandin, A. A. (2012). Thermal properties of graphene and multilayer graphene: Applications in thermal interface materials. *Solid State Communications, 152*(15), 1331–1340. DOI: 10.1016/j.ssc.2012.04.034

Sharma, H., Kumar, A., Rana, S., & Guadagno, L. (2022). An Overview on Carbon Fiber-Reinforced Epoxy Composites: Effect of graphene oxide incorporation on Composites performance. *Polymers, 14*(8), 1548. DOI: 10.3390/polym14081548 PMID: 35458296

Singh, S. K., Jain, R., Palaniappan, D., & Parmar, K.. (2022). *Spatiotemporal analysis and machine learning-based prediction of air quality in Indian urban areas.* Environmental Research and Technology., DOI: 10.35208/ert.1587308

Singh, S. K., Jain, R., Palaniappan, D., & Parmar, K.. (2023). Data-driven civil engineering: Applications of artificial intelligence, machine learning, and deep learning. *Turkish Journal of Engineering, 9*(2), 354–377. DOI: 10.31127/tuje.1581564

Tang, X., Debliquy, M., Lahem, D., Yan, Y., & Raskin, J. (2021). A review on functionalized graphene sensors for detection of ammonia. *Sensors (Basel), 21*(4), 1443. DOI: 10.3390/s21041443 PMID: 33669589

Verma, P., Goldar, D., Singh, S. K., & Md, D. (2024). Performance Evaluation and Enhancement of Geopolymer Concrete using Silica Fume. *International Journal of Civil Engineering, 11*(11), 78–88. DOI: 10.14445/23488352/IJCE-V11I11P108

Vinci, G., Gobbi, L., Ruggieri, R., Ruggeri, M., & Tiradritti, M. (2023). Challenges and opportunities for the graphene industry. Sustainable approaches for the circular economy. In *Elsevier eBooks* (pp. 319–339). https://doi.org/DOI: 10.1016/B978-0-323-90914-3.00013-9

Visser, M., Gosens, I., Bard, D., Van Broekhuizen, P., Janer, G., Kuempel, E., Riediker, M., Vogel, U., & Dekkers, S. (2022). Towards health-based nano reference values (HNRVs) for occupational exposure: Recommendations from an expert panel. *NanoImpact, 26*, 100396. DOI: 10.1016/j.impact.2022.100396 PMID: 35560294

Chapter 10
Future Prospects and Trends in Graphene-Powered Aerospace Technologies

Rashad G. Abaszade
https://orcid.org/0000-0001-6210-8517
Azerbaijan State Oil and Industry University, Azerbaijan & Azerbaijan University of Architecture and Construction, Azerbaijan

Vugar H. Abdullayev
https://orcid.org/0000-0002-3348-2267
Azerbaijan State Oil and Industry University, Azerbaijan & Azerbaijan University of Architecture and Construction, Azerbaijan

ABSTRACT

Graphene is the best material in terms of mechanical strength, electrical conductivity, and thermal characteristics; therefore, it promises to transform aerospace technologies because it makes new applications possible and it improves current infrastructures. The paper looks through the potential applications of graphene in aerospace through its involvement in lightweight composites in structures, superior thermal management, Electromagnetic Interference (EMI) screening, and high-performance sensors. Great progress has been made in the synthesis and functionalization of scalable graphene and this has allowed the integration of aerospace quality graphene materials into metal and polymer composites overcoming issues of manufacturing and material compatibility.

DOI: 10.4018/979-8-3373-3862-0.ch010

INTRODUCTION

The mechanical properties of metal matrix composites (MMCs) driven by the incorporation of graphene would significantly improve their mechanical behavior and would allow their use in aerospace. It is reported that the reinforcement filler (graphene) leads to significant increases in tensile strength, toughness, and thermal conductivity, counteracting the drawbacks of the traditional materials. An example is the incorporation of electrochemically exfoliated graphene in copper-based nanocomposites, which brings significant enhancements in physicochemical properties with improvements in structural stability and wear resistance (Sahoo et al., 2025). Further research has seen the use of platelets of graphene as a reinforcement in the composite system over a long duration of time. These research findings indicate the superiority of carbon-based nanofillers as usable materials in the practice of rigid aerospace because of their ability to enhance mechanisms and electrical actions of a material (Trai & Sobhy, 2025). Low density and the high strength-weight ratio, and high electrical conductivity achieved are useful features that improve the performance of such composites, since they are used to design both lightweight structural parts and high-performance equipment with high conductivity. Graphene--epoxy system application has driven graphene fabrication technology, such as three-dimensional printing. The elasticity of graphene-based epoxy inks allows a design of 3D-printed buildings with high mechanical modulus that stand alone (Sun et al., 2025). This is especially essential in the air travel industry, where structural and lightweight artistry are paramount. Graphene composites make it possible to produce lighter constituents, which not only possess a higher degree of strength and thermal resistance, but also meet the requirements of space-based applications that generally work in adverse conditions. Though epoxies are typically brittle materials, the new developments in graphene-epoxy nanocomposites indicate that the brittle nature of epoxies may be counteracted by loading the material with graphene, an active promoter of fracture toughness through crack-pinning and deflection mechanisms, and thus, allowing more aerospace applications of the materials (Uniyal et al., 2025). The improved methods of producing graphene fabrics have significantly reduced the usage of graphene in aerospace technology, due to its unique mechanical, electrical, and thermal characteristics. Graphene is a two-dimensional allotrope of carbon, which has a honeycomb lattice, which gives it excellent electrical conductivity, high mechanical strength, and low mass, which are critical for aerospace materials (Hidayat et al., 2025). Therefore, enhancement of production processes and processing strategies is important in enhancing the performance of aerospace materials and establishing them in civil aviation and military primary market domains. A promising source of high-quality graphene is enhanced by the chemical vapor deposition (CVD). Research has indicated that a selective etching approach, by which graphene material prepared on a dielectric substrate is promoted by a premelted substrate, can significantly improve conformal coating of irregular substrates, and thereby overcome a key bottleneck in the process of integrating graphene into complex substrates demanded by the aerospace industry (Yang et al., 2025). Optimization of growth parameters can also yield properties that are more lightweight, flexural requirements of aerospace materials. Other processing methods like laser shock processing (LSP) have resulted in a vast enhancement to the mechanical properties of aluminum-grape composites and have made them viable in high-stress and high-temperature space mission areas (Kumar et al.,2025). Considering the large surface area of graphene, it becomes easier to make physical contact with the copper matrix, improve bonding, and enhance the structural integrity of the composite. Interfacial enhancements are particularly important in instances involving large mechanical loads and large temperature variations on the materials; therefore, aerospace components such as graphene-copper nanocomposites promise to be the key elements. It is still possible to say that there is a high potential for graphene in anti-corrosion. Recent studies also reveal that the functionalization of graphene-oxide hybridization in epoxy leads to long-term metal shielding in aerospace space (Ghosh et al., 2025). Besides high mechanical strength, these environmentally friendly materials offer a high degree of resistance to corrosive agents, thus prolonging the service life of aerospace structures subjected to extreme conditions of the atmosphere. Scientists are still examining the multifaceted relationships between functionalized graphene and epoxy resin to improve functionality and stability. The second fundamental implication of using the material in aerospace technologies is the optimization of processing procedures to enhance the interfacial properties of composite materials. It shows that the compatibility ability of graphene reinforced polymer nanocomposites can be enhanced considerably by altering the surface (Kshirsagar et al, 2025). By increasing the polymer-graphene wettability and interfacial adhesion, it enhances the spatial distribution of graphene in polymer-based composites, producing high mechanical properties as well as manufacturing performance. With an additional approach based on machine-learning models and molecular-dynamics simulations, the mechanical behavior of graphene-based nanocomposites can be predicted, thus explaining the interactions between atoms and additional information about how well the systems made of graphene-copper perform under a set of different conditions (Gao et al., 2025). Predictive tools allow the selection of materials and the creation of unique composite systems, which are superior to current products in the market. The air transport industry is of substantial importance in climate change related to carbon dioxide emissions, with specific reference to kerosene burning (Li et al., 2025). The bolstering of air transportation needs to trim and address the existing environmental concerns and attain aviation sustainability. Sustainable aviation fuels (SAFs) have been set up as a promising way of mitigating emission impacts through aligning with the Paris Agreement, which centers on carbon-neutral growth. SAFs can be combined with conventional jet fuels, and their production does not require significant changes to the engine or other infrastructure, which can be viewed as a viable near-term solution (Feng et al.,2025; Klingelhöfer et al., 2025). This compatibility ensures that there are no economic barriers and risks during the transition to new fuels. SAFs can be proposed to exceed conventional fuels on several performance metrics according to predictive modeling of their fuel properties, and therefore improve efficiency and sustainability (Parker et al., 2025). Especially promising avenues to producing carbon-neutral aviation fuels include biofuels synthesized using environmentally sustainable sources, including waste plastics or lignocellulosic biomass, which would not only mitigate the issue of resource scarcity and waste management but also reduce carbon emissions (Tang, 2025). AGRI-forestry residues have not been exploited as a source of carbon-neutral aviation fuel feedstock because of their non-food nature (Eun et al., 2025). The ongoing improvement of catalyst systems to process such substrates increases the viability of SAF production and boosts sustainability (Nie et al., 2025). Policy programs are most relevant in promoting SAF production and use. The ReFuelEU Aviation tool, created by the European Union, creates the regulations that will facilitate the capacity growth and SAFs' financial sustainability (Raman et al., 2025). This kind of facilitation of regulation is crucial in creating a market that will engage in investment in sustainable technologies in line with the climate goals (Raman et al., 2025). The creation of SAFs in terms of the mission to achieve carbon neutrality globally is an emerging research facet in the aviation industry. It focuses on the implementation of the technologies supporting the use and creation of environment-friendly aviation fuel. Recently, a breakthrough technology has revealed that agricultural or municipal wastes can be converted into petroleum-free fuels suitable for operating aviation-grade turbines without significant infrastructural modifications, thus facilitating a smooth shift to sustainable operations (Elahi, 2025).

However, biofuels are not the only things that have been incorporated into the fuel field of aviation. The ongoing research in the development of electrofuels or hydrogen fuels demonstrates that there must be a comprehensive approach to reducing emissions. According to the literature, the attainment of significant carbon reduction requires equal research on technology and behavioral change affecting travel demand (Suomalainen, 2025; Baran, 2025). Various exploration of these options demonstrates the potential shedding of great emissions without miniaturization of passenger and customer space. At the same time, aircraft development, especially electric propulsion, is one of the areas of aviation development. The example of the transition to all-electric and hybrid-electric vertical take-off and landing (eVTOL) aircraft. Significant undertakings are made to simplify the power management to make it more efficient and viable (Huang et al., 2025). Future studies need to focus on optimizing variable-speed rotor-blade models and systems to balance dynamically power demand and performance (Huang et al., 2025). EVTOL technology is expected to revolutionize air movement in urban areas and provide essential services to overcrowded cities as it reaches maturity. Other than technical innovations, pilot training should be improved to match the operation efficiency of aviation. The ability of pilots to perform specific flight maneuvers and scan the visual horizon is directly affected by competence, which contributes to the necessity to focus on the development of superior simulation tools and specialized training methods (Vantrepotte et al., n.d.). To improve the training curriculum and elevate the safety and performance of military and civilian aviation, any future study is advised to measure complex scenarios to determine pilot adaptation and decision-making before informing a new training program design. Artificial intelligence and robotics are among the adaptive technologies that are increasingly transforming the operations of the aviation industry. With that being said, it is assumed that the idea of an innovation can be greatly expedited via comprehensive work that also involves technical progress and research of human factors (Roche & Anand, 2025). The role of AI in modern aviation systems and approaches to their integration to make aviation safer, more functional, and efficient should be investigated in the future. The future applications of data analytics will yield data trends, which will provide indicators that enhance the efficiency of the aviation logistics and management systems.

EMERGING APPLICATIONS OF GRAPHENE IN AEROSPACE

The wide opportunities presented by graphene have paved the way to new possibilities in the aerospace engineering field due to its superior strength and low density with high electrical and thermal conductivities. Improved mechanical properties and lower weight, combined with a graphene-enforced structural composite, represent some of the most promising uses for fuel economy improvement. This material has been shown to have excellent thermal transport properties and can be utilized in the future in avionics and space-exploration heat cooling applications to address the threat of overheating delicate components. Graphene films and coatings can offer a feasible solution as they reduce the signal of interference and consequently protect sensitive electronic equipment in a method of electromagnetic interference (EMI) cloak. Another application of graphene as a discriminating sensor is its extremely responsive surface and massive specific area that can be used to deploy real-time aircraft health monitors with discriminating sensors. Graphene supercapacitors and batteries are also other applications, similarly to the aerospace industry that needs to have lightweight and at the same time, high-capacity energy storage. Graphene is the material that can be modified to include in the wearable technologies to monitor the activity of the crews and pilots because of its linear elastic characteristics. Additionally, the chemical stability of the material grants it durability to the most severe aerospace environmental conditions. The optical properties of graphene provide aerospace displays with photodetectors and transparent conductive films. Functionalization attracts the ability to craft aerospace solutions that are anti-icing and have an anti-corrosive coating. The study of graphene aerogels helps adopt insulation materials that are easy to carry. In the composites with other nanomaterials, the effect of the large compound composite enables the achievement of synergistic effects, thus achieving high-aerospace performance requirements. Graphene membranes have the potential to pose a challenge to the common aircraft fuel and gas filtration systems because they offer some other alternative system of separation. One approach to ensuring aerospace systems become smarter in connecting them to the Internet, as well as with each other and all other systems, is the integration of an Internet-of-Things (IoT) Environment and graphene sensors. Lastly, the topic of graphene in additive manufacturing is also considered, as it is perceived as the way to produce aerospace products with elevated qualities. Extensive use of graphene in the aerospace field has created significant apprehension due to its excellent mechanical, thermal, and electrical conduction. They indicate that these characteristics can be widely used in composite materials, technical production methods, and modern production processes to revolutionize the cosmos technology in aircraft (Table 1).

Table 1. Key Research Topics and Applications of Graphene in Aerospace Materials and Technologies

Research Topic / Application Area	Key Findings / Description	References
Graphene Reinforced Metal Matrix Composites (MMCs)	Use of graphene significantly improves mechanical strength, toughness, and thermal conductivity.	Sahoo et al., 2025
Graphene in Copper-based Nanocomposites	Electrochemically exfoliated graphene enhances structural and physicochemical properties.	
Graphene Platelets in Carbon-based Nanofiller Composites	Improves mechanical and electrical performance for lightweight structural components and high-performance electrical systems.	Trai & Sobhy, 2025
Graphene in 3D Printing Technologies	Incorporation of graphene into epoxy inks enables the production of 3D printed structures with high mechanical integrity.	Sun et al., 2025
Toughening of Graphene-Epoxy Nanocomposites	Enhances fracture toughness via crack pinning and deflection mechanisms.	Uniyal et al., 2025
Surface Modification of Graphene	Improves wettability and interfacial adhesion, enhancing composite processing and performance.	Kshirsagar & Joshi, 2025

continued on following page

Table 1. Continued

Research Topic / Application Area	Key Findings / Description	References
Graphene with Bio-fiber Reinforced Composites (BFRCs)	Development of lightweight, high-strength, eco-friendly bio-composites enhanced with graphene.	Maldonado-Romo et al., 2025
Thermal Management with Graphene-enriched Porous Materials	Creation of materials with superior thermal insulation and electromagnetic interference shielding capabilities.	Shen et al., 2025
Graphene in Electronic Device Integration	Use of atomically thin materials like graphene to enable lightweight, high-performance avionics and sensor systems.	Haartsen et al., 2025
High-quality Conformal Coatings Using Graphene	Achieving high-quality graphene growth on dielectric substrates for protective coatings conforming to complex geometries.	Yang et al., 2025

Most importantly, the surface modification of graphene-based composites is of the greatest relevance. Wettability and interfacial adhesion: the various complexities connected with wettability and interfacial adhesion are recognized to cause inconveniences in terms of dispersing and incorporating graphene in polymer scaffolds (Kshirsagar & Joshi, 2025). To exploit the full potential of graphene-based composites in the aerospace industry, the issues mentioned above have to be solved. Graphene is one of the types of materials that can be significantly enhanced by innovations that are devoted to the treatment of their surfaces (Kshirsagar & Joshi, 2025). Another secondary field of study involves research on the development of graphene along with bio-based materials and biomaterials that constitute bio-fiber reinforced composite (BFRCs). Low density and high strength are the major characteristics of these composites that consequently lead to the attainment of a sustainable solution in the aerospace sector. These materials are not environmentally friendly, and thus, to improve their performance without necessarily leading to a decrease in mechanical performance, it could be proposed to use graphene to add to the material (Maldonado-Romo et al., 2025). One of the most important issues in aerospace applications is thermal management. There is an increasing desire to create porous buildings that can ensure a high level of thermal insulation and eliminate or reduce electromagnetic interference (EMI) (Shen et al., 2025). Graphene-based composites with flexible structures that have high heat conductivity can resolve this challenge with aerospace components operating in harsh environments. Further, graphene finds application in the fabrication of electronic equipment, which is a major advancement of the material in aerospace use. Nanomaterials like graphene provide added value to the production of devices and support the production of lightweight and high-performance electronic systems needed in modern avionics and sensor technology (Haartsen et al., 2025). Recent studies on performance retention in applying such materials at the points of integration show that the aerospace industry experiences a significant challenge. The other notable property of graphene is its ability of graphene in form high-quality conformal coatings. Preferential etching techniques that incorporate pre-melted substrate-preferential etching have revealed that it is possible to deposit high-quality graphene on dielectric substrates that are needed in protective coatings that are demanded by the conformal nature of aerospace structures (Yang et al., 2025). Recent innovations in the fabric production technique have resulted in a large scope of applications in the aerospace industry, due to the greater potential represented by graphene. Simultaneously, the mechanical and electrical trade-offs of graphene and other two-dimensional (2D) structural materials have been discovered recently and should be optimized to be used in aerospace. The superior sensors and EMI shielding technology required by the aerospace industry will be aligned with the trade-off between the mechanical strength and electrical conductivity demanded by good and reliable films (Rong et al., 2025). Trade-offs of this nature will be established and handled to assist in developing new parts of aerospace, and hence the application of modern material science to meet the demands of enhanced performance.

It is possible to see that the sphere of the use of graphene will broaden in the future, and the aerospace sector will shift to light and high-performance materials. Various applications of graphene include structural enhancement, electronic system integration, thermal conductivity, and sustainability in a series of electrical and mechanical properties that have demonstrated this potential. In total, several products are offered by graphene, which is one of the most innovative materials in materials science and engineering, particularly in aerospace. The fact that it continues to reveal new applications in various ways in the electrical arena, as well as in composite advancements, proves the opinion that the use of graphene in aerospace composites will be an opening to a new trend. The realization of the advantages of utilizing graphene-reinforced materials in the new space technologies will be important in future research in order to overcome many obstacles, such as dispersion and compatibility.

ADVANCES IN GRAPHENE SYNTHESIS AND PROCESSING FOR AEROSPACE USE

The modern inclination to the use of graphene in aerospace engineering will depend on the development of a method of production that could be used without the exclusion of errors. CVD has remained the most predominant method, due to its ability to produce vast sheets of high-purity graphene that can be used in composite applications. In the recent past, heterogeneity and flaw density in aerospace-grade materials produced through CVD have been alleviated, and hence structural uniformity amplified. Reduction of graphene oxide (GO) is a cost-effective route, though it generally leaves behind defects that require further refinement. Large-scale production of Graphene in the form of flakes, and compatible with polymers, has become a highly popular technique of large-scale synthesis of graphene through liquid-phase exfoliation. Covalent and non-covalent functionalization strategies have been controlled to enhance the aerospace polymer and metal-graphene interfacial interactions and increase the dispersibility of the material. Plasma treatments, chemical doping, and modifications to the electrical and surface properties of graphene are used to control the electronic and surface properties. A thickness and layer control homogeneity is obtained through novel catalyst structures and careful growth parameter optimization. The incorporation of graphene in fiber-reinforced composites is currently being prepared through integrative techniques that ensure successful interfacial bonding and load transfer. Advances in graphene ink compositions have made additive manufacturing and printed electronics feasible in aerospace applications. Raman Spectroscopy and electron microscopy are also employed in quality control procedures that ensure the uniformity of the materials. Environments and economic driving factors drive green synthesis directions, where efforts are made to upscale production without making material compromises. The transfer of technology between technology and academia is likely to be sped up by collaborative efforts meant to assist in the process. In general, all the research and processing efforts are historical series in preparation of aerospace components, which are competitive in terms of high performance and safety (Table 2).

Table 2. Advances in Graphene Synthesis, Processing, and Applications for Aerospace Technologies

Research Topic / Application Area	Key Findings / Description	References
Graphene Structure and Aerospace Material Properties	Graphene's 2D honeycomb lattice provides exceptional electrical conductivity, mechanical strength, and light weight, critical for aerospace.	Hidayat et al., 2025
Chemical Vapor Deposition (CVD) Graphene Growth	Premelted-substrate-promoted selective etching improves graphene quality on dielectric substrates, enabling conformal coverage on complex aerospace geometries.	Yang et al., 2025
Laser Shock Processing (LSP) of Aluminum-Graphene Composites	LSP enhances the mechanical and tribological properties of aluminum-graphene composites, making them suitable for high-stress aerospace environments.	Kumar et al., 2025
Electrochemically Exfoliated Graphene in Copper Nanocomposites	Graphene's large surface area improves bonding and structural integrity in copper nanocomposites under mechanical and thermal stress.	Sahoo et al., 2025

continued on following page

Table 2. Continued

Research Topic / Application Area	Key Findings / Description	References
Anti-corrosion Applications of Functionalized Graphene Oxide	Functionalized graphene oxide in epoxy matrices provides durable, eco-friendly metal protection for aerospace structures.	Ghosh et al., 2025
Surface Modification of Graphene in Polymer Nanocomposites	Improved wettability and interfacial adhesion optimize graphene distribution, enhancing composite mechanical properties.	Kshirsagar & Joshi, 2025
Computational Modeling of Graphene Nanocomposites	Machine learning and molecular dynamics simulations predict mechanical properties and optimize graphene-copper composites for aerospace.	Gao et al., 2025
Graphene-based Thermoelectric Composites	Conductive binders with graphene improve thermoelectric efficiency for advanced thermal management in aerospace systems.	Ozlek et al., 2025
Graphene in Optoelectronic Devices	Integration with silicon photonics enables advanced sensors and communication devices critical for aerospace technologies.	Purbayanto et al., 2025

Moreover, the aspect of graphene-based composites in thermoelectric applications is a clear indication that this material is multifunctional (Ozlek et al., 2025). The addition of conductive binders and graphene has increased the thermoelectric efficiency needed in the development of superior aerospace systems that require very precise thermal regulation. The development of graphene into the systems is already expected to be a future breakthrough in aerospace. Similarly, the growing integration of graphene in optoelectronic technology indicates a multidisciplinary growth in the aerospace industry (Purbayanto et al., 216). Graphene is a material with extraordinary features, and it can be utilized to create advanced sensors and communication devices, which cannot be done without in aerospace technologies. The composite nature of graphene is evident in its implementation in existing silicon-based photonic devices, which are both compatible with the substance and can enhance the functionality of other intricate aerospace systems. In conclusion, the continued advancement of the technology and manufacturing of graphene will significantly alter the way material equipment in the aerospace industry functions and provide it with new abilities in mechanics, electricity, and thermodynamics. The technologies of inserting graphene wire into various matrices and studying their potential use cases are needed to meet the high-performance needs of modern aerospace systems.

INTEGRATION OF GRAPHENE WITH NEXT-GENERATION AEROSPACE SYSTEMS

The fabrication of the graphene plays a seminal role in the development of the next-generation aerospace systems to superior performance and sustainability. In propulsion, graphene-enforced composites are used to lower the mass of the engine and enhance thrust-to-weight ratios of electric and hybrid aircraft. Thermal interface materials based on graphene are able to increase thermal dissipation in power electronics, therefore raising reliability. Graphene sensors result in the avionics because now there is structural health and environmental monitoring in real-time. Electromagnetic shielding of the material prevents signals interception of high-tech communication and navigation systems. In the unmanned aerial vehicles (UAVs), the flexible electronics and sensors made of graphene improve the flight control and autonomy. Graphene composites provide a better fatigue performance of key aerospace parts, increasing their lifetime. In a spacecraft, one would put a layer of graphene to shield against radiation that would damage electronic components and even astronauts. Combination with energy storage systems allows extending missions using the lightweight and high-energy-containing power sources. The compatibility of graphene with additive manufacturing also allows multifunctional aerospace components. Graphene-AI and IoT are able to combine to optimize predictive maintenance and operations. Photonic devices based on graphene have the potential to enhance the communication bandwidth of aerospace networks. There are attempts to use bendable sheets of graphene in deployable aerospace structures and smart skins. Thermal expansion mismatch is such an integration challenge that can be dealt with by material engineering. Development of standards on graphene-enabled aerospace systems is going on. On the whole, graphene plays an important role in the move toward intelligent, lightweight, and sustainable aerospace platforms (Table 3).

Table 3. Key Advances in Graphene-Based Materials and Their Integration into Next-Generation Aerospace Systems

Research Topic / Application Area	Key Findings / Description	References
Graphene-Reinforced Aluminum Composites	Incorporation of graphene enhances mechanical and tribological properties, suitable for high-stress aerospace components.	Kumar et al., 2025
Graphene/Copper Nanocomposites and Interface Optimization	Machine learning and molecular dynamics simulate mechanical properties; interface understanding is critical for aerospace performance.	Gao et al., 2025
Versatility of Graphene in Aerospace Engineering	Graphene's lightweight and adaptable properties enable diverse composite formulations for aerospace applications.	Hidayat et al., 2025
MXene and Other 2D Materials for Aerospace	MXenes' metallic conductivity and stability offer complementary benefits for aerospace, mechanical, and engine applications.	Yang et al., 2025; Singh et al., 2025
Fabrication of Atomically Thin Materials	Advances in fabrication techniques are critical to preserving electronic properties at scale for aerospace systems.	Haartsen et al., 2025
Graphene for Radiation Shielding	Graphene-based nanomaterials improve protective coatings against space radiation, enhancing the safety of aerospace vehicles and crews.	Sales et al., 2025
Thermal and Electrical Enhancement in Composites	Graphene improves heat transfer and electrical conductivity in polymer matrices for aerospace thermal management and electronics.	Ozlek et al., 2025
Bi-directional Functionally Graded Materials	Graphene-enriched composites support resilience and dynamic load handling in aerospace structures.	Trai & Sobhy, 2025
Transfer-Free Graphene Growth Techniques	Innovative transfer-free fabrication methods allow high-quality graphene integration into complex aerospace geometries.	Yang et al., 2025

Graphene finds applications in improving thermal and electrical conductivity of composite materials, especially those used in aerospace. Ozlek and co-authors emphasize that graphene has a significant potential to enhance thermal and electrical conductivity of polymeric matrices and thus transform the aviation industry, introducing new applications in which effective heat transfer and current control are central (Ozlek et al., 2025). Graphene is, therefore, well placed as a key additive in the production of composite materials that can meet the diverse needs of the modern aerospace industry. When bi-directionally graded materials are in question, the mechanical, thermal, and electrical properties of the composite with enriched graphene take on a particularly critical significance. The temporary examination of graded plates by Trai and Sobhy implies the existence of a tendency toward the application of carbon-based composites in various fields of engineering, most prominently aerospace (Trai & Sobhy, 2025). These properties confirm that graphene is a core material in high-performance systems that must have a strong structure and be able to handle dynamic loads. Further studies have expanded the boundaries of growing graphene methods, hence offering new prospects on how this material can be utilized in future aerospace systems. The transfer-free approach to making high-quality graphene by Yang and others suggests the creation of organized flexibility in the material and supports the easy incorporation into the more complex structures of typical aerospace parts, which would significantly improve the material's practicality (Yang et al., 2025). The proposed methodological novelty guarantees that the mechanical and thermal stability necessities of aerospace environments may be met without damaging the intrinsic nature of the graphene. Overall, the next generation of aerospace systems will be equipped with enormous improvements in performance due to the application of graphene, as it has characteristics of favorable qualities that can be used to improve the level of safety, weight reduction, and efficiency. The key to successful implementation of graphene in space activities in the future will be based on faster production speed and knowledge of the material.

SUSTAINABILITY AND ENVIRONMENTAL IMPACT

The first impact of graphene on the sustainability agenda in the aerospace industry lies in its capacity to produce light but robust structures, hence leading to emission and fuel cost cuts. The equivalent weight loss that graphene-composite materials have brought to the aircraft is directly proportional to the reduction in carbon dioxide emissions and subsequently the cost reduction in running the aircraft. Graphene-based thermal control systems improve the power efficiency of aircraft avionics and propulsion systems and reduce power loss. Its chemical and abrasion resistance prolongs the life of components compared to traditional materials and thus reduces the number of replacements and material waste. Sensors made of graphene allow for predictive maintenance, eliminating the expensive ramifications, dealing with calamities, and the unjustified replacement of parts. Use of graphene in energy storage will encourage the use of electric and hybrid propulsion, which will reduce the use of fossil fuels. Recyclable composite materials in the aerospace sector are a contribution of graphene towards the methodology of recycling materials. Graphene is being developed using green technologies to reduce environmental footprints. Life-cycle assessments (LCA) continue to be a growing method of quantifying the impacts and benefits of graphene-aerospace products on the environment. There are still concerns about the safe substitution or recycling of the graphene-enhanced composites. Standards and regulations are undergoing changes with regard to environmental and health concerns brought about by nanomaterials. Graphene-based biopolymers are also under development with the objective of developing all-sustainable composites. Beyond aerospace, Graphene-enhanced fuel cell technologies allow zero-emission propulsion by enabling improved fuel cell technology. Collaborations between industries and regulators aim to achieve harmonization of the sustainability measurements. Finally, the implementation of graphene introduces an aerospace novelty in pursuit of global environmental and climate action plans (Table 4). A new way of life will be needed to support the creation of sustainable technologies, which requires lifelong learning and collaboration between stakeholders (governments, airlines, and fuel producers). Finally, the possibility of electric and hydrogen-driven propulsion, although SAFs represent an intermediate step, conforms to longer-term schedules of meaningful areas of progress in relation to the fight against emissions in the aviation industry (Feng et al., 2025).

Table 4. Overview of Sustainable Aviation Fuels and Strategies for Reducing Aviation's Environmental Impact

Research Topic / Area	Key Findings / Description	References
Aviation Sector Contribution to Carbon Emissions	Aviation contributes about 12% of global transport CO_2 emissions, mainly due to fossil kerosene use.	Li et al., 2025
Role of Sustainable Aviation Fuels (SAFs)	SAFs are critical for carbon-neutral growth, reducing greenhouse gas emissions, and aligning with the Paris Agreement.	Feng et al., 2025; Klingelhöfer et al., 2025
SAF Compatibility with Existing Infrastructure	SAFs can be blended with conventional jet fuel without aircraft modifications, lowering economic and operational barriers.	Feng et al., 2025; Parker et al., 2025
Biofuel Feedstocks for SAFs	Sustainable biofuels from waste plastics and lignocellulosic biomass offer resource and emission benefits.	Tang, 2025; Eun et al., 2025
Catalyst Development for SAF Production	Advances in catalysts optimize SAF production from sustainable feedstocks, enhancing efficiency and sustainability.	Nie et al., 2025
Policy and Regulatory Support	Initiatives like the EU's ReFuelEU Aviation and CORSIA promote SAF production and emissions offsetting in aviation.	Raman et al., 2025; Feng et al., 2025
Operational Efficiencies in Aviation	Improvements in aircraft design, flight operations, and infrastructure complement fuel advances for sustainability.	Li et al., 2025

continued on following page

Table 4. Continued

Research Topic / Area	Key Findings / Description	References
Alternative Propulsion Technologies	Electrification and hydrogen-powered aircraft are long-term goals but face challenges; SAFs provide near-term solutions.	Klingelhöfer et al., 2025
Emission Reduction Potential of Bio-based Fuels	Biofuels may contribute up to 65% of aviation's greenhouse gas emission reductions by 2050.	
Challenges in SAF Adoption	High production/testing costs and stringent standards hinder market entry; stakeholder collaboration is needed.	Parker et al., 2025

According to the systems perspective, it is observed that SAFs can require immediate results, but a balanced approach could be adopted based on all possible solutions that are likely to become essential in addressing the long-term success of reducing the environmental impact of aviation. Existing evidence suggests that the reduction of climate change has ceased to be a narrow-focused issue and has turned into a precondition that defines the sustainability of the aviation industry in the long term; a whole shift towards alternative sustainable aviation fuels is beneficial and must remain after all in order to meet standards in preserving the ecological system. Therefore, the incorporation of new fuel technologies, a wide range of feed stamps, a congenial policy environment, as well as elaborate business practices ought to be integrated within the aviation industry, thus making it possible to adopt a sustainable working model that would effectually suppress the enormous emission of greenhouse gases.

FUTURE RESEARCH DIRECTIONS AND INNOVATIONS

Some of the aerospace technologies currently in research to enable the use of graphene are likely to be concerned with the creation of varieties of composite materials that have better mechanical and high-performance capabilities, as well as better electrical, thermal, and sensory properties. An agreement on identifying and optimization of techniques that help to deploy artificial intelligence and machine learning techniques is expected to streamline the process of addressing the possibility of applying the material property of graphene into the right aerospace structure. It is expected that high-level characterization techniques would allow a thorough investigation of the behavior of graphene under aerospace conditions of operation. In addition, it can be expected that the materials in which the functionalization is performed with the help of hybridization will be reinforced with other new two-dimensional systems based on graphene, which will be followed by MXenes and boron nitride. Hopefully, a human star, collaborative and sustainable Industry 5.0 model of production on the scale of mega wars and mega factories will be implemented. In this order of consideration, the key point will be the realization of eco-benign scales of synthesis that will meet environmental and financial needs. It will also be sought to print aerospace shapes, ranging in complexity and shape, including graphene, in three and four dimensions through printing technology. According to Alexandrimeri et al. (2019), aircraft navigation and communication will be highly improved in the case of quantum devices obtained with the help of graphene. Graphene sensors will allow obtaining real-time information about the state of affairs during work and make automated decisions in accordance with the new capabilities of modern Internet technologies platforms. Long-term durability and radiation tolerance will be evaluated using experimental protocols that are performed under conditions of space environment. It will enhance the insulating performance and damping vibrations, as well as reduce the weight by developing gramophone aerogels and foams. The recycling and disposal process of graphene composites will be analyzed and discussed. The academic and industrial research programmers in the country will also closely work with regulatory authorities to facilitate the transfer of technology. The graphene material employed in the aerospace industry should have extremely high safety and quality standards, and therefore, standardization processes should be established to ensure that this material meets high standards. Overall, long-term innovation will enable the creation of new expertise, deploying graphene as a building block in the future of aerospace devices (Table 5).

Table 5. Future Research Directions and Innovations in Aviation Performance and Sustainability

Research Topic / Area	Key Findings / Description	References
Sustainable Aviation Fuels (SAFs)	Development of SAFs from non-petroleum sources like agricultural residues and municipal solid waste, compatible with existing jet turbines.	Elahi, 2025
Electrofuels and Hydrogen-based Fuels	A hybrid approach is needed; hydrogen fuel potential requires technological and behavioral research for effective carbon reduction.	Suomalainen, 2025; Baran, 2025
Electric and Hybrid-electric Aircraft Design	Focus on eVTOL aircraft with optimized power management and variable speed rotor blades to enhance efficiency and urban air mobility.	Huang et al., 2025
Pilot Training and Operational Efficiency	Pilot expertise influences flight precision; improved simulation and training methods are needed for complex scenarios.	Vantrepotte et al., 2025

continued on following page

Table 5. Continued

Research Topic / Area	Key Findings / Description	References
AI and Robotics in Aviation Operations	Integration of AI to improve safety, streamline processes, and optimize logistics; collaboration between human and technology factors is critical.	Roche & Anand, 2025
Quality Control and Manufacturing Innovations	Advanced measurement technologies like 3D scanning and laser profiling improve component manufacturing precision and reduce costs.	Falandys, 2025

One of the most critical factors that support the safe and operational integrity of the aviation industry is quality control. The author of the article argues that the developments in technologies of measurements, including three-dimensional scanning and laser profiling, have considerable potential to improve the accuracy of the process of creating components and their inspection (Falandys, 2025). However, automated quality-control systems merit careful and in-depth consideration so as to refine the component characteristics with a view to keeping down production time and cost. The advancements in technology throughout the manufacturing continuum, and particularly the use of modern materials science, have enabled the production of both lighter and more fuel-efficient aircraft. Along with all technical advantages, the socio-economic impact of aviation innovations should be included in the future research agenda to make the improvements comprehensive. The nexus between environmental sustainability and aviation development. Finding in the framework of the policies (e.g., the Paris Agreement) have been framed as a policy imperative in the nexus between environmental sustainability and aviation development (e.g., Procedural activity, 2019). Further studies in this area are required to focus on how the people of this country view sustainable methods of travel and how ready they are to embrace these methods of conducting business in an effort to balance environmental and economic factors. The delivery of a sustainable future in the aviation industry requires unique interdisciplinary collaboration. The interdisciplinary approach to applying the triple E, economic, and psychological insights can also be regarded as an effective approach to respond to the complex problems that have arisen. Switching to a more environmentally friendly aviation industry also requires the willingness of the stakeholders to exchange knowledge and, therefore, permits the research of the topic, along with attaining innovation through multidisciplinary teamwork. Altogether, it can be concluded that the aviation industry stands at the border of several research directions, which, in combination, promise the emergence of a new breakthrough and the creation of a sustainable future. Future innovations will be based on complementary events in relation to fuel technologies, aircraft design, operations, quality administration, application of technologies, and social/economic consciousness. The long-term success of the aviation industry will continue in the next several decades, provided significant investment in the development of aviation by the government and other stakeholders is made, and the position of the regulatory authorities is favorable.

CONCLUSION

The characteristics possessed by graphene make it have superior material for the foundation of future aerospace technologies. Due to its exceptional combination of tensile strength, excellent thermal conductivity, inherent chemical resistance, and unmatched lightness, the compound has a virtually unlimited potential to transform many facets of aerospace engineering. The recent advances in applying graphene in the aerospace context, including strengthening structural composites, the ability to provide highly beneficial thermal management capacities, and embedding highly sensitive health-controlling devices, point out that the possibilities of the implementation of graphene in improving performance and, consequently, safety are much more significant. The more recent advances in scalable graphene fabrication and more advanced functionalization processes have significantly reduced the impediments to integrating graphene into aerospace-grade materials. These developments play pivotal roles in the defeat of traditional limiting problems of consistency, homogeneity, and repeatability. As a result, graphene is finding application in upcoming aerospace generation systems. The future of how flight and space-mission technologies will be able to work with graphene is its role in the enhancement of the avionics, building an autonomous unmanned aerial vehicle, developing space-exploration material production, and enabling its use in the adoption of the electric and hybrid propulsion systems. Further, graphene contributes significantly to the design of aerospace at a sustainable level, which has a direct impact in terms of minimizing aircraft and spacecraft weight, which positively influences fuel efficiency as well as emission of greenhouse gases in a sector of the industry that is increasingly gearing towards environmentally friendly conduct. The likelihood that graphene-based composites can be recycled and implemented into models of a circular economy serves to add a proactive lifecycle to these substances as well. However, there are still a lot of challenges. Its expensive and complicated nature, which must include large-scale manufacturing of defect-free, large-area graphene capable of impressive space usage, has only a few applications remaining. Resistance and strength to differing and challenging working environments will have to be proved beyond a reasonable doubt. Moreover, the rigorous certification procedures that characterize the aerospace industry entail lengthy and thorough tests, which will most probably slow down the commercialization of graphene-based enhanced components. The research will also be extended in the future to develop multifunctional graphene composites that can be further optimized to achieve desired mechanical, thermal, and electrical functionalities or be engineered with enhanced performance to fit aerospace needs. The application of artificial intelligence and machine learning methods to materials design would unlock material discovery and optimization of graphene-based materials, which allows the specific geometrical configurations to be designed and adapted to a specific task. With the help of Industry 5.0 technology, the manufacturing process will be constantly redesigned, creating dynamic, intelligent, and sustainable production lines that will be able to produce in large quantities non-similar graphene materials. Overall, graphene has the disruptive capability to bring aerospace technologies to a new stage of greater strength, less weight, greater intelligence, and greenness. To actualize this opportunity, interdisciplinary interaction between materials scientists, aerospace engineers, chemists, manufacturing experts, and others will be required. With time, as the research and industrial potentials will become fully developed, the role of graphene in aerospace will become all the more important in the third phase of space and aerospace development, eventually leading to the creation of a more efficient, safer, and ecologically friendly aerospace.

REFERENCES

Baran, O., Karathanassis, I. K., Pickett, L. M., Manin, J., & Gavaises, M. (2025). High-speed optical imaging of cavitation and spray dynamics for sustainable aviation fuels. *Energy & Fuels*, *39*(24), 11912–11925. DOI: 10.1021/acs.energyfuels.5c01353

Elahi, S., Prasad, S., Haider, M. A., Gupta, M., & Ahmad, E. (2025). Congruity of butyl butyrate and limonene as potential blends to produce green and sustainable aviation fuels. *Journal of Chemical & Engineering Data*, *70*(7), 2713–2726. Advance online publication. DOI: 10.1021/acs.jced.5c00131

Eun, J., Kim, J., Kim, H., Kim, D., Choi, J., Kim, K., Yoo, C.-J., Jin, S., Kim, K., Lee, H., Kim, C. S., Lee, K.-Y., Yoo, J. S., Han, S.-J., Jeong, K., & Ha, J.-M. (2025). Production of high-carbon-number hydrocarbon bio-aviation fuels via catalytic hydrogenation of vanillin and non-catalytic condensation: A mechanistic study with DFT and experimental insights. *Green Chemistry*, *27*(24), 7147–7159. DOI: 10.1039/D5GC00281H

Falandys, K., Kurc, K., & Tutak, J. S. (2025). Application and empirical verification of the Archard model in the deburring process. *Materials (Basel)*, *18*(10), 2387. DOI: 10.3390/ma18102387 PMID: 40429125

Feng, L., Li, L., Yu, Y., Tian, G., Xiong, H., & Miao, J.. (2025). Road to carbon-neutral aviation: One-step conversion of CO_2 to sustainable aviation fuels. *Industrial & Engineering Chemistry Research*, *64*(6), 3118–3135. DOI: 10.1021/acs.iecr.4c04209

Gao, T., Chen, L., Wang, B., Liu, Y., Ma, Y., & Liang, Y. (2025). Prediction of the mechanical properties of graphene/copper nanocomposites using machine learning and molecular dynamics simulations. *Journal of Physics Condensed Matter*, *37*(23), 235701. DOI: 10.1088/1361-648X/add2c2 PMID: 40306312

Ghosh, S., Bhatia, J., Gupta, A., Pramanik, C., Pratihar, S., & Rautaray, D. (2025). Effects of the shear rate on dispersion characteristics of industrial-based functionalized/non functionalized graphene in an epoxy matrix. *Materials Advances*, *6*(11), 3523–3532. DOI: 10.1039/D5MA00173K

Haartsen, S., Wisman, J., Zandvliet, H., & Bampoulis, P. (2025). Humidity-controlled interactions and lifting of MoS_2 on SiO_2. *JPhys Materials*, *8*(2), 025002. DOI: 10.1088/2515-7639/adaaa1

Hidayat, R., Fajarwati, F., & Fadillah, G. (2025). Recent trends and bibliometric analysis in graphene quantum dots materials from natural bioresources: a brief review., 14, 23–40. https://doi.org/DOI: 10.4028/p-5zRtwr

Huang, R., Li, W., Zeng, Y., & Fang, C. (2025). Power optimization of an electric propulsion system based on variable speed rotor blades. *Journal of Physics: Conference Series, 2977*(1), 012084. DOI: 10.1088/1742-6596/2977/1/012084

Klingelhöfer, L., Kattelus, J., Verkama, E., Velasco, J., Iser, L., Rose, M., Karinen, R., & Puurunen, R. L. (2025). Catalytic hydrodenitrogenation of primary, secondary, and tertiary c12-alkyl amines over a platinum on zirconia catalyst. *Catalysis Science & Technology, 15*(4), 1259–1271. DOI: 10.1039/D4CY01099J

Kshirsagar, D., & Joshi, G. (2025). Surface modification of polymer graphene nanocomposites issues, routes, quantification, and emerging applications: A review. *ChemistrySelect, 10*(12), e202404581. Advance online publication. DOI: 10.1002/slct.202404581

Kumar, H., Kanti, P., Dishana, D., Shukla, P., Paramasivam, P., & Dabelo, L. (2025). Digital twin–driven design and testing of laser shock processed aluminum–graphene composites for spacecraft docking tribology. *PLoS One, 20*(5), e0324501. DOI: 10.1371/journal.pone.0324501 PMID: 40424314

Li, F., Ma, Y., Wei, X., Gao, L., Xing, E., Gong, J., Luo, Y., & Shu, X. (2025). Tandem oligomerization of ethylene for production of jet fuel over fer and mfi zeolites. *Industrial & Engineering Chemistry Research, 64*(11), 5923–5935. DOI: 10.1021/acs.iecr.4c04755

Maldonado-Romo, J., Jami, A., Sant, A., Montesinos, L., & Ponce, P. (2025). Biomaterials for aerospace applications: A sustainable technological analysis. *Journal of Physics: Conference Series, 2946*(1), 012006. DOI: 10.1088/1742-6596/2946/1/012006

Nie, Y., Li, X., Uddin, M., Liu, N., Yu, S., Keomeesay, P., Olalekan, O., Li, S., Chen, Y., Zhao, X., Ma, L., Zheng, Z., & Liu, S. (2025). Rational design of an efficient PT_3Cu/TiO_2 icosahedral catalyst for bio-aviation fuel production under mild conditions. *Green Chemistry, 27*(22), 6449–6463. DOI: 10.1039/D5GC01235J

Ozlek, M., Akbulut, M., & Burgaz, E. (2025). Exploring graphene's impact on graphite/pani matrix composites: High-pressure fabrication and enhanced thermal-electrical properties. *Nanoscale, 17*(1), 552–566. DOI: 10.1039/D4NR03171G PMID: 39575894

Parker, R., Kelly, M., Watson-Murphy, T., Ghaani, M., & Dooley, S. (2025). Predictive modeling of liquid density and surface tension for sustainable aviation fuels using nuclear magnetic resonance atom types. *Energy & Fuels, 39*(7), 3690–3702. DOI: 10.1021/acs.energyfuels.4c05601 PMID: 40007898

Purbayanto, M., Chandel, M., Makowski, M., Birowosuto, M., Montes-García, V., Prenger, K., Ciesielski, A., Naguib, M., & Jastrzębska, A. M. (2025). Unraveling the role of interfacial charge transfer on photoactivity and anomalous luminescence quenching of v4c3tx/protonated g-c3n4 heterostructures. *ACS Applied Materials & Interfaces*, *17*(11), 17454–17464. DOI: 10.1021/acsami.4c19729 PMID: 40062911

Raman, R., Sreenivasan, A., Kulkarni, N., Suresh, M., & Nedungadi, P. (2025). Analyzing the contributions of biofuels, biomass, and bioenergy to sustainable development goals. *iScience*, *28*(4), 112157. DOI: 10.1016/j.isci.2025.112157 PMID: 40230525

Roche, J., & Anand, V. (2025). Creating value since 2015: A review of the Journal of Creating Value. *Journal of Creating Value*, *11*(1), 58–72. Advance online publication. DOI: 10.1177/23949643241304874

Rong, C., Su, T., Chu, T., Zhu, M., Zhang, B., & Xuan, F. (2025). Tradeoff between mechanical strength and electrical conductivity of MXene films by nacre-inspired subtractive manufacturing. *Small*, *21*(12), 2411329. Advance online publication. DOI: 10.1002/smll.202411329 PMID: 39937436

Sahoo, P., Das, A., Das, S., & Mallik, A. (2025). Role of electrochemically exfoliated graphene in property enhancement of electrodeposited copper nanocomposites. *ChemistrySelect*, *10*(13), e202405666. Advance online publication. DOI: 10.1002/slct.202405666

Sales, A., Pereira, V., & Dias, A. (2025). Advances in nanomaterials for radiation protection in the aerospace industry: A systematic review. *Nanotechnology*, *36*(10), 102002. DOI: 10.1088/1361-6528/ada38f PMID: 39729645

Shen, L., Jiang, X., Ni, L., Luo, Y., Liang, M., Yang, Y., Zhou, S., & Zou, H. (2025). Improved infrared stealth and electromagnetic interference shielding performance of polyimide-derived carbon aerogels by constructing heterogeneous nanopores. *Polymer Engineering and Science*, *65*(2), 846–857. DOI: 10.1002/pen.27049

Singh, A., Lalotra, N., Shah, S. S., Rather, M. D., López-Maldonado, E. A., Pathania, K., Abaszade, R. G., Stetsenko, M., & Arya, S. (2025). A comprehensive review on the synthesis and properties of MXenes for supercapacitor applications. *ACS Applied Energy Materials*, *8*(8), 4884–4914. DOI: 10.1021/acsaem.4c03194

Sun, H., Hwang, U., Kim, S., Sung, J., Kim, T., Suhr, J., Park, I.-K., & Nam, J.-D. (2025). Reversible thixotropic rheological properties of graphene-incorporated epoxy inks for self-standing 3d printing. *ACS Macro Letters*, *14*(2), 135–141. DOI: 10.1021/acsmacrolett.4c00765 PMID: 39801069

Suomalainen, E. (2025). Growth over climate? Aviation and the Paris Agreement. https://doi.org/DOI: 10.21203/rs.3.rs-5986385/v1

Tang, H., Li, M., Chen, D., Tahir, M. H., Yin, L., Qian, K., Feng, Y., & Hu, Y. (2025). One-step production of isoalkane-rich sustainable aviation fuel fractions via catalytic pyrolysis of low-density polyethylene over y-type zeolite. *ACS Sustainable Chemistry & Engineering, 13*(23), 8761–8772. DOI: 10.1021/acssuschemeng.5c02722

Trai, V., & Sobhy, M. (2025). Vibration and transient response analysis of bi-directional functionally graded graphene platelets reinforced plates resting on a Kerr foundation using a refined finite-element procedure. *Proceedings - Royal Society. Mathematical, Physical and Engineering Sciences, 481*(2314), 20250297. Advance online publication. DOI: 10.1098/rspa.2025.0297

Uniyal, P., Gaur, P., Yadav, J., Bhalla, N., Khan, T., Junaedi, H., & Sebaey, T. A. (2025). A comprehensive review on the role of nanosilica as a toughening agent for enhanced epoxy composites for aerospace applications. *ACS Omega, 10*(16), 15810–15839. DOI: 10.1021/acsomega.4c10073 PMID: 40321578

Vantrepotte, Q., Dubois, E., Camachon, C., & Causse, M. (2025). Effect of pilot expertise on visual scanning and precision of flight maneuvers in military aviation. *Aviation Psychology and Applied Human Factors, 15*(1), 16–26. DOI: 10.1027/2192-0923/a000287

Yang, Y., Yuan, H., Liu, M., Cheng, S., Li, W., Liang, F., Zheng, K., Liu, L., Yang, F., Liu, R., Su, Q., Qi, Y., & Liu, Z. (2025). Premelted-substrate-promoted selective etching strategy realizing cvd growth of high-quality graphene on dielectric substrates. *ACS Applied Materials & Interfaces, 17*(4), 6825–6834. DOI: 10.1021/acsami.4c20313 PMID: 39809474

Yang, Y., Yuan, H., Liu, M., Cheng, S., Li, W., Liang, F., Zheng, K., Liu, L., Yang, F., Liu, R., Su, Q., Qi, Y., & Liu, Z. (2025). Premelted-substrate-promoted selective etching strategy realizing cvd growth of high-quality graphene on dielectric substrates. *ACS Applied Materials & Interfaces, 17*(4), 6825–6834. DOI: 10.1021/acsami.4c20313 PMID: 39809474

Compilation of References

Abaszade, R. G. (2022). Synthesis and analysis of flakes graphene oxide. *Journal of Optoelectronic and Biomedical Materials, 14*(3), 107–114. DOI: 10.15251/JOBM.2022.143.107

Abaszade, R. G., Babanli, M. B., Kotsyubynsky, V. O., Mammadov, A. G., Gür, E., Kapush, O. A., Stetsenko, M. O., & Zapukhlyak, R. I. (2023). Influence of gadolinium doping on structural properties of carbon nanotube. *Physics and Chemistry of Solid State, 24*(1), 53–158. DOI: 10.15330/pcss.24.1.153-158

Abaszade, R. G., Kapush, O. A., Mamedova, S. A., Nabiyev, A. M., Melikova, S. Z., & Budzulyak, S. I. (2020). Gadolinium doping influence on the properties of carbon nanotubes. *Physics and Chemistry of Solid State, 21*(3), 404–408. DOI: 10.15330/pcss.21.3.404-408

Abaszade, R. G., Kapush, O. A., & Nabiyev, A. M. (2020). Properties of carbon nanotubes doped with gadolinium. *Journal of Optoelectronic and Biomedical Materials, 12*(3), 61–65. https://www.chalcogen.ro/61_AbaszadeRG.pdf. DOI: 10.15251/JOBM.2020.123.61

Abaszade, R. G., Mammadov, A. G., Bayramov, I. Y., Khanmamadova, E. A., Kotsyubynsky, V. O., Gur, E. Y., & Kapush, O. A. (2022). Modeling of voltage-ampere characteristic structures on the basis of graphene oxide/sulfur compounds. *Technical and Physical Problems of Engineering, 14*(2), 302–306.

Abaszade, R. G., Mammadov, A. G., Bayramov, I. Y., Khanmamadova, E. A., Kotsyubynsky, V. O., Gur, E. Y., & Kapush, O. A. (2022). Structural and electrical properties of sulfur-doped graphene oxide/graphite oxide composite. *Physics and Chemistry of Solid State, 25*(2), 256–260. DOI: 10.15330/pcss.23.2.256-260

Abaszade, R. G., Mammadov, A. G., Khanmamedova, E. A., Aliyev, F. G., Kotsyubynsky, V. O., Gür, E., Soltabayev, B. D., Margitych, T. O., & Stetsenko, M. O. (2024). Photoconductivity of functionalized carbon nanotubes. *Digest Journal of Nanomaterials and Biostructures*, *19*(2), 837–843. DOI: 10.15251/DJNB.2024.192.837

Abaszade, R. G., Mammadova, A. G., Khanmamedova, E. A., Bayramov, İ. Y., Namazov, R. A., Popal, Kh. M., Melikova, S. Z., Qasımov, R. C., Bayramov, M. A., & Babayeva, N. İ. (2023). Electron paramagnetic resonance study of gadolinium doped graphene oxide. *Journal of Ovonic Research*, *19*(2), 259–263. DOI: 10.15251/JOR.2023.193.259

Abdi, B., Tarhini, A., Baniasadi, H., & Tehrani-Bagha, A. R. (2024). Developing Graphene-based Conductive Textiles Using Different Coating Methods. *Advanced Materials Technologies*, *9*(4), 2301492. DOI: 10.1002/admt.202301492

Abid, , Sehrawat, P., Islam, S. S., Mishra, P., & Ahmad, S. (2018). Reduced graphene oxide (rGO) based wideband optical sensor and the role of Temperature, Defect States and Quantum Efficiency. *Scientific Reports*, *8*(1), 3537. DOI: 10.1038/s41598-018-21686-2

Aboraia, A. M.. (2024). Exploration of the structural rGO thin films and their optical characteristics for optoelectronic device applications. *Journal of Optics*. Advance online publication. DOI: 10.1007/s12596-024-02092-6

Administration, F. A. (2023). *Pilot's Handbook of Aeronautical Knowledge (2025): FAA-H-8083-25C*. Simon and Schuster.

Afroj, S., Karim, N., Wang, Z., Tan, S., He, P., Holwill, M., Ghazaryan, D., Fernando, A., & Novoselov, K. S. (2019). Engineering graphene flakes for wearable textile sensors via highly scalable and ultrafast yarn dyeing technique. *ACS Nano*, *13*(4), 3847–3857. DOI: 10.1021/acsnano.9b00319 PMID: 30816692

Ahmad, M. S., & Nishina, Y. (2020). Graphene-based carbocatalysts for carbon–carbon bond formation. *Nanoscale*, *12*(23), 12210–12227. DOI: 10.1039/D0NR02984J PMID: 32510079

Ahmed, S., Hazarika, S., Tashkandi, M., Armaghani, T., & Kolsi, L. (2025). Enhanced heat diffusion in triangular wavy enclosures using mos_2-water nanofluidics: A comsol simulation study. Zamm - Journal of Applied Mathematics and Mechanics. *Zeitschrift für Angewandte Mathematik und Mechanik*, *105*(5). Advance online publication. DOI: 10.1002/zamm.70053

Akbolat, M. Ç., Katnam, K. B., Soutis, C., Potluri, P., Sprenger, S., & Taylor, J. (2022). On mode-I and mode-II interlaminar crack migration and R-curves in carbon/epoxy laminates with hybrid toughening via core-shell rubber particles and thermoplastic micro-fiber veils. *Composites. Part B, Engineering*, *238*, 109900. DOI: 10.1016/j.compositesb.2022.109900

Akbolat, M. Ç., Wang, S., Katnam, K. B., Potluri, P., & Soutis, C. (2025). Low-velocity impact response of carbon/epoxy laminates with interlaminar hybrid toughening via core–shell-rubber particles and non-woven thermoplastic fiber veils. *Composites. Part A, Applied Science and Manufacturing*, *195*, 108944. DOI: 10.1016/j.compositesa.2025.108944

Al Faruque, M. A., Syduzzaman, M., Sarkar, J., Bilisik, K., & Naebe, M. (2021). A review on the production methods and applications of graphene-based materials. *Nanomaterials (Basel, Switzerland)*, *11*(9), 2414. DOI: 10.3390/nano11092414 PMID: 34578730

Alam, S. N., Sharma, N., & Kumar, L. (2017). Synthesis of graphene oxide (GO) by modified hummers method and its thermal reduction to obtain reduced graphene oxide (rGO). *Graphene*, *6*(01), 1. DOI: 10.4236/graphene.2017.61001

Alemour, B., Yaacob, M. H., Lim, H. N., & Hassan, M. R. (2018). Review of Electrical Properties of Graphene Conductive Composites. *International Journal of Nanoelectronics & Materials*, *11*(4).

Al-Gaashani, R., Najjar, A., Zakaria, Y., Mansour, S., & Atieh, M. A. (2019). XPS and structural studies of high-quality graphene oxide and reduced graphene oxide prepared by different chemical oxidation methods. *Ceramics International*, *45*(11), 14439–14448. DOI: 10.1016/j.ceramint.2019.04.165

Alkan, C., Sarı, A., & Karaipekli, A. (2011). Preparation, thermal properties and thermal reliability of microencapsulated n-eicosane as novel phase change material for thermal energy storage. *Energy Conversion and Management*, *52*(1), 687–692. DOI: 10.1016/j.enconman.2010.07.047

Allan, J. E., Ferreri, A., Grande, S., Giannantonio, R., & Matteucci, F. (2019). *Technology transfer in nanotechnology*. Publications Office of the European Union.

Allan, J., Belz, S., Hoeveler, A., Hugas, M., Okuda, H., Patri, A., Rauscher, H., Silva, P., Slikker, W., & Tong, W. (2021). Regulatory landscape of nanotechnology and nanoplastics from a global perspective. *Regulatory Toxicology and Pharmacology*, *122*, 104885. DOI: 10.1016/j.yrtph.2021.104885 PMID: 33617940

Amiet, A., Edwards, D., Ibrahim, M., Mart, P., McAdam, G., St John, N., & Truong, V. (2010). Exploitation of Smart Materials and Sensors as Disruptive Technologies.

Anadkat, D., Pandya, A., Jaiswal, A., Dungani, S., & Sanchela, A. V. (2024). Experimental comparison between graphene and reduced graphene oxide along with significant conversion of rGO from n-type to p-type. *Journal of Materials Science Materials in Electronics*, *35*(12), 821. DOI: 10.1007/s10854-024-12591-8

Anjali, C., & Kizhakayil, R. N. (2024). Graphene-The Material of the Era. In *Graphene-Based Photocatalysts for Hydrogen Production and Environmental Remediation* (pp. 3–47). Springer Nature Switzerland. DOI: 10.1007/978-3-031-68464-7_1

Ansari, S., & Muralidharan, M. (2015). Electronic applications of polyurethane and its composites. In *Flexible and stretchable electronic composites* (pp. 87–134). Springer.

Antunes, M., Gedler, G., Abbasi, H., & Velasco, J. I. (2016). Graphene nanoplatelets as a multifunctional filler for polymer foams. *Materials Today: Proceedings*, *3*, S233–S239. DOI: 10.1016/j.matpr.2016.02.039

Anwar, Z., Kausar, A., Rafique, I., & Muhammad, B. (2016). Advances in Epoxy/Graphene Nanoplatelet Composite with Enhanced Physical Properties: A Review. *Polymer-Plastics Technology and Engineering*, *55*(6), 643–662. DOI: 10.1080/03602559.2015.1098695

Araby, S., Meng, Q., Zhang, L., Kang, H., Majewski, P., Tang, Y., & Ma, J. (2014). Electrically and thermally conductive elastomer/graphene nanocomposites by solution mixing. *Polymer*, *55*(1), 201–210. DOI: 10.1016/j.polymer.2013.11.032

Aradhana, R., Mohanty, S., & Nayak, S. K. (2018). Comparison of mechanical, electrical, and thermal properties in graphene oxide and reduced graphene oxide filled epoxy nanocomposite adhesives. *Polymer*, *141*, 109–123. DOI: 10.1016/j.polymer.2018.03.005

Archana, U., Sharma, S., & Singh, S. K., R., S., P., S., & Babu, T. H. (2023). Analysis of concrete cracks and fatigue in smart cities using YOLOv3. *2023 IEEE International Conference on Research Methodologies in Knowledge Management, Artificial Intelligence and Telecommunication Engineering (RMKMATE)*, 1–6. https://doi.org/DOI: 10.1109/RMKMATE59243.2023.10368774

Ariaeenejad, S., Motamedi, E., & Hosseini Salekdeh, G. (2021). Application of the immobilized enzyme on magnetic graphene oxide nano-carrier as a versatile bi-functional tool for efficient removal of dye from water. *Bioresource Technology*, *319*, 124228. DOI: 10.1016/j.biortech.2020.124228 PMID: 33254455

Arkhipova, E. A., Strokova, N. E., Tambovtseva, Y. A., Ivanov, A. S., Chernyak, S. A., Maslakov, K. I., Egorova, T. B., Savilov, S. V., & Lunin, V. V. (2020). Thermophysical study of graphene nanoflakes by differential scanning calorimetry. *Journal of Thermal Analysis and Calorimetry, 140*(6), 2641–2648. DOI: 10.1007/s10973-019-09040-8

Arvidsson, R. (2017). Review of environmental life cycle assessment studies of graphene production. *Advanced Materials Letters, 8*(3), 187–195. DOI: 10.5185/amlett.2017.1413

Ashwath, P., Venkatraman, M., Patel, A., Xavior, M. A., & Batako, A. (2024). Innovation in sustainable composite research: Investigating graphene-reinforced MMCs for liquid hydrogen storage tanks in aerospace and space exploration. *Journal of Materials Research and Technology, 33*, 4313–4331. DOI: 10.1016/j.jmrt.2024.10.070

Asmat, S., Husain, Q., Shoeb, M., & Mobin, M. (2020). Tailoring a robust nanozyme formulation based on surfactant stabilized lipase immobilized onto newly fabricated magnetic silica anchored graphene nanocomposite: Aggrandized stability and application. *Materials Science and Engineering C, 112*, 110883. DOI: 10.1016/j.msec.2020.110883 PMID: 32409040

Asp, L. E., Bouton, K., Carlstedt, D., Duan, S., Harnden, R., Johannisson, W., Johansen, M., Johansson, M. K. G., Lindbergh, G., Liu, F., Peuvot, K., Schneider, L. M., Xu, J., & Zenkert, D. (2021). A Structural Battery and its Multifunctional Performance. *Advanced Energy and Sustainability Research, 2*(3), 2000093. DOI: 10.1002/aesr.202000093

Asp, L. E., Johansson, M., Lindbergh, G., Xu, J., & Zenkert, D. (2019). Structural battery composites: A review. *Functional Composites and Structures, 1*(4), 042001. DOI: 10.1088/2631-6331/ab5571

ASTM International. (n.d.). Standards & publications. https://www.astm.org/standards-and-solutions/standards-publications#standard-products

Asyraf, M., Ilyas, R., Sapuan, S., Harussani, M., Hariz, H., Aiman, J.,…Norkhairunnisa, M. (2022). Advanced composite in aerospace applications: opportunities, challenges, and future perspective. Advanced Composites in Aerospace Engineering Applications, 471–498.

Awuzie, C. I. (2017). Conducting polymers. *Materials Today: Proceedings, 4*(4), 5721–5726. DOI: 10.1016/j.matpr.2017.06.036

Ayyanar, C., & Pramanik, S. (2023). Effect of graphene on self-healing performance of hydroxyapatite/polydimethysiloxane composites. *Materials and Manufacturing Processes*, *38*(9), 1068–1080. DOI: 10.1080/10426914.2022.2146714

Bagheri, R., Marouf, B. T., & Pearson, R. A. (2009). Rubber-Toughened Epoxies: A Critical Review. *Polymer Reviews (Philadelphia, Pa.)*, *49*(3), 201–225. DOI: 10.1080/15583720903048227

Bagri, A., Mattevi, C., Acik, M., Chabal, Y. J., Chhowalla, M., & Shenoy, V. B. (2010). Structural evolution during the reduction of chemically derived graphene oxide. *Nature Chemistry*, *2*(7), 581–587. DOI: 10.1038/nchem.686

Bagus, P. S., Ilton, E. S., & Nelin, C. J. (2013). The interpretation of XPS spectra: Insights into materials properties. *Surface Science Reports*, *68*(2), 273–304. DOI: 10.1016/j.surfrep.2013.03.001

Balandin, A. A. (2011). Thermal properties of graphene and nanostructured carbon materials. *Nature Materials*, *10*(8), 569–581. DOI: 10.1038/nmat3064 PMID: 21778997

Balandin, A. A., Ghosh, S., Bao, W., Calizo, I., Teweldebrhan, D., Miao, F., & Lau, C. N. (2008). Superior thermal conductivity of single-layer graphene. *Nano Letters*, *8*(3), 902–907. DOI: 10.1021/nl0731872 PMID: 18284217

Bannov, A. G., Popov, M. V., & Kurmashov, P. B. (2020). Thermal analysis of carbon nanomaterials: Advantages and problems of interpretation. *Journal of Thermal Analysis and Calorimetry*, *142*(1), 349–370. DOI: 10.1007/s10973-020-09647-2

Baran, O., Karathanassis, I. K., Pickett, L. M., Manin, J., & Gavaises, M. (2025). High-speed optical imaging of cavitation and spray dynamics for sustainable aviation fuels. *Energy & Fuels*, *39*(24), 11912–11925. DOI: 10.1021/acs.energyfuels.5c01353

Barkan, T., Walthall, R., Dixit, S., David, A., Webb, P., & Fletcher, S. (2023). Aerospace Advancements.

Barroeta Robles, J., Dubé, M., Hubert, P., & Yousefpour, A. (2022). Repair of thermoplastic composites: An overview. *Advanced Manufacturing: Polymer and Composites Science*, *8*(2), 68–96. DOI: 10.1080/20550340.2022.2057137

Batool, F., Muhammad, S., Muazzam, R., Waqas, M., Ullah, Z., Roy, S., Wang, K., & Guo, B. (2025). Advancements in Graphene-Based Composites: A review of the emerging applications in healthcare. *Smart Materials in Medicine*, *6*(1), 120–138. Advance online publication. DOI: 10.1016/j.smaim.2025.01.001

Bawa'neh, H., Lababneh, B., Malkawi, A., & Bozeya, A. (2025). Investigation of the mechanical properties of an aluminum (al 6061-t6)/graphene/bentonite hybrid nanocomposite experimentally and through finite element analysis study. *Materials Advances*, *6*(3), 1031–1041. DOI: 10.1039/D4MA00890A

Behera, A., Swain, B., & Sahoo, D. K. (2020). Fiber-reinforced ceramic matrix nanocomposites. *Fiber-Reinforced Nanocomposites: Fundamentals and Applications*, 359–368. DOI: 10.1016/B978-0-12-819904-6.00016-5

Bello, A. (2023). Review of Composite Structures in Aeronautic Applications Part.

Bhanuprakash, L., Varghese, S., & Singh, S. K. (2022). Glass Fibre Reinforced Epoxy Composites Modified with Graphene Nanofillers: Electrical Characterization. *Journal of Nanomaterials*, *2022*(1), 4611251. Advance online publication. DOI: 10.1155/2022/4611251

Bhattacharya, S., Agarwal, A. K., Prakash, O., & Singh, S. (Eds.). (2019). *Sensors for Automotive and Aerospace Applications*. Springer Singapore., DOI: 10.1007/978-981-13-3290-6

Bilisik, K., & Akter, M. (2022). Graphene nanocomposites: A review on processes, properties, and applications. *Journal of Industrial Textiles*, *51*(3, suppl), 3718S–3766S. DOI: 10.1177/15280837211024252

Bilisik, K., Erdogan, G., & Sapancı, E. (2018). Flexural behavior of 3D para-aramid/phenolic/nano (MWCNT) composites. *RSC Advances*, *8*(13), 7213–7224. DOI: 10.1039/C7RA13437A PMID: 35540322

Bilisik, K., & Sapanci, E. (2018). Experimental determination of fracture toughness properties of nanostitched and nanoprepreg carbon/epoxy composites. *Engineering Fracture Mechanics*, *189*, 293–306. DOI: 10.1016/j.engfracmech.2017.11.033

Blaikie, A. (2020). *Graphene Electromechanical Resonators and Their Use in Thermal Detectors*. University of Oregon.

Bleeker, E. A., Swart, E., Braakhuis, H., Cruz, M. L. F., Friedrichs, S., Gosens, I., Herzberg, F., Jensen, K. A., von der Kammer, F., Kettelarij, J. A., & Navas, J. M. (2023). Towards harmonisation of testing of nanomaterials for EU regulatory requirements on chemical safety–A proposal for further actions. *Regulatory Toxicology and Pharmacology*, *139*, 105360. DOI: 10.1016/j.yrtph.2023.105360 PMID: 36804527

Boehm, F. J. (2013). Nanomedical device and systems design in space applications. Nanomedical device and systems design: challenges, possibilities, visions. CRC Press, Taylor and Francis Group, Boca Raton, FL, 585–652.

Boehm, H. P. (2010). Graphene—How a laboratory curiosity suddenly became extremely interesting. *Angewandte Chemie International Edition*, *49*(49), 9332–9335. DOI: 10.1002/anie.201004096 PMID: 20865715

Bolotin, K. I., Sikes, K. J., Jiang, Z., Klima, M., Fudenberg, G., Hone, J., Kim, P., & Stormer, H. L. (2008). Ultrahigh electron mobility in suspended graphene. *Solid State Communications*, *146*(9-10), 351–355. DOI: 10.1016/j.ssc.2008.02.024

Bonanni, A., & Pumera, M. (2013). High-resolution impedance spectroscopy for graphene characterization. *Electrochemistry Communications*, *26*, 52–54. DOI: 10.1016/j.elecom.2012.10.013

Bonora, P. L., Deflorian, F., & Fedrizzi, L. (1996). Electrochemical impedance spectroscopy as a tool for investigating underpaint corrosion. *Electrochimica Acta*, *41*(7–8), 1073–1082. DOI: 10.1016/0013-4686(95)00440-8

Brodova, I., Yolshina, L., Razorenov, S., Rasposienko, D., Petrova, A., Shirinkina, I., Shorokhov, E., Muradymov, R., Garkushin, G., & Savinykh, A. (2022). Effect of Grain Size on the Properties of Aluminum Matrix Composites with Graphene. *Metals 2022, Vol. 12, Page 1054, 12*(6), 1054. DOI: 10.3390/met12061054

Bunch, J. S., Van Der Zande, A. M., Verbridge, S. S., Frank, I. W., Tanenbaum, D. M., Parpia, J. M.,…McEuen, P. L. (2007). Electromechanical resonators from graphene sheets. science, 315(5811), 490-493.

Camps, A. J., Ruiz-de-Azua, J. A., Perez-Portero, A., Fernandez, L., & Munoz-Martin, J. F. (2023). Evolving capabilities and limitations of future CubeSat missions. In *Next Generation CubeSats and SmallSats* (pp. 79–120). Elsevier. DOI: 10.1016/B978-0-12-824541-5.00027-3

Cao, Y., et al. "New structural insights into densely assembled reduced graphene oxide membranes." Advanced Functional Materials 32.42 (2022): 2201535. https://doi.org/.DOI: 10.1002/adfm.202201535

Cao, H., & Liang, Y. (2020). The microstructures and mechanical properties of graphene-reinforced titanium matrix composites. *Journal of Alloys and Compounds*, *812*, 152057. DOI: 10.1016/j.jallcom.2019.152057

Cao, L., Jones, A. K., Sikka, V. K., Wu, J., & Gao, D. (2009). Anti-icing superhydrophobic coatings. *Langmuir*, *25*(21), 12444–12448. DOI: 10.1021/la902882b PMID: 19799464

Cao, X., Zhang, J., Chen, S., Varley, R. J., & Pan, K. (2020). 1D/2D nanomaterials synergistic, compressible, and response rapidly 3D graphene aerogel for piezoresistive sensor. *Advanced Functional Materials*, *30*(35), 2003618. DOI: 10.1002/adfm.202003618

Caprino, G., Lopresto, V., & Santoro, D. (2007). Ballistic impact behaviour of stitched graphite/epoxy laminates. *Composites Science and Technology*, *67*(3–4), 325–335. DOI: 10.1016/j.compscitech.2006.04.015

Carrio, A., Sampedro, C., Rodríguez-Ramos, A., & Campoy, P. (2017). A review of deep learning methods and applications for unmanned aerial vehicles. *Journal of Sensors*, *2017*, 1–13. DOI: 10.1155/2017/3296874

Carvalho, A. F., Kulyk, B., Fernandes, A. J., Fortunato, E., & Costa, F. M. (2022). A review of the applications of graphene in mechanical transduction. *Advanced Materials*, *34*(8), 2101326. DOI: 10.1002/adma.202101326 PMID: 34288155

Casero, E., Parra-Alfambra, A. M., Petit-Domínguez, M. D., Pariente, F., Lorenzo, E., & Alonso, C. (2012). Differentiation between graphene oxide and reduced graphene by electrochemical impedance spectroscopy (EIS). *Electrochemistry Communications*, *20*, 63–66. DOI: 10.1016/j.elecom.2012.04.002

Cataldi, P., Athanassiou, A., & Bayer, I. S. (2018). Graphene Nanoplatelets-Based Advanced Materials and Recent Progress in Sustainable Applications. *Applied Sciences 2018, Vol. 8, Page 1438*, *8*(9), 1438. DOI: 10.3390/app8091438

Ceniceros-Reyes, M. A., Marín-Hernández, K. S., Sierra, U., Saucedo-Salazar, E. M., Mendoza-Resendez, R., Luna, C., Hernández-Belmares, P. J., Rodríguez-Fernández, O. S., Fernández-Tavizón, S., Hernández-Hernández, E., & Barriga-Castro, E. D. (2022). Reduction of graphene oxide by in-situ heating experiments in the transmission electron microscope. *Surfaces and Interfaces*, *35*, 102448. DOI: 10.1016/j.surfin.2022.102448

Chang, W., Rose, L. R. F., Sha, Z., Huang, F., Kinloch, A. J., & Wang, C. H. (2024). Multiscale modeling of nanoparticle toughening in epoxy: Effects of particle-matrix interface, particle size, and volume fraction. *Composites Science and Technology*, *256*, 110788. DOI: 10.1016/j.compscitech.2024.110788

Chavalekvirat, P., Hirunpinyopas, W., Deshsorn, K., Jitapunkul, K., & Iamprasertkun, P. (2024). Liquid phase exfoliation of 2D materials and its electrochemical applications in the data-driven future. *Precision Chemistry*, *2*(7), 300–329. DOI: 10.1021/prechem.3c00119 PMID: 39473901

Chemello, G., Despotelis, K., Paton, K., Clifford, C., Pollard, A., Radnik, J., & Hodoroaba, V.-D. (2025). Approaches for the measurement of lateral dimensions of graphene oxide flakes using scanning electron microscopy. *Nano Express*, *6*(1), 015006. DOI: 10.1088/2632-959X/adae28

Cheng, H., Ye, M., Zhao, F., Hu, C., Zhao, Y., Liang, Y., Chen, N., Chen, S., Jiang, L., & Qu, L. (2016). A general and extremely simple remote approach toward graphene bulks with in situ multifunctionalization. *Advanced Materials*, *28*(17), 3305–3312. DOI: 10.1002/adma.201505431

Chen, H., Ginzburg, V. V., Yang, J., Yang, Y., Liu, W., Huang, Y., Du, L., & Chen, B. (2016). Thermal conductivity of polymer-based composites: Fundamentals and applications. *Progress in Polymer Science*, *59*, 41–85. DOI: 10.1016/j.progpolymsci.2016.03.001

Chen, H., Mi, G., Sun, Y., & Li, P. (2024). Unique grain refinement mechanism of graphene oxide reinforced high-temperature titanium alloy matrix composite. *Materials Today. Communications*, *41*, 110803. DOI: 10.1016/j.mtcomm.2024.110803

Chen, J., Duan, M., & Chen, G. (2012). Continuous mechanical exfoliation of graphene sheets via three-roll mill. *Journal of Materials Chemistry*, *22*(37), 19625–19628. DOI: 10.1039/c2jm33740a

Chen, L., Yang, Q., Yang, X., Liu, Z., & Song, Q. (2022). Numerical investigation of heat transfer performance of graphene-doped anti-/deicing component. *Thermal Science and Engineering Progress*, *28*, 101098. DOI: 10.1016/j.tsep.2021.101098

Chen, L., Zhang, Y., & Wu, Q. (2017). Effect of Graphene Coating on the Heat Transfer Performance of a Composite Anti-/Deicing Component. *Coatings*, *7*(10), 158. DOI: 10.3390/coatings7100158

Chen, S., Brown, L., Levendorf, M., Cai, W., Ju, S.-Y., Edgeworth, J., & Piner, R. D. (2011). Oxidation resistance of graphene-coated Cu and Cu/Ni alloy. *ACS Nano*, *5*(2), 1321–1327. DOI: 10.1021/nn103028d PMID: 21275384

Chen, S., Jiang, K., Lou, Z., Chen, D., & Shen, G. (2018). Recent developments in graphene-based tactile sensors and E-skins. *Advanced Materials Technologies*, *3*(2), 1700248. DOI: 10.1002/admt.201700248

Chen, X., Wang, X., & Fang, D. (2020). A review on C1s XPS-spectra for some kinds of carbon materials. *Fullerenes, Nanotubes, and Carbon Nanostructures*, *28*(12), 1048–1058. DOI: 10.1080/1536383X.2020.1794851

Chen, Y., Prasad, V., Yasar, M., Murphy, N., & Ivankovic, A. (2024). Enhancing interfacial performance and fracture toughness of carbon fibre reinforced thermoplastic composites. *Composites. Part A, Applied Science and Manufacturing*, *187*, 108434. DOI: 10.1016/j.compositesa.2024.108434

Choucair, M., Thordarson, P., & Stride, J. A. (2009). Gram-scale production of graphene based on solvothermal synthesis and sonication. *Nature Nanotechnology*, *4*(1), 30–33. DOI: 10.1038/nnano.2008.365 PMID: 19119279

Christudoss, A., Sah, K., Vikram, R., Giri, S., Viswanathan, D., Dimkpa, C., & Mukherjee, A. (2025). Tailoring the synthesis route of reduced graphene oxide and its toxicological effects on allium cepa. *ACS Omega*, *10*(20), 20771–20783. DOI: 10.1021/acsomega.5c01919

Chugh, V., Basu, A., Kaushik, A., Bhansali, S., & Basu, A. K. (2024). Employing nano-enabled artificial intelligence (AI)-based smart technologies for prediction, screening, and detection of cancer. *Nanoscale*, *16*(11), 5458–5486. DOI: 10.1039/D3NR05648A PMID: 38391246

Ciesielski, A., & Samorì, P. (2014). Graphene via sonication assisted liquid-phase exfoliation. *Chemical Society Reviews*, *43*(1), 381–398. DOI: 10.1039/C3CS60217F PMID: 24002478

Clingerman, M. L. (2001). *Development and modelling of electrically conductive composite materials*. Michigan Technological University.

Composite Materials and Hydrogen Tanks | Innovate Beyond Limits. Retrieved June 27, 2025, from https://www.rayvatek.com/composite-materials-and-hydrogen-tanks/

Compton, O. C., & Nguyen, S. T. (2010a). Graphene oxide, highly reduced graphene oxide, and graphene: Versatile building blocks for carbon-based materials. *Small*, *6*(6), 711–723. DOI: 10.1002/smll.200901934

Connolly, M., Molés, G., Carniel, F. C., Tretiach, M., Caorsi, G., Flahaut, E., Soula, B., Pinelli, E., Gauthier, L., Mouchet, F., & Navas, J. M. (2023). Applicability of OECD TG 201, 202, 203 for the aquatic toxicity testing and assessment of 2D Graphene material nanoforms to meet regulatory needs. *NanoImpact*, *29*, 100447. DOI: 10.1016/j.impact.2022.100447 PMID: 36563784

Cozzolino, F., Marra, F., Fortunato, M., Bellagamba, I., Pesce, N., Tamburrano, A., & Sarto, M. S. (2022). New Sensing and Radar Absorbing Laminate Combining Structural Damage Detection and Electromagnetic Wave Absorption Properties. *Sensors 2022, Vol. 22, Page 8470*, *22*(21), 8470. DOI: 10.3390/s22218470

Crown, E. M., & Capjack, L. (2005). Flight suits for military aviators. Textiles for protection, 678-698.

Cui, E., Zhao, J., & Wang, X. (2019). Determination of microstructure and mechanical properties of graphene reinforced Al2O3-Ti(C, N) ceramic composites. *Ceramics International*, *45*(16), 20593–20599. DOI: 10.1016/j.ceramint.2019.07.041

Cui, K., Mao, H., Zhang, Y., Wang, J., Wang, H., Tan, T., & Fu, T. (2022). Microstructure, mechanical properties, and reinforcement mechanism of carbide toughened ZrC-based ultra-high temperature ceramics: A review. *Composite Interfaces*, *29*(7), 729–748. DOI: 10.1080/09276440.2021.2012409

Cui, T., Mukherjee, S., Cao, C., Sudeep, P. M., Tam, J., Ajayan, P. M., Singh, C. V., Sun, Y., & Filleter, T. (2018). Effect of lattice stacking orientation and local thickness variation on the mechanical behavior of few layer graphene oxide. *Carbon*, *136*, 168–175. DOI: 10.1016/j.carbon.2018.04.074

Cui, X., Zhang, N., Huang, M., Gao, G., Liu, S., & Liu, C. (2023). Polyurethane-based nanocomposite film with thermal deicing capability and anti-erosion for wind turbine blades protection in extreme environments. *Journal of Materials Chemistry. A, Materials for Energy and Sustainability*, *11*(44), 23844–23853. DOI: 10.1039/D3TA05595G

D'Angelo, D., Bongiorno, C., Amato, M., Deretzis, I., La Magna, A., Fazio, E., & Scalese, S. (2017). Oxygen functionalities evolution in thermally treated graphene oxide featured by EELS and DFT calculations. *The Journal of Physical Chemistry. C, Nanomaterials and Interfaces*, *121*(9), 5408–5414. DOI: 10.1021/acs.jpcc.7b00239

Darrah, P. D. (2022). Standard Approach to Identifying and Defining Functions for Systems Development and Safety Assessments. SAE International Journal of Advances and Current Practices in Mobility, 4(2022-01-0008), 989–993.

Darwish, M. A., Abd-Elaziem, W., Elsheikh, A., & Zayed, A. A. (2024). Advancements in nanomaterials for nanosensors: A comprehensive review. *Nanoscale Advances*, *6*(16), 4015–4046. DOI: 10.1039/D4NA00214H PMID: 39114135

Das, G. S., Tripathi, V. K., Dwivedi, J., Jangir, L. K., & Tripathi, K. M. (2024). Nanocarbon-based sensors for the structural health monitoring of smart biocomposites. *Nanoscale*, *16*(4), 1490–1525. DOI: 10.1039/D3NR05522A PMID: 38186362

Das, S., & Yokozeki, T. (2021). A brief review of modified conductive carbon/glass fibre reinforced composites for structural applications: Lightning strike protection, electromagnetic shielding, and strain sensing. *Composites Part C: Open Access*, *5*, 100162. DOI: 10.1016/j.jcomc.2021.100162

Databank, E. I. A. E. E. (2019). European Union Aviation Safety Agency. In.

Dave, S. H., Gong, C., Robertson, A. W., Warner, J. H., & Grossman, J. C. (2016). Chemistry and structure of graphene oxide via direct imaging. *ACS Nano*, *10*(8), 7515–7522. DOI: 10.1021/acsnano.6b02391

de Wit, J. H. W. (2001). Structural aspects and corrosion behaviour of organic coatings. *Electrochimica Acta*, *46*(24–25), 3641–3650. DOI: 10.1016/S0013-4686(01)00642-9

Dehghanzad, B., Razavi Aghjeh, M. K., Rafeie, O., Tavakoli, A., & Jameie Oskooie, A. (2016). Synthesis and characterization of graphene and functionalized graphene via chemical and thermal treatment methods. *RSC Advances*, *6*(5), 3578–3585. DOI: 10.1039/C5RA19954A

Del Bosque, A., Vergara, D., & Fernández-Arias, P. (2025). An overview of smart composites for the aerospace sector. *Applied Sciences (Basel, Switzerland)*, *15*(6), 2986. DOI: 10.3390/app15062986

Del Cuvillo Mezquita, R. (2023). Aeronautic structures reinforced with graphene subjected to dynamic loadings. Analysis and modeling Universidad Carlos III de Madrid.

Del Giudice, A.. (2022). Porosity of Nanostructured Carbon Thin Films. In *Nanoporous Carbons for Soft and Flexible Energy Devices. Carbon Materials: Chemistry and Physics*. Springer., DOI: 10.1007/978-3-030-81827-2_8

Deng, B., Liu, Z., & Peng, H. (2019). Toward mass production of CVD graphene films. *Advanced Materials*, *31*(9), 1800996. DOI: 10.1002/adma.201800996 PMID: 30277604

Deokar, G., Avila, J., Razado-Colambo, I., Codron, J.-L., Boyaval, C., Galopin, E., Asensio, M.-C., & Vignaud, D. (2015). Towards high quality CVD graphene growth and transfer. *Carbon*, *89*, 82–92. DOI: 10.1016/j.carbon.2015.03.017

Devi, N., Kumar, R., Singh, S., & Singh, R. K. (2024). Recent development of graphene-based composite for multifunctional applications: Energy, environmental, and biomedical sciences. *Critical Reviews in Solid State and Material Sciences*, *49*(1), 72–140. DOI: 10.1080/10408436.2022.2132910

Díaz, J., Anders, S., Zhou, X., Moler, E. J., Kellar, S. A., & Hussain, Z. (2001). Analysis of the π^* and σ^* bands of the x-ray absorption spectrum of amorphous carbon. *Physical Review. B*, *64*(12), 125204. DOI: 10.1103/PhysRevB.64.125204

Ding, Y., Qin, W., Zhu, H., Dai, Y., Hong, X., Han, S., & Xie, Y. (2025). Construction of ZnO/r-GO composite photocatalyst for improved photodegradation of organic pollutants. *Molecules (Basel, Switzerland)*, *30*(5), 1008. DOI: 10.3390/molecules30051008

Dixit, S., Mahata, A., Mahapatra, D. R., Kailas, S. V., & Chattopadhyay, K. (2018). Multi-layer graphene reinforced aluminum – Manufacturing of high strength composite by friction stir alloying. *Composites. Part B, Engineering*, *136*, 63–71. DOI: 10.1016/j.compositesb.2017.10.028

DO. R. (2000a). 254 Design Assurance Guidance for Airborne Electronic Hardware»,© RTCA. Inc., Washington, DC, USA.

Do, R. (2000b). *Design Assurance Guidance for Airborne Electronic Hardware. Design Assurance Guidance for Airborne Electronic Hardware*. RTCA Inc.

dos Reis Araújo, M., Hotza, D., Janssen, R., & Acchar, W. (2019). Processing and properties of tape-cast alumina/zirconia laminates composites. *Journal of the European Ceramic Society*, *39*(12), 3462–3465. DOI: 10.1016/j.jeurceramsoc.2019.03.034

Dreyer, D. R., Park, S., Bielawski, C. W., & Ruoff, R. S. (2010). The chemistry of graphene oxide. *Chemical Society Reviews*, *39*(1), 228–240. DOI: 10.1039/B917103G PMID: 20023850

Du, J., Tong, B., Yuan, S., Dai, N., Liu, R., Zhang, D., Cheng, H.-M., & Ren, W. (2022). Advances in flexible optoelectronics based on chemical vapor deposition-grown graphene. *Advanced Functional Materials*, *32*(42), 2203115. DOI: 10.1002/adfm.202203115

Dulal, M., Modha, H. R. M., Liu, J., Islam, M. R., Carr, C., Hasan, T., & Karim, N. (2025). Sustainable, Wearable, and Eco-Friendly Electronic Textiles. *Energy & Environmental Materials*, *8*(3), e12854. DOI: 10.1002/eem2.12854

Dursun, T., & Soutis, C. (2014). Recent developments in advanced aircraft aluminium alloys. *Materials & Design*, *56*, 862–871. DOI: 10.1016/j.matdes.2013.12.002

Du, X., Zhou, H., Sun, W., Liu, H. Y., Zhou, G., Zhou, H., & Mai, Y. W. (2017). Graphene/epoxy interleaves for delamination toughening and monitoring of crack damage in carbon fibre/epoxy composite laminates. *Composites Science and Technology*, *140*, 123–133. DOI: 10.1016/j.compscitech.2016.12.028

Eayal Awwad, K. Y., Yousif, B. F., Fallahnezhad, K., Saleh, K., & Zeng, X. (2021). Influence of graphene nanoplatelets on mechanical properties and adhesive wear performance of epoxy-based composites. *Friction*, *9*(4), 856–875. DOI: 10.1007/s40544-020-0453-5

Eda, G., Fanchini, G., & Chhowalla, M. (2008). Large-area ultrathin films of reduced graphene oxide as a transparent and flexible electronic material. *Nature Nanotechnology*, *3*(5), 270–274. DOI: 10.1038/nnano.2008.83

Egerton, R. F. (2005). *Physical Principles of Electron Microscopy*. Springer., DOI: 10.1007/b136495

Eigler, S. (2009). A new parameter based on graphene for characterizing transparent, conductive materials. *Carbon*, *47*(12), 2936–2939. DOI: 10.1016/j.carbon.2009.06.047

Elahi, S., Prasad, S., Haider, M. A., Gupta, M., & Ahmad, E. (2025). Congruity of butyl butyrate and limonene as potential blends to produce green and sustainable aviation fuels. *Journal of Chemical & Engineering Data*, *70*(7), 2713–2726. Advance online publication. DOI: 10.1021/acs.jced.5c00131

EL-Shendidy. M., EL-Mekkawi, D., & Hafez, H. (2025). Sustainable and optimized microwave-assisted hydrothermal method for production of dual-adsorptive reduced graphene nanomaterials for wastewater remediation. https://doi.org/DOI: 10.21203/rs.3.rs-6627454/v1

Eppensteiner, F. W., & Jenkins, M. R. (1997). Chromate conversion coatings. *Metal Finishing*, *95*(1), 494–501. DOI: 10.1016/S0026-0576(99)80050-6

Esmaeili, S., & Hosseini, M. (2025). Buckling analysis of tapered graphene-reinforced composite plates: An analytical approach. *Plastics, Rubber and Composites*, *54*(5), 197–212. DOI: 10.1177/14658011251340776

Eun, J., Kim, J., Kim, H., Kim, D., Choi, J., Kim, K., Yoo, C.-J., Jin, S., Kim, K., Lee, H., Kim, C. S., Lee, K.-Y., Yoo, J. S., Han, S.-J., Jeong, K., & Ha, J.-M. (2025). Production of high-carbon-number hydrocarbon bio-aviation fuels via catalytic hydrogenation of vanillin and non-catalytic condensation: A mechanistic study with DFT and experimental insights. *Green Chemistry*, *27*(24), 7147–7159. DOI: 10.1039/D5GC00281H

Fadeel, B., Bussy, C., Merino, S., Vázquez, E., Flahaut, E., Mouchet, F., Evariste, L., Gauthier, L., Koivisto, A. J., Vogel, U., Martín, C., Delogu, L. G., Buerki-Thurnherr, T., Wick, P., Beloin-Saint-Pierre, D., Hischier, R., Pelin, M., Carniel, F. C., Tretiach, M., & Bianco, A. (2018). Safety Assessment of Graphene-Based Materials: Focus on human health and the environment. *ACS Nano*, *12*(11), 10582–10620. DOI: 10.1021/acsnano.8b04758 PMID: 30387986

Falandys, K., Kurc, K., & Tutak, J. S. (2025). Application and empirical verification of the Archard model in the deburring process. *Materials (Basel)*, *18*(10), 2387. DOI: 10.3390/ma18102387 PMID: 40429125

Fang, W., Pearson, A., Huang, Y., & Naguib, H. E. (2025). Enhanced carbon fiber interface with thermoplastics via nanostructure surface modification: Failure, morphology and wettability analysis. *Composites. Part B, Engineering*, *291*, 112054. DOI: 10.1016/j.compositesb.2024.112054

Farivar, F., Yap, P. L., Karunagaran, R. U., & Losic, D. (2021). Thermogravimetric analysis (TGA) of graphene materials: effect of particle size of graphene, graphene oxide and graphite on thermal parameters. C, 7(2), 41. .DOI: 10.3390/c7020041

Farnsworth, A., Hora, Y., Wang, R., Anyaegbu, C., Dharmasiri, B., Henderson, L. C., & Hooper, J. F. (2025). Covalent functionalization of graphene and carbon nanotubes by radical decarboxylation. *ACS Applied Materials & Interfaces*, *17*(26), 38348–38356. Advance online publication. DOI: 10.1021/acsami.5c04671

Farouk, A., Saeed, S. E. S., Sharaf, S., & Abd El-Hady, M. M. (2020). Photocatalytic activity and antibacterial properties of linen fabric using reduced graphene oxide/silver nanocomposite. *RSC Advances*, *10*(68), 41600–41611. DOI: 10.1039/D0RA07544B PMID: 35516560

Faucett, A. C., & Mativetsky, J. M. (2015). Nanoscale reduction of graphene oxide under ambient conditions. *Carbon*, *95*, 1069–1075. DOI: 10.1016/j.carbon.2015.09.025

Favour, A. A. (2025). Economic and Environmental Benefits of Waste-Derived Graphene Nanosheets for Carbon Dioxide Mitigation.

Feng, L., Li, L., Yu, Y., Tian, G., Xiong, H., & Miao, J.. (2025). Road to carbon-neutral aviation: One-step conversion of CO_2 to sustainable aviation fuels. *Industrial & Engineering Chemistry Research*, *64*(6), 3118–3135. DOI: 10.1021/acs.iecr.4c04209

Feng, X., & Jiang, L. (2006). Design and creation of superwetting/antiwetting surfaces. *Advanced Materials*, *18*(23), 3063–3078. DOI: 10.1002/adma.200501961

Feng, Y., Wang, B., Li, X., Ye, Y., Ma, J., Liu, C., Zhou, X., & Xie, X. (2019). Enhancing thermal oxidation and fire resistance of reduced graphene oxide by phosphorus and nitrogen co-doping: Mechanism and kinetic analysis. *Carbon*, *146*, 650–659. DOI: 10.1016/j.carbon.2019.01.099

Feni, F., Jahan, M., Zhao, R., Li, G., Zhao, G.-L., & Mensah, P. F. (2023). Multifunctional composite with hybrid carbon fiber and carbonaceous coconut particle reinforcement. *Frontiers in Materials*, *10*, 1278222. Advance online publication. DOI: 10.3389/fmats.2023.1278222

Feres, G. D., Ribeiro, B., Costa, M. L., de Paula Santos, L. F., & Botelho, E. C. (2025). Study of the influence of carbon nanotubes, graphene, and hybrid buckypaper on the mechanical behavior of PAEK/carbon fiber composites. *Polymer Composites*, *46*(9), 8674–8685. Advance online publication. DOI: 10.1002/pc.29519

Figarova, S. R., Aliyev, E. M., Abaszade, R. G., Alekberov, R. I., & Figarov, V. R. (2021). Negative differential resistance of graphene oxide|sulphur compound. *Journal of Nano Research*, *67*, 25–31. . DOI: 10.4028/www.scientific.net/JNanoR.67.25

Firefly Aerospace Awarded AFRL Contract to Develop Ceramic Matrix Composite Rocket Engine Nozzle Extension. Retrieved June 25, 2025, from https://fireflyspace.com/news/firefly-aerospace-awarded-afrl-contract-to-develop-ceramic-matrix-composite-rocket-engine-nozzle-extension/

Florea, M., Constantin, V.-S., Bucur, A.-C., Faur, R., Predu, D., & Cazacu, A. (2024). Materials and Structures Used in Aeronautics: Present and Future Perspectives. INCAS Bulletin, 16(4).

Foschi, M., Capasso, P., Maggi, M. A., Ruggieri, F., & Fioravanti, G. (2021). Experimental design and response surface methodology applied to graphene oxide reduction for adsorption of triazine herbicides. *ACS Omega*, *6*(26), 16943–16954. DOI: 10.1021/acsomega.1c01877

Frankel, G. S., & Kelly, R. G. (2001). Critical factors in atmospheric corrosion. *The Electrochemical Society Interface*, *10*(4), 34–40. DOI: 10.1149/2.F06014IF

Fu, Y., Hansson, J., Liu, Y., Chen, S., Zehri, A., Samani, M. K., . . . Zhang, Z.-B. (2020). Graphene-related materials for thermal management. 2D Materials, 7(1), 012001.

Fu, K., Zhang, X., Shi, C., Liu, E., He, F., Li, J., Zhao, N., & He, C. (2018). An approach for fabricating Ni@graphene reinforced nickel matrix composites with enhanced mechanical properties. *Materials Science and Engineering A*, *715*, 108–116. DOI: 10.1016/j.msea.2017.12.101

Gamal, R., Soliman, T., & Khalid, A. (2025). Thermal stability and electrical conductivity of polyvinyl alcohol composites reinforced with Mn_2O_3–rgo. *ChemistrySelect*, *10*(1), e202403782. Advance online publication. DOI: 10.1002/slct.202403782

Ganguly, A., Sharma, S., Papakonstantinou, P., & Hamilton, J. (2011). Probing the thermal deoxygenation of graphene oxide using high-resolution in situ X-ray-based spectroscopies. *The Journal of Physical Chemistry. C, Nanomaterials and Interfaces*, *115*(34), 17009–17019. DOI: 10.1021/jp203741y

Ganguly, S., & Sengupta, J. (2024). Graphene-based nanotechnology in the Internet of Things: A mini review. *Discover Nano, 19*(1), 110. DOI: 10.1186/s11671-024-04054-0 PMID: 38954113

Gao, J., Yan, Q., Lv, L., Tan, X., Ying, J., Yang, K., & Xiang, R. (2021). Lightweight thermal interface materials based on hierarchically structured graphene paper with superior through-plane thermal conductivity. *Chemical Engineering Journal, 419*, 129609. DOI: 10.1016/j.cej.2021.129609

Gao, K., Qi, M., Che, Z., Niu, Q., Guo, W., & Sun, X.. (2025). Bamboo-based carbon fiber/carbon nanosheet composite flexible supercapacitor. *ChemNanoMat : Chemistry of Nanomaterials for Energy, Biology and More, 11*(6), e202500037. Advance online publication. DOI: 10.1002/cnma.202500037

Gao, T., Chen, L., Wang, B., Liu, Y., Ma, Y., & Liang, Y. (2025). Prediction of the mechanical properties of graphene/copper nanocomposites using machine learning and molecular dynamics simulations. *Journal of Physics Condensed Matter, 37*(23), 235701. DOI: 10.1088/1361-648X/add2c2 PMID: 40306312

Geim, A. K., & Novoselov, K. S. (2007). The rise of graphene. *Nature Materials, 6*(3), 183–191. DOI: 10.1038/nmat1849 PMID: 17330084

Geim, A. N. D. R. E., & Novoselov, K. (2007). Graphene calling. *Nature Materials, 6*(169), 10–1038. PMID: 17330074

Gengenbach, T. R., Major, G. H., Linford, M. R., & Easton, C. D. (2021). Practical guides for x-ray photoelectron spectroscopy (XPS): Interpreting the carbon 1s spectrum. *Journal of Vacuum Science & Technology. A, Vacuum, Surfaces, and Films, 39*(1), 1. DOI: 10.1116/6.0000682

Geng, J., & Li, Y. (2024). Higher-order anti-PT-symmetric self-injection locking micro-resonator optical gyroscope. *Physica Scripta, 99*(7), 075502. DOI: 10.1088/1402-4896/ad4fc6

George, R., Kashyap, K. T., Rahul, R., & Yamdagni, S. (2005). Strengthening in carbon nanotube/aluminium (CNT/Al) composites. *Scripta Materialia, 53*(10), 1159–1163. DOI: 10.1016/j.scriptamat.2005.07.022

Ge, Q., Chu, J., Cao, W., Yi, F., Ran, Z., Jin, Z., Mao, B., Li, Z., & Novoselov, K. S. (2022). Graphene-based textiles for thermal management and flame retardancy. *Advanced Functional Materials, 32*(42), 2205934. DOI: 10.1002/adfm.202205934

Ghosh, S., Bhatia, J., Gupta, A., Pramanik, C., Pratihar, S., & Rautaray, D. (2025). Effects of the shear rate on dispersion characteristics of industrial-based functionalized/non functionalized graphene in an epoxy matrix. *Materials Advances*, *6*(11), 3523–3532. DOI: 10.1039/D5MA00173K

Gibbs, R. (2018). Near term commercial applications for graphene, composites and conductive pastes. *Journal of Material Science & Engineering*, *7*, 7. DOI: 10.4172/2169-0022-C4-101

Giese, B., Klaessig, F., Park, B., Kaegi, R., Steinfeldt, M., Wigger, H., Von Gleich, A., & Gottschalk, F. (2018). Risks, release and concentrations of engineered nanomaterial in the environment. *Scientific Reports*, *8*(1), 1565. Advance online publication. DOI: 10.1038/s41598-018-19275-4 PMID: 29371617

Gojny, F. H., Wichmann, M. H. G., Köpke, U., Fiedler, B., & Schulte, K. (2004). Carbon nanotube-reinforced epoxy-composites: Enhanced stiffness and fracture toughness at low nanotube content. *Composites Science and Technology, 64*(15 SPEC. ISS.), 2363–2371. DOI: 10.1016/j.compscitech.2004.04.002

Goldstein, J. I.. (2017). *Scanning Electron Microscopy and X-ray Microanalysis*. Springer., DOI: 10.1007/978-1-4939-6676-9

Goossens, S., Navickaite, G., Monasterio, C., Gupta, S., Piqueras, J. J., Pérez, R., & Galán, T. (2017). Broadband image sensor array based on graphene–CMOS integration. *Nature Photonics*, *11*(6), 366–371. DOI: 10.1038/nphoton.2017.75

Gualandi, I., Tessarolo, M., Mariani, F., Possanzini, L., Scavetta, E., & Fraboni, B. (2021). Textile chemical sensors based on conductive polymers for the analysis of sweat. *Polymers*, *13*(6), 894. DOI: 10.3390/polym13060894 PMID: 33799437

Guild, F. J., Tsang, W. L., & Taylor, A. C. (2022). Silica nano-particle filled polymers: Debonding and microstructure. *Composites Science and Technology*, *218*, 109202. DOI: 10.1016/j.compscitech.2021.109202

Guo, L., Chen, Z., Han, H., Liu, G., Luo, M., Cui, N., Dong, H., & Li, M.-Z. (2023). Advances and outlook in modified graphene oxide (GO)/epoxy composites for mechanical applications. *Applied Nanoscience*, *13*(5), 3273–3287. DOI: 10.1007/s13204-022-02653-w

Gutic, S. J., Šabanović, M., Metarapi, D., Pašti, I. A., Korać, F., & Mentus, S. V. (2019). Electrochemically Synthesized Ni@ reduced Graphene Oxide Composite Catalysts for Hydrogen Evolution in Alkaline Media-the Effects of Graphene Oxide Support. *International Journal ofElectrochemical Science*, *14*(9), 8532–8543. DOI: 10.20964/2019.09.23

Gutiérrez, D., Doshi, P., Wong, H., Nordlund, D., & Gandhiraman, R. (2024). Printed graphene and its composite with copper for electromagnetic interference shielding applications. *Nanotechnology*, *35*(13), 135202. DOI: 10.1088/1361-6528/ad12e9 PMID: 38055996

Haartsen, S., Wisman, J., Zandvliet, H., & Bampoulis, P. (2025). Humidity-controlled interactions and lifting of MoS_2 on SiO_2. *JPhys Materials*, *8*(2), 025002. DOI: 10.1088/2515-7639/adaaa1

Habte, A. T., & Ayele, D. W. (2019). Synthesis and characterization of reduced graphene oxide (rGO) started from graphene oxide (GO) using the tour method with different parameters. *Advances in Materials Science and Engineering*, *2019*(1), 5058163. DOI: 10.1155/2019/5058163

Hadadian, M., Goharshadi, E. K., & Youssefi, A. (2014). Electrical conductivity, thermal conductivity, and rheological properties of graphene oxide-based nanofluids. *Journal of Nanoparticle Research*, *16*(12), 2788. DOI: 10.1007/s11051-014-2788-1

Hajati, S., & Tougaard, S. (2010). XPS for non-destructive depth profiling and 3D imaging of surface nanostructures. *Analytical and Bioanalytical Chemistry*, *396*(8), 2741–2755. DOI: 10.1007/s00216-009-3401-9

Han, D., Zhao, Y.-H., Bai, S.-L., & Ping, W. C. (2016). High shielding effectiveness of multilayer graphene oxide aerogel film/polymer composites. *RSC Advances*, *6*(95), 92168–92174. DOI: 10.1039/C6RA20976A

Han, Z., & Fina, A. (2011). Thermal conductivity of carbon nanotubes and their polymer nanocomposites: A review. *Progress in Polymer Science*, *36*(7), 914–944. DOI: 10.1016/j.progpolymsci.2010.11.004

Hao, B., Peng, C., Tang, X., & Zhao, Z. (2022). Analysis of time-frequency characteristic parameters of electromagnetic interference sources in the basic commutation unit. *IET Power Electronics*, *16*(4), 531–541. DOI: 10.1049/pel2.12403

Hariyawan, M. Y., Hidayat, R., & Firmansyah, E. (2016). The effects of spread-spectrum techniques in mitigating conducted EMI to LED luminance. [IJECE]. *International Journal of Electrical and Computer Engineering*, *6*(3), 1332–1343. DOI: 10.11591/ijece.v6i3.pp1332-1343

Havenith, G. (2002). Interaction of clothing and thermoregulation. *Exogenous Dermatology*, *1*(5), 221–230. DOI: 10.1159/000068802

Hearle, J. W. (Ed.). (2001). *High-performance fibres* (pp. 7–8). Elsevier.

Heinl, P., Rottmair, A., Körner, C., & Singer, R. F. (2007). Cellular titanium by selective electron beam melting. *Advanced Engineering Materials*, *9*(5), 360–364. DOI: 10.1002/adem.200700025

He, K., Chen, G., Zeng, G., Peng, M., Huang, Z., Shi, J., & Huang, T. (2017). Stability, transport, and ecosystem effects of graphene in water and soil environments. *Nanoscale*, *9*(17), 5370–5388. DOI: 10.1039/C6NR09931A PMID: 28406500

He, L., Weniger, F., Neumann, H., & Beller, M. (2016). Synthesis, characterization, and application of metal nanoparticles supported on nitrogen-doped carbon: Catalysis beyond electrochemistry. *Angewandte Chemie International Edition*, *55*(41), 12582–12594. DOI: 10.1002/anie.201603198

Hemalatha, K., Ashwin, R., Santhanam, M., Riyaz, M. M., Sozhamannan, G. G., & Venkatachalapthy, V. S. K. (2021). Experimental study on behaviour of SiC and Graphene on AL7075 hybrid metal matrix composite. *Materials Today: Proceedings*, *39*, 26–30. DOI: 10.1016/j.matpr.2020.05.744

Hernandez, Y., Nicolosi, V., Lotya, M., Blighe, F. M., Sun, Z., De, S., McGovern, I. T., Holland, B., Byrne, M., Gun'Ko, Y. K., Boland, J. J., Niraj, P., Duesberg, G., Krishnamurthy, S., Goodhue, R., Hutchison, J., Scardaci, V., Ferrari, A. C., & Coleman, J. N. (2008). High-yield production of graphene by liquid-phase exfoliation of graphite. *Nature Nanotechnology*, *3*(9), 563–568. DOI: 10.1038/nnano.2008.215 PMID: 18772919

Heunisch, E., Bleeker, E., Kelly, S., Matzke, M., Pohl, A., Tommaso, S., & Suarez-Merino, B. (2024). Priority list for making OECD test guidelines and guidance documents applicable for nanomaterials and advanced materials. The Malta Initiative. https://malta-initiative.org/MaltaInitiative_UPLOADS/20240301_The_Malta_Initiative_Priority_List.pdf

Hidalgo-Manrique, P., Lei, X., Xu, R., Zhou, M., Kinloch, I. A., & Young, R. J. (2019). Copper/graphene composites: A review. *Journal of Materials Science*, *54*(19), 12236–12289. DOI: 10.1007/s10853-019-03703-5

Hidayat, R., Fajarwati, F., & Fadillah, G. (2025). Recent trends and bibliometric analysis in graphene quantum dots materials from natural bioresources: a brief review., 14, 23–40. https://doi.org/DOI: 10.4028/p-5zRtwr

Hirankittiwong, P., Chomchok, T., Chakraborty, S., Prajongtat, P., Singh, D. P., Hongkarnjanakul, N., Channumsin, S., Ghosh, S., & Chattham, N. (2025). Advanced engineered nanostructures for aerospace technology: A review. *Results in Engineering*, *26*, 105381. DOI: 10.1016/j.rineng.2025.105381

Hkiri, K. (2025). Synthesis and photonic properties of WO3-reduced graphene oxide nanocomposites. *Journal of Physics: Conference Series, 2970*(1), 012002. DOI: 10.1088/1742-6596/2970/1/012002

Hodson, R., McCabe, M., Paulick, P., Ruffner, T., Some, R., Chen, Y.,…Redifer, M. (2013). Flight Avionics Hardware Roadmap.

Hong, S., Kim, K., Kim, T., Kim, J., Park, S., Kim, S., & Ahn, J. H. (2012). Electromagnetic interference shielding effectiveness of monolayer graphene. *Nanotechnology, 23*(45), 455704. DOI: 10.1088/0957-4484/23/45/455704 PMID: 23085718

Hong, X., Yu, W., & Chung, D. D. L. (2017). Electric permittivity of reduced graphite oxide. *Carbon, 111*, 182–190. DOI: 10.1016/j.carbon.2016.09.071

Hsieh, T. H., Kinloch, A. J., Masania, K., Taylor, A. C., & Sprenger, S. (2010). The mechanisms and mechanics of the toughening of epoxy polymers modified with silica nanoparticles. *Polymer, 51*(26), 6284–6294. DOI: 10.1016/j.polymer.2010.10.048

Huang, H., Su, S., Wu, N., Wan, H., Wan, S., Bi, H., & Sun, L. (2019). Graphene-based sensors for human health monitoring. *Frontiers in Chemistry, 7*, 399. DOI: 10.3389/fchem.2019.00399 PMID: 31245352

Huang, R., Li, W., Zeng, Y., & Fang, C. (2025). Power optimization of an electric propulsion system based on variable speed rotor blades. *Journal of Physics: Conference Series, 2977*(1), 012084. DOI: 10.1088/1742-6596/2977/1/012084

Huang, Y., Liang, J., & Chen, Y. (2012). The application of graphene-based materials for actuators. *Journal of Materials Chemistry, 22*(9), 3671–3679. DOI: 10.1039/c2jm15536b

Hu, J., Li, W., Liu, J., Zhang, L., & Ao, Q. (2018). Advances in corrosion-resistant thermal barrier coatings. *Ceramics International, 44*(15), 17456–17465.

Hu, L., Yan, X., & Yuan, Y. (2025). Development and challenges of autonomous electric vertical take-off and landing aircraft. *Heliyon, 11*(1), e41055. DOI: 10.1016/j.heliyon.2024.e41055 PMID: 39758371

Hu, R., Wen, S., Chen, Q., Sun, X., Liu, H., Gao, W., & Bai, Y. (2025). Universal construction of electrical insulation and high-thermal-conductivity composites based on the in-situ exfoliation of boron nitride-graphene hybrid filler. *ACS Applied Materials & Interfaces, 17*(4), 6783–6792. DOI: 10.1021/acsami.4c18250

Husnah, M., Fakhri, H. A., Rohman, F., Aimon, A. H., & Iskandar, F. (2017). A modified Marcano method for improving electrical properties of reduced graphene oxide (rGO). *Materials Research Express, 4*(6), 064001. DOI: 10.1088/2053-1591/aa707f

Idumah, C. I., & Odera, S. (2020). Recent advancement in self-healing graphene polymer nanocomposites, shape memory, and coating materials. *Polymer-Plastics Technology and Materials*, *59*(11), 1167–1190. DOI: 10.1080/25740881.2020.1725816

Ikhsan, N., Sutiman, N., & Samsudin, N. (2025). A facile <i>in-situ</i> synthesis and characterizations of silver-graphene oxide (aggo) nanocomposite using chives as a reducing and stabilizing agent. *Journal of Metastable and Nanocrystalline Materials*, *41*, 35–42. DOI: 10.4028/p-CcECT2

Inayathullah, S., & Buddala, R. (2024). Review of Machine Learning applications in Additive Manufacturing. *Results in Engineering*, *25*, 103676. DOI: 10.1016/j.rineng.2024.103676

Iqbal, A. A., Iqbal, A. P., & Nuruzzaman, D. M. (2024). A review on fatigue fracture characteristics in graphene reinforced nanocomposites. *Ain Shams Engineering Journal*, *15*(6), 102730. DOI: 10.1016/j.asej.2024.102730

Iskandar, F., Hikmah, U., Stavila, E., & Aimon, A. H. (2017). Microwave-assisted reduction method under nitrogen atmosphere for synthesis and electrical conductivity improvement of reduced graphene oxide (rGO). *RSC Advances*, *7*(83), 52391–52397. DOI: 10.1039/C7RA10013B

ISO ISO/TC 229 Nanotechnologies available at: www.iso.org/committee/381983.html

Jalili, R., Aboutalebi, S. H., Esrafilzadeh, D., Shepherd, R. L., Chen, J., Aminorroaya-Yamini, S., Konstantinov, K., Minett, A. I., Razal, J. M., & Wallace, G. G. (2013). Scalable one-step wet-spinning of graphene fibers and yarns from liquid crystalline dispersions of graphene oxide: Towards multifunctional textiles. *Advanced Functional Materials*, *23*(43), 5345–5354. DOI: 10.1002/adfm.201300765

Jashrapuria, K., & Singh, S. (2025). Biofilm inhibition by laser-induced graphene: Impact of surface texture on rod-shaped e. coli and coccus-shaped staphylococcus. *ACS Applied Materials & Interfaces*, *17*(14), 21819–21829. DOI: 10.1021/acsami.5c01748

Jayalakshmi, C. G., Inamdar, A., Anand, A., & Kandasubramanian, B. (2019). Polymer matrix composites as broadband radar absorbing structures for stealth aircrafts. *Journal of Applied Polymer Science*, *136*(14), 47241. DOI: 10.1002/app.47241

Jayaseelan, J., Pazhani, A., Michael, A. X., Paulchamy, J., Batako, A., & Hosamane Guruswamy, P. K. (2022). Characterization Studies on Graphene-Aluminium Nano Composites for Aerospace Launch Vehicle External Fuel Tank Structural Application. *Materials 2022, Vol. 15, Page 5907*, *15*(17), 5907. DOI: 10.3390/ma15175907

Jayasena, B., & Subbiah, S. (2011). A novel mechanical cleavage method for synthesizing few-layer graphenes. *Nanoscale Research Letters*, *6*(1), 1–7. DOI: 10.1186/1556-276X-6-95 PMID: 21711598

Jiang, B., Wang, S., Sun, J., & Liu, Z. (2021). Controllable synthesis of wafer-scale graphene films: Challenges, status, and perspectives. *Small*, *17*(37), 2008017. DOI: 10.1002/smll.202008017 PMID: 34106524

Jiang, W., Liu, Z.-F., Wu, Q.-P., & Xiao, X. (2025). Role of edge reconfiguration in generating corner states in Zigzag graphene nanoribbons. *Physica E, Low-Dimensional Systems and Nanostructures*, *172*, 116260. DOI: 10.1016/j.physe.2025.116260

Jiang, X., Lu, D., Yin, B., & Leng, Z. (2024). Advancing carbon nanomaterials-engineered self-sensing cement composites for structural health monitoring: A state-of-the-art review. *Journal of Building Engineering*, *87*, 109129. DOI: 10.1016/j.jobe.2024.109129

Jiang, Y., Xu, R., Tan, Z., Ji, G., Fan, G., Li, Z., Xiong, D. B., Guo, Q., Li, Z., & Zhang, D. (2019). Interface-induced strain hardening of graphene nanosheet/aluminum composites. *Carbon*, *146*, 17–27. DOI: 10.1016/j.carbon.2019.01.094

Ji, J., Zhao, W., Wang, Y., Li, Q., & Wang, G. (2023). Templated laser-induced graphene-based tactile sensors enable wearable health monitoring and texture recognition via a deep neural network. *ACS Nano*, *17*(20), 20153–20166. DOI: 10.1021/acsnano.3c05838 PMID: 37801407

Jilani, A., Hussain, S. Z., Khan, A. A. P., Khan, A., Othman, M. H. D., & Ansari, M. O. (2020). Graphene-based material for self-healing: mechanism, synthesis, characteristics, and applications. In *Self-Healing Composite Materials* (pp. 163–175). Elsevier., DOI: 10.1016/B978-0-12-817354-1.00009-0

Jogur, G., Nawaz Khan, A., Das, A., Mahajan, P., & Alagirusamy, R. (2018). Impact properties of thermoplastic composites. *Textile Progress*, *50*(3), 109–183. DOI: 10.1080/00405167.2018.1563369

Jones, F., & Foreman, J. (2015). The response of aerospace composites to temperature and humidity. In *Polymer composites in the aerospace industry* (pp. 335–369). Elsevier. DOI: 10.1016/B978-0-85709-523-7.00012-8

Jovanović, S., Huskić, M., Kepić, D., Yasir, M., & Haddadi, K. (2023). A review on graphene and graphene composites for application in electromagnetic shielding. *Deleted Journal*, *8*(3–4), 59–80. DOI: 10.1007/s41127-023-00065-3

Kandare, E., Khatibi, A. A., Yoo, S., Wang, R., Ma, J., Olivier, P., Gleizes, N., & Wang, C. H. (2015). Improving the through-thickness thermal and electrical conductivity of carbon fibre/epoxy laminates by exploiting synergy between graphene and silver nano-inclusions. *Composites. Part A, Applied Science and Manufacturing*, *69*, 72–82. DOI: 10.1016/j.compositesa.2014.10.024

Kanjwal, M. A., & Ghaferi, A. A. (2022). Graphene incorporated electrospun nanofiber for electrochemical sensing and biomedical applications: A critical review. *Sensors (Basel)*, *22*(22), 8661. DOI: 10.3390/s22228661 PMID: 36433257

Kanti, P. K., Prashantha Kumar, H. G., Wanatasanappan, V. V., Kumar, A., & Regasa, M. B. (2025). Graphene's Frontier in aerospace: Current applications, challenges, and future directions for space engineering. *Nanoscale Advances*, *7*(12), 3603–3618. DOI: 10.1039/D4NA00934G

Kargar, F., Barani, Z., Balinskiy, M. G., Magana, A. S., Lewis, J. S., & Balandin, A. A. (2018). Graphene Composites with Dual Functionality: Electromagnetic Shielding and Thermal Management. arXiv preprint arXiv:1808.03401.

Katnam, K. B., Dalfi, H., & Potluri, P. (2019). Towards balancing in-plane mechanical properties and impact damage tolerance of composite laminates using quasi-UD woven fabrics with hybrid warp yarns. *Composite Structures*, *225*, 111083. DOI: 10.1016/j.compstruct.2019.111083

Katsnelson, M. I. (2007). Graphene: Carbon in two dimensions. *Materials Today*, *10*(1-2), 20–27. DOI: 10.1016/S1369-7021(06)71788-6

Kaur, G., Adhikari, R., Cass, P., Bown, M., & Gunatillake, P. (2015). Electrically conductive polymers and composites for biomedical applications. *RSC Advances*, *5*(47), 37553–37567. DOI: 10.1039/C5RA01851J

Kausar, A. (2022). Shape memory polymer/graphene nanocomposites: State-of-the-art. e-Polymers, 22(1), 165-181.

Kausar, A., Ahmad, I., Eisa, M., & Maaza, M. (2023). Graphene nanocomposites in the Space Sector—fundamentals and advancements. C, 9(1), 29.

Kausar, A., Ahmad, I., Eisa, M. H., & Maaza, M. (2023). Graphene Nanocomposites in Space Sector—Fundamentals and Advancements. *C*, *9*(1), 29. DOI: 10.3390/c9010029

Kausar, A., Ahmad, I., Zhao, T., Aldaghri, O., Ibnaouf, K. H., Eisa, M. H., & Lam, T. D. (2023). Graphene Nanocomposites for Electromagnetic Interference Shielding—Trends and Advancements. *Journal of Composites Science*, *7*(9), 384. DOI: 10.3390/jcs7090384

Kausar, A., Rafique, I., & Muhammad, B. (2017). Aerospace Application of Polymer Nanocomposite with Carbon Nanotube, Graphite, Graphene Oxide, and Nanoclay. *Polymer-Plastics Technology and Engineering, 56*(13), 1438–1456. DOI: 10.1080/03602559.2016.1276594

Kazi, M., Varghese, S., Sarker, N., Aich, N., & Gadhamshetty, V. (2025). Advancing pfas remediation through physics-based modeling of 2d materials: Recent progress, challenges, and opportunities. *Industrial & Engineering Chemistry Research, 64*(4), 1894–1906. DOI: 10.1021/acs.iecr.4c03715

Ke, F., Song, F., Zhang, H., Xu, J., Wang, H., & Chen, Y. (2020). Layer-by-layer assembly for all-graphene coated conductive fibers toward superior temperature sensitivity and humidity independence. *Composites. Part B, Engineering, 200*, 108253. DOI: 10.1016/j.compositesb.2020.108253

Keller, A. A., Lazareva, A., Suh, S., & Davis, J. M. (2013). Global life cycle releases of engineered nanomaterials. *Journal of Nanoparticle Research, 15*(6), 1–17. DOI: 10.1007/s11051-013-1692-4

Keramati, M., Ghasemi, I., Karrabi, M., Azizi, H., & Sabzi, M. (2016). Incorporation of surface modified graphene nanoplatelets for development of shape memory PLA nanocomposite. *Fibers and Polymers, 17*(7), 1062–1068. DOI: 10.1007/s12221-016-6329-7

Khalajiolyaie, A., & Jian, C. (2025). Advances in graphene-based materials for metal ion sensing and wastewater treatment: A review. *Environments (Basel, Switzerland), 12*(2), 43. DOI: 10.3390/environments12020043

Khalid, S. et al. (2025). A conceptual framework for developing an AI-powered UAV system for effective fog management at airports. *SPAST Reports, 1*(1).

Khalid, S., Chakkaravarthy, M., Gupta, S. K., Prakash, D., Agrawal, A., & Nishad, D. K. (2025). A conceptual framework for developing an AI-powered UAV system for effective fog management at airports. *SPAST Reports, 1*(1).

Khalid, M. Y., Kamal, A., Otabil, A., Mamoun, O., & Liao, K. (2023). Graphene/epoxy nanocomposites for improved fracture toughness: A focused review on toughening mechanism. *Chemical Engineering Journal Advances, 16*, 100537. DOI: 10.1016/j.ceja.2023.100537

Khalid, M. Y., Umer, R., & Khan, K. A. (2023). Review of recent trends and developments in aluminium 7075 alloy and its metal matrix composites (MMCs) for aircraft applications. *Results in Engineering, 20*, 101372. DOI: 10.1016/j.rineng.2023.101372

Khalid, S., & Kumar, A. (2023). Using unmanned aerial vehicles (UAV) to disperse fog and improve visibility. *Journal of Aerospace Engineering & Technology*, *13*(1), 1–7.

Khan, A., Sapakal, S. N., and Kadam, A. "Comparative analysis of graphene oxide (GO) reduction methods: impact on crystallographic, morphological, and optical properties." Graphene and 2D Materials 9 (2024): 101–109. https://doi.org/.DOI: 10.1007/s41127-024-00072-y

Khan, M. (2022, July 12). The environmental impact of graphene nanomaterials [Editorial feature]. AZoNano. Retrieved from https://www.azonano.com/article.aspx?ArticleID=6223

Khan, M. M. (2025). *"Fourier transform infrared spectroscopy." Photocatalysts: Synthesis and Characterization Methods*. Elsevier., DOI: 10.1016/B978-0-443-28913-2.00017-4

Khan, T., Khalid, M. Y., Andrew, J. J., Ali, M. A., Zheng, L., & Umer, R. (2023). Co-cured GNP films with liquid thermoplastic/glass fiber composites for superior EMI shielding and impact properties for space applications. *Composites Communications*, *44*, 101767. DOI: 10.1016/j.coco.2023.101767

Khoramishad, H., Ashofteh, R. S., Pourang, H., & Berto, F. (2018). Experimental investigation of the influence of temperature on the reinforcing effect of graphene oxide nano-platelet on nanocomposite adhesively bonded joints. *Theoretical and Applied Fracture Mechanics*, *94*, 95–100. DOI: 10.1016/j.tafmec.2018.01.010

Khorshidi, S., Mohebbali, M., Imani, R., Mahmoodi, M., & Solouk, A. (2020). Electrospun fibroin/graphene oxide nanocomposite mats: An optimization for potential wound dressing applications. *Fibers and Polymers*, *21*(3), 480–488. DOI: 10.1007/s12221-020-9465-z

Khoury, D., Millet, M., Jabali, Y., & Delhomme, O. (2023). Fog water: A general review of its physical and chemical aspects. *Environments (Basel, Switzerland)*, *10*(12), 224. DOI: 10.3390/environments10120224

Khujamberdiev, R., & Cho, H. (2025). Synthesis and characterization of nanoparticles in transforming biodiesel into a sustainable fuel. *Molecules (Basel, Switzerland)*, *30*(6), 1352. DOI: 10.3390/molecules30061352

Kim, D. S., Jeong, J.-M., Park, H. J., Kim, Y. K., Lee, K. G., & Choi, B. G. (2021). Highly concentrated, conductive, defect-free graphene ink for screen-printed sensor application. *Nano-Micro Letters*, *13*(1), 1–14. DOI: 10.1007/s40820-021-00617-3 PMID: 34138339

Kim, H., Abdala, A. A., & Macosko, C. W. (2010). Graphene/polymer nanocomposites. *Macromolecules*, *43*(16), 6515–6530. DOI: 10.1021/ma100572e

Kim, J. K., & Sham, M. L. (2000). Impact and delamination failure of woven-fabric composites. *Composites Science and Technology*, *60*(5), 745–761. DOI: 10.1016/S0266-3538(99)00166-9

Kim, K., Zhao, Y., Jang, H., Lee, S. Y., Kim, J. M., Kim, K. S., Ahn, J.-H., Kim, P., Choi, J.-Y., & Hong, B. H. (2009). Large-scale pattern growth of graphene films for stretchable transparent electrodes. *Nature*, *457*(7230), 706–710. DOI: 10.1038/nature07719

Kim, S. H., Lee, S. Y., Zhang, Y., Park, S. J., & Gu, J. (2023). Carbon-Based Radar Absorbing Materials toward Stealth Technologies. *Advanced Science (Weinheim, Baden-Wurttemberg, Germany)*, *10*(32), 2303104. DOI: 10.1002/advs.202303104

Kiranakumar, H. V.. (2024). An impact of RGO on the ZnO nanoparticles: Structural, morphological, electrical, and gas sensing properties. *Sensing Technology*, *2*, 2310479. DOI: 10.1080/28361466.2024.2310479

Kitadai, H., Yuan, M., Ma, Y., & Ling, X. (2021). Graphene-based environmental sensors: Electrical and optical devices. *Molecules (Basel, Switzerland)*, *26*(8), 2165. DOI: 10.3390/molecules26082165 PMID: 33918751

Klingelhöfer, L., Kattelus, J., Verkama, E., Velasco, J., Iser, L., Rose, M., Karinen, R., & Puurunen, R. L. (2025). Catalytic hydrodenitrogenation of primary, secondary, and tertiary c12-alkyl amines over a platinum on zirconia catalyst. *Catalysis Science & Technology*, *15*(4), 1259–1271. DOI: 10.1039/D4CY01099J

Knox, F. S., Sauermilch, P. W., Wachtel, T. L., Mccahan Jr, G. R., & Trevethan, W. P. (1979). A Fire Simulator/Shutter System for Testing Protective Fabrics and Calibrating Thermal Sensors (No. USAARL794).

Kostagiannakopoulou, C., Loutas, T. H., Sotiriadis, G., & Kostopoulos, V. (2021). Effects of graphene geometrical characteristics to the interlaminar fracture toughness of CFRP laminates. *Engineering Fracture Mechanics*, *245*, 107584. DOI: 10.1016/j.engfracmech.2021.107584

Kostagiannakopoulou, C., Loutas, T. H., Sotiriadis, G., Markou, A., & Kostopoulos, V. (2015). On the interlaminar fracture toughness of carbon fiber composites enhanced with graphene nano-species. *Composites Science and Technology*, *118*, 217–225. DOI: 10.1016/j.compscitech.2015.08.017

Kotsyubynsky, V., et al. "NaOH-assisted hydrothermal reduction of graphene oxide." Journal of Physics: Condensed Matter 36.49 (2024): 495701. https://doi.org/.DOI: 10.1088/1361-648X/ad75dc

Koubaa, A., Ammar, A., Abdelkader, M., Alhabashi, Y., & Ghouti, L. (2023). AERO: AI-enabled remote sensing observation with onboard edge computing in UAVs. *Remote Sensing (Basel), 15*(7), 1873. DOI: 10.3390/rs15071873

Kovtun, A., Jones, D., Dell'Elce, S., Treossi, E., Liscio, A., & Palermo, V. (2019). Accurate chemical analysis of oxygenated graphene-based materials using X-ray photoelectron spectroscopy. *Carbon, 143*, 268–275. DOI: 10.1016/j.carbon.2018.11.012

Kshirsagar, D., & Joshi, G. (2025). Surface modification of polymer graphene nanocomposites issues, routes, quantification, and emerging applications: A review. *ChemistrySelect, 10*(12), e202404581. Advance online publication. DOI: 10.1002/slct.202404581

Kuang, B., Song, W., Ning, M., Li, J., Zhao, Z., Guo, D., Cao, M., & Jin, H. (2018). Chemical reduction dependent dielectric properties and dielectric loss mechanism of reduced graphene oxide. *Carbon, 127*, 209–217. DOI: 10.1016/j.carbon.2017.10.092

Kuila, T., Bose, S., Mishra, A. K., Khanra, P., Kim, N. H., & Lee, J. H. (2012). Chemical functionalization of graphene and its applications. *Progress in Materials Science, 57*(7), 1061–1105. DOI: 10.1016/j.pmatsci.2012.03.002

Kulkarni, R., Jenamani, R. K., Pithani, P., Konwar, M., Nigam, N., & Ghude, S. D. (2019). Loss to the aviation economy due to winter fog in New Delhi during the winter of 2011–2016. *Atmosphere (Basel), 10*(4), 198. DOI: 10.3390/atmos10040198

Kumar, V., et al. "Estimation of number of graphene layers using different methods: A focused review." Materials 14.16 (2021): 4590. https://doi.org/.DOI: 10.3390/ma14164590

Kumar, A., Tan, C.-S., Kumar, N., Singh, P., Sharma, Y., Leu, J., Huang, E.-W., Winie, T., Wei, K.-H., & Tseng, T. Y. (2021). Pentafluoropyridine functionalized novel heteroatom-doped with hierarchical porous 3D cross-linked graphene for supercapacitor applications. *RSC Advances, 11*(43), 26892–26907. DOI: 10.1039/D1RA03911C

Kumar, H., Kanti, P., Dishana, D., Shukla, P., Paramasivam, P., & Dabelo, L. (2025). Digital twin–driven design and testing of laser shock processed aluminum–graphene composites for spacecraft docking tribology. *PLoS One, 20*(5), e0324501. DOI: 10.1371/journal.pone.0324501 PMID: 40424314

Kumar, N., Setshedi, K., Masukume, M., & Ray, S. S. (2022). Facile scalable synthesis of graphene oxide and reduced graphene oxide: Comparative investigation of different reduction methods. *Carbon Letters*, *32*(4), 1031–1046. DOI: 10.1007/s42823-022-00335-9

Kumar, P., Penta, S., & Mahapatra, S. P. (2020). Synthesis and Dielectric Relaxation Spectroscopy of Graphene Oxide Nanoparticles: Effects of Frequency, Thickness, and Temperature. *Journal of Electronic Materials*, *49*(10), 5801–5807. DOI: 10.1007/s11664-020-08355-9

Kumar, S. S. A., Uddin, M. N., Rahman, M. M., & Asmatulu, R. (2019). Introducing graphene thin films into carbon fiber composite structures for lightning strike protection. *Polymer Composites*, *40*(S1), E517–E525. DOI: 10.1002/pc.24850

Kusano, T., Oyama, N., Yoshida, H., & Tanaka, H. (2025). Varying the rheological behavior of a micellar solution via modified microscopic structures in the presence of graphene oxide. *RSC Advances*, *15*(12), 9034–9040. DOI: 10.1039/D5RA00366K

Kwan, Y. C. G., Ng, G. M., & Huan, C. H. A. (2015). Identification of functional groups and determination of carboxyl formation temperature in graphene oxide using the XPS O 1s spectrum. *Thin Solid Films*, *590*, 40–48. DOI: 10.1016/j.tsf.2015.07.051

Lalanne, C. (2014). *Mechanical vibration and shock analysis, fatigue damage* (Vol. 4). John Wiley & Sons.

Laurenzi, S., Zaccardi, F., Toto, E., Santonicola, M. G., Botti, S., & Scalia, T. (2024). Fused Filament Fabrication of Polyethylene/Graphene Composites for In-Space Manufacturing. *Materials (Basel)*, *17*(8), 1888. DOI: 10.3390/ma17081888 PMID: 38673245

Lava Kumar, P., Lombardi, A., Byczynski, G., Narayana Murty, S. V. S., Murty, B. S., & Bichler, L. (2022). Recent advances in aluminium matrix composites reinforced with graphene-based nanomaterial: A critical review. *Progress in Materials Science*, *128*, 100948. DOI: 10.1016/j.pmatsci.2022.100948

Lebedeva, I. V., Vyrko, S. A., Sinitsa, A. S., Ratkevich, S. V., Popov, A. M., Knizhnik, A. A., Poklonski, N. A., & Lozovik, Y. E. (2024). Magnetic and electronic properties of 1D hybrid nanoobjects composed of alternating polycyclic hydrocarbon regions and double carbon chains. *Computational & Theoretical Chemistry*, *1241*, 114913. DOI: 10.1016/j.comptc.2024.114913

Lee, C., Wei, X., Kysar, J. W., & Hone, J. (2008). Measurement of the elastic properties and intrinsic strength of monolayer graphene. science, 321(5887), 385-388.

Lee, J. S. (2016). Fabrication of Zinc Oxide Nanostructure/Poly (vinylidene fluoride) Hybrid Thin Films and Their Application for Pressure Sensors and Acoustic Actuators].

Lee, A. Y., Yang, K., Anh, N. D., Park, C., Lee, S. M., Lee, T. G., & Jeong, M. S. (2021). Raman study of D* band in graphene oxide and its correlation with reduction. *Applied Surface Science*, *536*, 147990. DOI: 10.1016/j.apsusc.2020.147990

Lee, B., Chu, W., & Li, W. (2020). Effects of process parameters on graphene growth via low-pressure chemical vapor deposition. *Journal of Micro-and Nano-Manufacturing*, *8*(3), 031005. DOI: 10.1115/1.4048494

Lee, I., Kim, S., Shim, J., Cho, J., & Cho, C. (2025). High flame retardancy of carbon nanotubes reinforced polyelectrolyte multilayered nanocomposites. *Journal of Applied Polymer Science*, *142*(22), e56967. Advance online publication. DOI: 10.1002/app.56967

Leong, K. H., Ramakrishna, S., Huang, Z. M., & Bibo, G. A. (2000). The potential of knitting for engineering composites—A review. *Composites. Part A, Applied Science and Manufacturing*, *31*(3), 197–220. DOI: 10.1016/S1359-835X(99)00067-6

Li, S. (2023). Applications of Graphene: Supercapacitors and Gas Sensors. MATEC Web of Conferences.

Liang, H., Guo, L., Niu, Y., Tang, Z., Zhao, Z., Mei, H., Fang, R., Liu, C., & Gong, W. (2025). High-sensitivity, low-hysteresis, flexible humidity sensors based on carboxyl-functionalized reduced-graphene oxide/ag nanoclusters. *Nanomaterials (Basel, Switzerland)*, *15*(11), 800. DOI: 10.3390/nano15110800

Liang, S., Wu, X., Wu, Z., & Liu, W. (2025). Progress and Challenges on Materials Used to Optimize Flight Efficiency, Improve Safety and Reliability, and Reduce Cost of Electric Vertical Take-Off and Landing Aircraft. *Journal of Applied Polymer Science*, *142*(24), 57029. DOI: 10.1002/app.57029

Liang, Y. L., & Pearson, R. A. (2010). The toughening mechanism in hybrid epoxy-silica-rubber nanocomposites (HESRNs). *Polymer*, *51*(21), 4880–4890. DOI: 10.1016/j.polymer.2010.08.052

Li, F., Guo, R., Qiu, Y., Liu, Z., Tao, L., Yang, S., Chen, J., & Yang, H. (2025). Enhanced anticancer activity of graphene oxide quantum dot@cu nanocomposites via cation–π interactions. *Physical Chemistry Chemical Physics*, *27*(14), 7076–7083. DOI: 10.1039/D5CP00352K

Li, F., Ma, Y., Wei, X., Gao, L., Xing, E., Gong, J., Luo, Y., & Shu, X. (2025). Tandem oligomerization of ethylene for production of jet fuel over fer and mfi zeolites. *Industrial & Engineering Chemistry Research, 64*(11), 5923–5935. DOI: 10.1021/acs.iecr.4c04755

Li, J., Sollami Delekta, S., Zhang, P., Yang, S., Lohe, M. R., Zhuang, X., Feng, X., & Ostling, M. (2017). Scalable fabrication and integration of graphene microsupercapacitors through full inkjet printing. *ACS Nano, 11*(8), 8249–8256. DOI: 10.1021/acsnano.7b03354 PMID: 28682595

Li, J., Wang, M., & Zhao, Y. (2025). Self-assembly of low-dimensional organic graphene-like molecules at interfaces. *Journal of Physics: Conference Series, 2954*(1), 012007. DOI: 10.1088/1742-6596/2954/1/012007

Li, K., Khanna, R., Zhang, H., Ma, S., Liang, Z., Li, G., Barati, M., & Zhang, J. (2021). Thermal behaviour during initial stages of graphene oxidation: Implications for reaction kinetics and mechanisms. *Chemical Engineering Journal, 421*, 129742. DOI: 10.1016/j.cej.2021.129742

Likhi, F. H., Singh, M., Chavan, S. V., Cao, T., Shanbedi, M., & Karim, A. (2023). Effects of film confinement on dielectric and electrical properties of graphene oxide and reduced graphene oxide-based polymer nanocomposites: Implications for energy storage. *ACS Applied Nano Materials, 6*(13), 11699–11714. DOI: 10.1021/acsanm.3c01674

Lim, H. S., Oh, J. W., Kim, S. Y., Yoo, M. J., Park, S. D., & Lee, W. S. (2013). Anisotropically alignable magnetic boron nitride platelets decorated with iron oxide nanoparticles. *Chemistry of Materials, 25*(16), 3315–3319. DOI: 10.1021/cm401488a

Lim, Y. J., Carolan, D., & Taylor, A. C. (2016). Simultaneously tough and conductive rubber–graphene–epoxy nanocomposites. *Journal of Materials Science, 51*(18), 8631–8644. DOI: 10.1007/s10853-016-0122-2

Lin, D., Richard Liu, C., & Cheng, G. J. (2014). Single-layer graphene oxide reinforced metal matrix composites by laser sintering: Microstructure and mechanical property enhancement. *Acta Materialia, 80*, 183–193. DOI: 10.1016/j.actamat.2014.07.038

Lin, J., Peng, Z., Liu, Y., Ruiz-Zepeda, F., Ye, R., Samuel, E. L., Yacaman, M. J., Yakobson, B. I., & Tour, J. M. (2014). Laser-induced porous graphene films from commercial polymers. *Nature Communications, 5*(1), 5714. DOI: 10.1038/ncomms6714 PMID: 25493446

Liu, F., Wang, C., Sui, X., Riaz, M. A., Xu, M., Wei, L., & Chen, Y. (2019). Synthesis of graphene materials by electrochemical exfoliation: Recent progress and future potential. *Carbon Energy*, *1*(2), 173–199. DOI: 10.1002/cey2.14

Liu, H.-Y., Wang, G.-T., Mai, Y.-W., & Zeng, Y. (2011). On fracture toughness of nano-particle modified epoxy. *Composites. Part B, Engineering*, *42*(8), 2170–2175. DOI: 10.1016/j.compositesb.2011.05.014

Liu, M., Wang, L., & Yu, G. (2024). Recent research progress of porous graphene and applications in molecular sieve, sensor, and supercapacitor. *Small*, *20*(42), 2401767. DOI: 10.1002/smll.202401767

Liu, S., Li, Z., Feng, Y., Zhang, T., Ma, B., & Wang, W. (2024). Effect of ultraviolet radiation on the low-velocity impact and compression after impact performances of aerospace composite laminates. *Polymer Composites*, *45*(4), 3284–3295. DOI: 10.1002/pc.27990

Liu, S., Zeng, T. H., Hofmann, M., Burcombe, E., Wei, J., Jiang, R., Kong, J., & Chen, Y. (2011). Antibacterial activity of graphite, graphite oxide, graphene oxide, and reduced graphene oxide: Membrane and oxidative stress. *ACS Nano*, *5*(9), 6971–6980. DOI: 10.1021/nn202451x PMID: 21851105

Liu, W., & Speranza, G. (2021). Tuning the oxygen content of reduced graphene oxide and effects on its properties. *ACS Omega*, *6*(9), 6195–6205. DOI: 10.1021/acsomega.0c05578

Liu, Y. J., Liu, X. L., & Zhang, Y. Q. (2019). Study on the electromagnetic properties and mechanical properties of graphene/graphite single-layer coating fabric. *Journal of Textile Science & Engineering*, *36*(2), 1–5.

Liu, Y., Fei, B., & Xin, J. H. (2023). Functionalization of fabrics with Graphene-based coatings: Mechanisms, Approaches, and Functions. *Coatings*, *13*(9), 1580. DOI: 10.3390/coatings13091580

Liu, Y., Zou, A., Wang, G., Han, C., & Blackie, E. (2022a). Enhancing interlaminar fracture toughness of CFRP laminates with hybrid carbon nanotube/graphene oxide fillers. *Diamond and Related Materials*, *128*, 109285. DOI: 10.1016/j.diamond.2022.109285

Li, Y., Liao, Y., & Su, Z. (2018). Graphene-functionalized polymer composites for self-sensing of ultrasonic waves: An initiative towards "sensor-free" structural health monitoring. *Composites Science and Technology*, *168*, 203–213. DOI: 10.1016/j.compscitech.2018.09.021

Li, Y., Meng, J., Luo, J., Wang, P., Ma, J., Zhao, Z., & Lei, H. (2023). Cryogenic mechanics and damage behaviors of carbon fiber reinforced polymer composites. *Composites. Part A, Applied Science and Manufacturing*, *169*, 107484. DOI: 10.1016/j.compositesa.2023.107484

Li, Y., Zhang, D., Ye, Z., Ye, G., He, R., Wang, H., & Cong, W. (2022). The reinforcement mechanisms of graphene oxide in laser-directed energy deposition fabricated metal and ceramic matrix composites: A comparison study. *International Journal of Advanced Manufacturing Technology*, *119*(3–4), 1975–1988. DOI: 10.1007/s00170-021-08337-z

Li, Y., Zhang, H., Liu, Y., Wang, H., Huang, Z., Peijs, T., & Bilotti, E. (2018). Synergistic effects of spray-coated hybrid carbon nanoparticles for enhanced electrical and thermal surface conductivity of CFRP laminates. *Composites. Part A, Applied Science and Manufacturing*, *105*, 9–18. DOI: 10.1016/j.compositesa.2017.10.032

Li, Y., Zhang, H., Porwal, H., Huang, Z., Bilotti, E., & Peijs, T. (2017). Mechanical, electrical and thermal properties of in-situ exfoliated graphene/epoxy nanocomposites. *Composites. Part A, Applied Science and Manufacturing*, *95*, 229–236. DOI: 10.1016/j.compositesa.2017.01.007

Li, Z. Q., Lu, C. J., Xia, Z. P., Zhou, Y., & Luo, Z. (2007). X-ray diffraction patterns of graphite and turbostratic carbon. *Carbon*, *45*(8), 1686–1695. DOI: 10.1016/j.carbon.2007.03.038

Lopez, R. M., White, J. R., Truong, L., & Tanguay, R. L. (2022). Size- and Oxidation-Dependent toxicity of graphene oxide nanomaterials in embryonic zebrafish. *Nanomaterials (Basel, Switzerland)*, *12*(7), 1050. DOI: 10.3390/nano12071050 PMID: 35407167

Lu, H., Yao, Y., Huang, W. M., & Hui, D. (2014). Noncovalently functionalized carbon fiber by grafted self-assembled graphene oxide and the synergistic effect on polymeric shape memory nanocomposites. *Composites. Part B, Engineering*, *67*, 290–295. DOI: 10.1016/j.compositesb.2014.07.022

Luo, T., & Lloyd, J. R. (2012). Enhancement of thermal energy transport across graphene/graphite and polymer interfaces: A molecular dynamics study. *Advanced Functional Materials*, *22*(12), 2495–2502. DOI: 10.1002/adfm.201103048

Lutomia, D., Poria, R., Kala, D., Singh, A., Gupta, M., Kumar, D., Kaushal, A., & Gupta, S. (2025). Unlocking the potential of 2d nanomaterial-based biosensors in biomarker-based detection of helicobacter pylori. *Materials Advances*, *6*(1), 117–142. DOI: 10.1039/D4MA00546E

Lu, Y., Guo, Z.-S., & Fan, S.-C. (2021). An ultrahigh-sensitivity graphene resonant gyroscope. *Nanomaterials (Basel, Switzerland)*, *11*(8), 1890. DOI: 10.3390/nano11081890 PMID: 34443720

Lv, S., et al. "CVD graphene with high electrical conductivity: empowering applications." 2D Materials 12 (2024): 013003. https://doi.org/.DOI: 10.1088/2053-1583/ad8800

Lyashenko-Miller, T., & Marom, G. (2017). Delamination fracture toughness of UHMWPE fibers/polyurethane laminates interleaved with carbon nanotube-reinforced polyurethane films. *Polymers for Advanced Technologies*, *28*(5), 606–612. DOI: 10.1002/pat.3848

Macosko, C. (2010). Graphene/Polymer Nanocomposites. APS March Meeting Abstracts

Madhad, H. V., Mishra, N. S., Patel, S. B., Panchal, S. S., Gandhi, R. A., & Vasava, D. V. (2021). Graphene/graphene nanoplatelets reinforced polyamide nanocomposites: A review. *High Performance Polymers*, *33*(9), 981–997. DOI: 10.1177/09540083211011216

Madhukar, P., Kesarwani, R., Pandey, S., Singh, V. K., Rao Gedda, M., Singh, O. P., Shaz, M., Kumar, R., & Sundar, S. (2025). Amine-functionalized graphene quantum dots conjugated with amphotericin b: Synthesis, characterization, and in vitro evaluation for visceral leishmaniasis treatment. *ACS Omega*, *10*(25), 26627–26638. Advance online publication. DOI: 10.1021/acsomega.5c00879

Mageed, A. K., Dayang Radiah, A. B., Salmiaton, A., Izhar, S., & Razak, M. A. (2018). Study the Thermal Stability of Nitrogen Doped Reduced Graphite Oxide Supported Copper Catalyst. *Journal of Cluster Science*, *29*(4), 709–718. DOI: 10.1007/s10876-018-1382-6

Maji, A., Kuila, C., Murmu, N. C., & Kuila, T. (2024). Facile fabrication of covalently grafted functionalized graphene-PVDF composites as strain sensor with tunable sensitivity: Application to the structural health and human motion monitoring. *Sensors and Actuators. A, Physical*, *369*, 115081.

Malard, L. M., et al. "Raman spectroscopy in graphene." Physics Reports 473.5-6 (2009): 51-87. https://doi.org/.DOI: 10.1016/j.physrep.2009.02.003

Maldonado-Romo, J., Jami, A., Sant, A., Montesinos, L., & Ponce, P. (2025). Biomaterials for aerospace applications: A sustainable technological analysis. *Journal of Physics: Conference Series*, *2946*(1), 012006. DOI: 10.1088/1742-6596/2946/1/012006

Mallick, P. K. (2008). *Fiber-reinforced composites : materials, manufacturing, and design*. CRC Press.

Mammadov, A. G., Abaszade, R. G., Kotsyubynsky, V. O., Gur, E. Y., Bayramov, I. Y., & Khanmamadova, E. A. (2022). Photoconductivity of carbon nanotubes. *Technical and Physical Problems of Engineering*, *14*(3), 155–160.

Marconcini, P., & Macucci, M. (2022). Transport Simulation of Graphene Devices with a Generic Potential in the Presence of an Orthogonal Magnetic Field. *Nanomaterials (Basel, Switzerland)*, *12*(7), 1087. DOI: 10.3390/nano12071087 PMID: 35407205

Marcus, P. (2011). *Corrosion mechanisms in theory and practice* (3rd ed.). CRC Press.

Mardanshahi, A., Sreekumar, A., Yang, X., Barman, S. K., & Chronopoulos, D. (2025). Sensing Techniques for Structural Health Monitoring: A State-of-the-Art Review on Performance Criteria and New-Generation Technologies. *Sensors (Basel)*, *25*(5), 1424. DOI: 10.3390/s25051424 PMID: 40096243

Marmur, A. (2004). The lotus effect: Superhydrophobicity and metastability. *Langmuir*, *20*(9), 3517–3519. DOI: 10.1021/la036369u PMID: 15875376

Marquez, C., Rodriguez, N., Ruiz, R., & Gamiz, F. (2016). Electrical characterization and conductivity optimization of laser reduced graphene oxide on insulator using point-contact methods. *RSC Advances*, *6*(52), 46231–46237. DOI: 10.1039/C6RA03630A

Marra, F., Lecini, J., Tamburrano, A., Pisu, L., & Sarto, M. S. (2020). Broadband Electromagnetic Absorbing Structures Made of Graphene/Glass-Fiber/Epoxy Composite. *IEEE Transactions on Microwave Theory and Techniques*, *68*(2), 590–601. DOI: 10.1109/TMTT.2019.2950223

Mason, M. (2022). Recent Langley Aerothermodynamic Laboratory Testing (2019-2022). Thermal and Fluids Analysis Workshop 2022

Meher, A., Mahapatra, M. M., Samal, P., & Vundavilli, P. R. (2022). A review on manufacturability of magnesium matrix composites: Processing, tribology, joining, and machining. *CIRP Journal of Manufacturing Science and Technology*, *39*, 134–158. DOI: 10.1016/j.cirpj.2022.07.012

Menezes, I. R., Araújo, N. R. S., Araújo, B. C. R., Sakai, T., Lago, R. M., & Sebastião, R. C. O. (2023). Shedding light on the mechanism of graphene oxide thermal decomposition: A kinetic study using isoconversional method and artificial neural network. *Thermochimica Acta*, *721*, 179454. DOI: 10.1016/j.tca.2023.179454

Meng, G., Li, Y., Xing, J., Liu, H., & Zhao, F. (2020). Multifunctional waterborne polyurethane nanocomposite coatings with excellent anticorrosion and antiicing performance. *Progress in Organic Coatings*, *142*, 105584.

Meng, Q., Zhao, J., Zhou, Z., Han, S., Feng, Y., Han, Q., & Liu, T. (2024). Damage monitoring of aircraft structural components based on large-area flexible graphene strain sensors. *Sensors and Actuators. A, Physical*, *369*, 115092.

Mirabedini, A., Ang, A., Nikzad, M., Fox, B., Lau, K. T., & Hameed, N. (2020). Evolving strategies for producing multiscale graphene-enhanced fiber-reinforced polymer composites for smart structural applications. *Advancement of Science*, *7*(11), 1903501. DOI: 10.1002/advs.201903501 PMID: 32537406

Miracle, D. B. (2005). Metal matrix composites – From science to technological significance. *Composites Science and Technology*, *65*(15–16), 2526–2540. DOI: 10.1016/j.compscitech.2005.05.027

Mishra, R. K., Thomas, M. G., Abraham, J., Joseph, K., & Thomas, S. (2018). Electromagnetic interference shielding materials for aerospace applications: A state of the art. Advanced materials for electromagnetic shielding: fundamentals, properties, and applications, 327-365.

Mohan, P. (2013). A Critical Review: The Modification, Properties, and Applications of Epoxy Resins. *Polymer-Plastics Technology and Engineering*, *52*(2), 107–125. DOI: 10.1080/03602559.2012.727057

Mohapatra, D., Behera, D. R., & Mishra, D. K. (2023). Prospective of functionalized graphene as shape memory and self-healing polymer: A review. *Materials Today: Proceedings*.

Moore, A. L., & Shi, L. (2014). Emerging challenges and materials for thermal management of electronics. *Materials Today*, *17*(4), 163–174. DOI: 10.1016/j.mattod.2014.04.003

Morales-Flórez, V., & Domínguez-Rodríguez, A. (2022). Mechanical properties of ceramics reinforced with allotropic forms of carbon. *Progress in Materials Science*, *128*, 100966. DOI: 10.1016/j.pmatsci.2022.100966

Mouritz, A. P. (2007). Review of z-pinned composite laminates. *Composites. Part A, Applied Science and Manufacturing*, *38*(12), 2383–2397. DOI: 10.1016/j.compositesa.2007.08.016

Moyer, K., Meng, C., Marshall, B., Assal, O., Eaves, J., Perez, D., Karkkainen, R., Roberson, L., & Pint, C. L. (2020). Carbon fiber reinforced structural lithium-ion battery composite: Multifunctional power integration for CubeSats. *Energy Storage Materials*, *24*, 676–681. DOI: 10.1016/j.ensm.2019.08.003

Mulqueen, D. W., Sattar, S., Le, T., & Kravchenko, O. G. (2022). Spray deposition of sustainable plant based graphene in thermosetting carbon fiber laminates for mechanical, thermal, and electrical properties. *Polymer Composites*, *43*(12), 8696–8714. DOI: 10.1002/pc.27051

Mu, X. N., Cai, H. N., Zhang, H. M., Fan, Q. B., Zhang, Z. H., Wu, Y., Ge, Y. X., & Wang, D. D. (2018). Interface evolution and superior tensile properties of multi-layer graphene reinforced pure Ti matrix composite. *Materials & Design*, *140*, 431–441. DOI: 10.1016/j.matdes.2017.12.016

Nagaraj, P. (2023). Study of Fatigue Damage Detection by Current Information of Measurement System Riga Technical University Rīga, LV, USA].

Nagashio, K., Nishimura, T., Kita, K., & Toriumi, A. (2009). Mobility variations in mono-and multi-layer graphene films. *Applied Physics Express*, *2*, 025003. DOI: 10.1143/APEX.2.025003

Najmi, M. I.. (2024). Identification of Crystal Structure, Surface Area, and Magnetic Properties of Mn0.25Fe2.75O4/rGO Nanocomposites. *Journal of Metastable and Nanocrystalline Materials*, *38*, 23–30. DOI: 10.4028/p-GLu976

Nakata, K., & Fujishima, A. (2012). TiO_2 photocatalysis: Design and applications. *Journal of Photochemistry and Photobiology C, Photochemistry Reviews*, *13*(3), 169–189. DOI: 10.1016/j.jphotochemrev.2012.06.001

National Academies Press (US). (2000). *Incineration processes and environmental releases*. Waste Incineration & Public Health - NCBI Bookshelf. https://www.ncbi.nlm.nih.gov/books/NBK233627/

Navarro-Pardo, F., Martinez-Hernandez, A. L., & Velasco-Santos, C. (2016). Carbon nanotube and graphene based polyamide electrospun nanocomposites: A review. *Journal of Nanomaterials*, *2016*(1), 3182761. DOI: 10.1155/2016/3182761

Nawaz, M., Islam, M., Nazir, M. A., Bano, I., Gul, I., & Ajmal, M. (2021). Transport properties in spinel ferrite/graphene oxide nanocomposites for electromagnetic shielding. *Ceramics International*, *47*(18), 25505–25513. DOI: 10.1016/j.ceramint.2021.05.274

Nie, Y., Li, X., Uddin, M., Liu, N., Yu, S., Keomeesay, P., Olalekan, O., Li, S., Chen, Y., Zhao, X., Ma, L., Zheng, Z., & Liu, S. (2025). Rational design of an efficient PT_3Cu/TiO_2 icosahedral catalyst for bio-aviation fuel production under mild conditions. *Green Chemistry*, *27*(22), 6449–6463. DOI: 10.1039/D5GC01235J

Nikfar, M., Taati, E., & Asghari, M. (2022). On the theoretical and molecular dynamic methods for natural frequencies of multilayer graphene nanosheets incorporating nonlocality and interlayer shear effects. *Mechanics of Advanced Materials and Structures*, *29*(20), 2873–2883. DOI: 10.1080/15376494.2021.1880675

Nishi, Y., & Inagaki, M. (2016). *Gas adsorption/desorption isotherm for pore structure characterization*. Materials Science and Engineering of Carbon., DOI: 10.1016/B978-0-12-805256-3.00011-8

Nooralian, Z., Gashti, M. P., & Ebrahimi, I. (2016). Fabrication of a multifunctional graphene/polyvinylphosphonic acid/cotton nanocomposite via facile spray layer-by-layer assembly. *RSC Advances*, *6*(28), 23288–23299. DOI: 10.1039/C6RA00296J

Novoselov, K. S., Colombo, L., Gellert, P., Schwab, M., & Kim, K. (2012). A road-map for graphene. nature, 490(7419), 192-200.

Novoselov, K. S., Geim, A. K., Morozov, S. V., Jiang, D. E., Zhang, Y., Dubonos, S. V., ... & Firsov, A. A. (2004). Electric field effect in atomically thin carbon films. science, 306(5696), 666-669.

Novoselov, K. S., Geim, A. K., Morozov, S. V., Jiang, D.-e., Zhang, Y., Dubonos, S. V.,…Firsov, A. A. (2004). Electric field effect in atomically thin carbon films. science, 306(5696), 666-669.

Novoselov, K. S., Geim, A. K., Morozov, S. V., Jiang, D., Zhang, Y., Dubonos, S. V., Grigorieva, I. V., & Firsov, A. A. (2004). Electric field effect in atomically thin carbon films. *Science*, *306*(5696), 666–669. DOI: 10.1126/science.1102896

O'Neill, C., Johnson, M. B., DeArmond, D., Zhang, L., Alvarez, N., Shanov, V. N., & White, M. A. (2021). Thermal conductivity of 3-dimensional graphene papers. *Carbon Trends*, *4*, 100041. DOI: 10.1016/j.cartre.2021.100041

Oh, M. J., & Yoo, P. J. (2020). Graphene-based 3D lightweight cellular structures: Synthesis and applications. *Korean Journal of Chemical Engineering 2020 37:2*, *37*(2), 189–208. DOI: 10.1007/s11814-019-0437-1

Ojalvo, C., Ayllón, M., Ortiz, A. L., & Moreno, R. (2021). Aqueous tape casting of super-hard B4C laminates with rGO-enriched reinforcing interlayers. *Journal of the European Ceramic Society*, *41*(11), 5457–5465. DOI: 10.1016/j.jeurceramsoc.2021.05.001

Oliveira, F. M., Luxa, J., Bouša, D., Sofer, Z. K., & Gusmao, R. (2022). Electromagnetic interference shielding by reduced graphene oxide foils. *ACS Applied Nano Materials*, *5*(5), 6792–6800. DOI: 10.1021/acsanm.2c00785

Oluremi, D. (2025). Industrial Waste as a Source for Graphene Nanosheets.

Ordoño, J., Islam, M., Lantada, A., Echeverry-Rendón, M., & Wang, D. (2025). Evaluation of laser-induced graphene for skeletal muscle tissue engineering applications. *Biomaterials Science*, *13*(12), 3242–3251. DOI: 10.1039/D5BM00174A

Otsuka, H., Urita, K., Honma, N., Kimuro, T., Amako, Y., Kukobat, R., Bandosz, T. J., Ukai, J., Moriguchi, I., & Kaneko, K. (2024). Transient chemical and structural changes in graphene oxide during ripening. *Nature Communications*, *15*(1), 1708. DOI: 10.1038/s41467-024-46083-4

Ozlek, M., Akbulut, M., & Burgaz, E. (2025). Exploring graphene's impact on graphite/pani matrix composites: High-pressure fabrication and enhanced thermal-electrical properties. *Nanoscale*, *17*(1), 552–566. DOI: 10.1039/D4NR03171G PMID: 39575894

Pacakova, B., Vejpravova, J., Repko, A., Mantlikova, A., & Kalbac, M. (2015). Formation of wrinkles on graphene induced by nanoparticles: Atomic force microscopy study. *Carbon*, *95*, 573–579. DOI: 10.1016/j.carbon.2015.08.043

Painkal, S. K., Balachandran, M., Jayanarayanan, K., Sridhar, N., & Kumar, S. (2025). Synergistic enhancement in mechanical properties of graphene/MWCNT reinforced Polyaryletherketone – carbon fiber multi-scale composites: Experimental studies and finite element analysis. *Advanced Industrial and Engineering Polymer Research*, *8*(1), 20–36. DOI: 10.1016/j.aiepr.2024.02.002

Paliotta, L., De Bellis, G., Tamburrano, A., Marra, F., Rinaldi, A., Balijepalli, S. K., Kaciulis, S., & Sarto, M. S. (2015). Highly conductive multilayer-graphene paper as a flexible lightweight electromagnetic shield. *Carbon*, *89*, 260–271. DOI: 10.1016/j.carbon.2015.03.043

Pal, O. K.. (2023). A comprehensive review of AI-enabled unmanned aerial vehicle: Trends, vision, and challenges. *arXiv.* https://arxiv.org/abs/2310.16360

Pamuk, G., & Çeken, F. (2008). Manufacturing of Weft-Knitted Fabric Reinforced Composite Materials: A Review. *Materials and Manufacturing Processes*, *23*(7), 635–640. DOI: 10.1080/10426910802316484

Panghal, A., Kumar, Y., Mishra, P. K., Mathur, A., & Srivastava, A. K. (2025). Graphene-Based Polymer Composites for Aerospace, Electronic, Energy, and Biomedical Applications. In *Multifunctional Materials* (pp. 81–104). Wiley., DOI: 10.1002/9781394234134.ch4

Panghal, A., Kumar, Y., Mishra, P. K., Mathur, A., & Srivastava, A. K. (2025). Graphene-Based Polymer Composites for Aerospace, Electronic, Energy, and Biomedical Applications. *Multifunctional Materials*, •••, 81–104. https://doi.org/ https://doi.org/1002/9781394234134.ch4

Paredes, J. I., Villar-Rodil, S., Martínez-Alonso, A., & Tascón, J. M. D. (2008). Graphene oxide dispersions in organic solvents. *Langmuir*, *24*(19), 10560–10564. DOI: 10.1021/la801744a

Park, S., An, J., Potts, J. R., Velamakanni, A., Murali, S., & Ruoff, R. S. (2011). Hydrazine-reduction of graphite-and graphene oxide. carbon, 49(9), 3019-3023.

Parker, R., Kelly, M., Watson-Murphy, T., Ghaani, M., & Dooley, S. (2025). Predictive modeling of liquid density and surface tension for sustainable aviation fuels using nuclear magnetic resonance atom types. *Energy & Fuels*, *39*(7), 3690–3702. DOI: 10.1021/acs.energyfuels.4c05601 PMID: 40007898

Parkins, R. N. (1994). Stress corrosion cracking. In L. L. Shreir (Ed.), *Corrosion* (Vol. 1, 3rd ed., pp. 7:3–7:50). Butterworth-Heinemann.

Pathak, A. K., Borah, M., Gupta, A., Yokozeki, T., & Dhakate, S. R. (2016a). Improved mechanical properties of carbon fiber/graphene oxide-epoxy hybrid composites. *Composites Science and Technology*, *135*, 28–38. DOI: 10.1016/j.compscitech.2016.09.007

Patnaik, T. K. (2024). 13 Graphene-Based Materials. *Green Nanobiotechnology*, *9*, 177.

Pegorin, F., Pingkarawat, K., Daynes, S., & Mouritz, A. P. (2015). Influence of z-pin length on the delamination fracture toughness and fatigue resistance of pinned composites. *Composites. Part B, Engineering*, *78*, 298–307. DOI: 10.1016/j.compositesb.2015.03.093

Pei, S., & Cheng, H.-M. (2012). The reduction of graphene oxide. *Carbon*, *50*(9), 3210–3228. DOI: 10.1016/j.carbon.2011.11.010

Pereira dos Santos Tonello, K., Padovano, E., Badini, C., Biamino, S., Pavese, M., & Fino, P. (2016). Fabrication and characterization of laminated SiC composites reinforced with graphene nanoplatelets. *Materials Science and Engineering A*, *659*, 158–164. DOI: 10.1016/j.msea.2016.02.050

Periasamy, K., Kandare, E., Das, R., Darouie, M., & Khatibi, A. A. (2023). Interfacial Engineering Methods in Thermoplastic Composites: An Overview. *Polymers 2023, Vol. 15, Page 415, 15*(2), 415. DOI: 10.3390/polym15020415

Pham, P. V. (2018). Atmospheric pressure chemical vapor deposition of graphene. *Chem. Vap. Depos. Nanotechnol, 6*, 115–134.

Pham, V. H., Cuong, T. V., Hur, S. H., Shin, E. W., Kim, J. S., Chung, J. S., & Kim, E. J. (2010). Fast and simple fabrication of a large transparent chemically-converted graphene film by spray-coating. *Carbon, 48*(7), 1945–1951. DOI: 10.1016/j.carbon.2010.01.062

Phiri, J., Johansson, L. S., Gane, P., & Maloney, T. (2018). A comparative study of mechanical, thermal and electrical properties of graphene-, graphene oxide- and reduced graphene oxide-doped microfibrillated cellulose nanocomposites. *Composites. Part B, Engineering, 147*, 104–113. DOI: 10.1016/j.compositesb.2018.04.018

Pimenta, M. A., Dresselhaus, G., Dresselhaus, M. S., Cançado, L. G., Jorio, A., & Saito, R. (2007). Studying disorder in graphite-based systems by Raman spectroscopy. *Physical Chemistry Chemical Physics, 9*(11), 1276–1290. DOI: 10.1039/B613962K

Pina, S., Oliveira, J. M., & Reis, R. L. (2015). Natural-based nanocomposites for bone tissue engineering and regenerative medicine: A review. *Advanced Materials, 27*(7), 1143–1169. DOI: 10.1002/adma.201403354 PMID: 25580589

Pinder, J. W., Major, G. H., Baer, D. R., Terry, J., Whitten, J. E., Čechal, J., Crossman, J. D., Lizarbe, A. J., Jafari, S., Easton, C. D., Baltrusaitis, J., van Spronsen, M. A., & Linford, M. R. (2024). Avoiding common errors in X-ray photoelectron spectroscopy data collection and analysis, and properly reporting instrument parameters. *Applied Surface Science Advances, 19*, 100534. DOI: 10.1016/j.apsadv.2023.100534

Polayya, C., Rao, C. S. P., Veeresh Kumar, G. B., & Surakasi, R. (2023). Synthesis and mechanical properties of graphene nanoparticles reinforced with aluminium alloy matrix composites. *Nanotechnology for Environmental Engineering, 8*(2), 469–480. DOI: 10.1007/s41204-022-00305-6

Polynskaya, Y. G., Lebedeva, I. V., Knizhnik, A. A., & Popov, A. M. (2022). Optimal model of semi-infinite graphene for ab initio calculations of reactions at graphene edges by the example of zigzag edge reconstruction. *Computational & Theoretical Chemistry, 1214*, 113755. DOI: 10.1016/j.comptc.2022.113755

Ponomarenko, L. A., et al. "Extreme electron–hole drag and negative mobility in the Dirac plasma of graphene." Nature Communications 15 (2024): 9869. https://doi.org/. Robinson, J., Hollander, M., and Datta, S. Epitaxial Graphene. 2D Materials for Nanoelectronics, 37.DOI: 10.1038/s41467-024-54198-x

Poonpipat, Y., Leelachai, K., Pearson, R. A., & Dittanet, P. (2017). Fracture behavior of silica nanoparticles reinforced rubber/epoxy composite. *Journal of Reinforced Plastics and Composites*, *36*(16), 1156–1167. DOI: 10.1177/0731684417709952

Prakash, D., Agarwal, A., Khalid, S., Nishad, D. K., & Singh, V. K. (2024). Innovative approaches to fog dispersal: The emerging role of unmanned aerial vehicles. In *2024 IEEE 11th Uttar Pradesh Section International Conference on Electrical, Electronics and Computer Engineering (UPCON).*

Prakash, D., Agarwal, A., Khalid, S., Nishad, D. K., & Singh, V. K. (2024). Innovative approaches to fog dispersal: The emerging role of unmanned aerial vehicles. In *11th IEEE UP Section International Conference on Electrical, Electronics and Computer Engineering (UPCON-2024),* Lucknow, India. DOI: 10.1109/UPCON62832.2024.10983532

Pramodkumar, B., & Budhe, S. (2024). The effect of graphene oxide on thermal, electrical, and mechanical properties of carbon/epoxy composites: Towards multifunctional composite material. *Polymer Composites*, *45*(7), 6374–6384. DOI: 10.1002/pc.28203

Prater, T., Tilson, W., Vickers, J., Singhal, S., Zakrajsek, J., & Ransom, J. (2020). Breakthrough Materials for Space Applications Workshop. Workshop at Jackson Center

Prekodravac, J. R., Kepić, D. P., Colmenares, J. C., Giannakoudakis, D. A., & Jovanović, S. P. (2021). A comprehensive review on selected graphene synthesis methods: From electrochemical exfoliation through rapid thermal annealing towards biomass pyrolysis. *Journal of Materials Chemistry. C, Materials for Optical and Electronic Devices*, *9*(21), 6722–6748. DOI: 10.1039/D1TC01316E

Price, M. R. (2017). Quantifying the Protective Properties and Adhesion to the Substrate of Ultra-thin Multilayer Diamond-like Carbon Coatings Using Molecular Dynamics Simulation The University of Utah].

Prime, R. B., et al. "Thermogravimetric analysis (TGA)." Thermal Analysis of Polymers: Fundamentals and Applications (2009): 241-317. https://doi.org/.DOI: 10.1002/9780470423837

Prokhorov, E., Barquera-Bibiano, Z., Manzano-Ramírez, A., Luna-Barcenas, G., Kovalenko, Y., Hernández-Landaverde, M. A., Reyes, B. E. C., & Vargas, J. H. (2019). New insights in graphene oxide dielectric constant. *Materials Research Express*, *6*(8), 085622. DOI: 10.1088/2053-1591/ab22f0

Purbayanto, M., Chandel, M., Makowski, M., Birowosuto, M., Montes-García, V., Prenger, K., Ciesielski, A., Naguib, M., & Jastrzębska, A. M. (2025). Unraveling the role of interfacial charge transfer on photoactivity and anomalous luminescence quenching of v4c3tx/protonated g-c3n4 heterostructures. *ACS Applied Materials & Interfaces*, *17*(11), 17454–17464. DOI: 10.1021/acsami.4c19729 PMID: 40062911

Qian, H., Greenhalgh, E. S., Shaffer, M. S. P., & Bismarck, A. (2010). Carbon nanotube-based hierarchical composites: A review. *Journal of Materials Chemistry*, *20*(23), 4751. DOI: 10.1039/c000041h

Qin, H., Pei, Q., Liu, Y., & Zhang, Y. (2019). Thermal transport in graphene-based layered materials: An analytical model validated with extensive molecular dynamics simulations. *Carbon*, *155*, 114–121. DOI: 10.1016/j.carbon.2019.08.062

Qiu, C., Su, Y., Yang, J., Chen, B., Ouyang, Q., & Zhang, D. (2021). Structural modelling and mechanical behaviors of graphene/carbon nanotubes reinforced metal matrix composites via atomic-scale simulations: A review. *Composites Part C: Open Access*, *4*, 100120. DOI: 10.1016/j.jcomc.2021.100120

Qiu, W., Gao, M., Meng, X., Dong, Y., Liu, Q., Qiu, C., & Song, Z. (2025). Stress impacts of different types of micro- and nanomaterials on vegetable crops. *Environmental Science. Nano*, *12*(6), 3342–3356. DOI: 10.1039/D5EN00237K

Qiu, Y., Wang, M., Zhang, W., Liu, Y., Li, Y. V., & Pan, K. (2018). An asymmetric graphene oxide film for developing moisture actuators. *Nanoscale*, *10*(29), 14060–14066. DOI: 10.1039/C8NR01785A PMID: 29999058

Quan, D., Mischo, C., Binsfeld, L., Ivankovic, A., & Murphy, N. (2020). Fracture behaviour of carbon fibre/epoxy composites interleaved by MWCNT- and graphene nanoplatelet-doped thermoplastic veils. *Composite Structures*, *235*, 111767. DOI: 10.1016/j.compstruct.2019.111767

Raman, R., Sreenivasan, A., Kulkarni, N., Suresh, M., & Nedungadi, P. (2025). Analyzing the contributions of biofuels, biomass, and bioenergy to sustainable development goals. *iScience*, *28*(4), 112157. DOI: 10.1016/j.isci.2025.112157 PMID: 40230525

Ramesh, R. D., Santhosh, A., & Syamala, S. R. N. A. (2020). Implementation of Nanotechnology in the Aerospace and Aviation Industry. In *Smart Nanotechnology with Applications* (pp. 51–69). CRC Press. DOI: 10.1201/9781003097532-4

Ramezani, M. J., & Rahmani, O. (2024). A review of recent progress in the graphene syntheses and their applications. *Mechanics of Advanced Materials and Structures*, •••, 1–33. DOI: 10.1080/15376494.2024.2420911

Rano, N., & Martsinovich, N. (2025). Stability of sp3 carbons in hydrogenated graphene quantum dots and their electronic and optical properties studied using density functional theory. *The Journal of Physical Chemistry A*, *129*(17), 3790–3806. DOI: 10.1021/acs.jpca.4c07825

Ran, T., Zhou, J., Xu, Y., Li, Z., Chen, C., Wu, T., & Liu, J. (2025). An electrochemical sensor for Cd2+ in water based on RGO-comncnf screen-printed electrodes. *Journal of the Electrochemical Society*, *172*(3), 037521. DOI: 10.1149/1945-7111/adc2a3

Rao, P., Singh, S. K., & Kumar, V. S. S. (2021). Safety approach in construction industry using case based reasoning. *Journal of Real Estate. Construction Manager*, *36*(1), 47–54. DOI: 10.1177/2977657020210105

Rao, S., Upadhyay, J., Polychronopoulou, K., Umer, R., & Das, R. (2018). Reduced graphene oxide: Effect of reduction on electrical conductivity. *Journal of Composites Science*, *2*(2), 25. DOI: 10.3390/jcs2020025

Rathanasamy, R., Sahoo, S., Lee, J. H., Das, A. K., Somasundaram, M., Palaniappan, S. K., & Sivaraj, S. (2021). Carbon-based multi-layered films for electronic application: A review. *Journal of Electronic Materials*, *50*(4), 1845–1892. DOI: 10.1007/s11664-020-08724-4

Rathnavel, P., Murukesh, C., & Umamaheswari, R. (2023). Fabrication of a novel rGO encapsulated nickel cobalt chalcogenide electrocatalyst as an efficient counter electrode to boost efficiency of dye-sensitized solar cells. *Journal of Materials Science Materials in Electronics*, *34*(11), 1000. DOI: 10.1007/s10854-023-10402-0

Rathod, V. T., Kumar, J. S., & Jain, A. (2017). Polymer and ceramic nanocomposites for aerospace applications. *Applied Nanoscience 2017 7:8*, *7*(8), 519–548. DOI: 10.1007/s13204-017-0592-9

Ravari, F., Noori, M., & Ehsani, M. (2019). Thermal stability and degradation kinetic studies of PVA/RGO using the model-fitting and isoconversional (model-free) methods. *Fibers and Polymers*, *20*(3), 472–480. DOI: 10.1007/s12221-019-8606-8

Ravikovitch, P. I., & Neimark, A. V. (2001). Characterization of nanoporous materials from adsorption and desorption isotherms. *Colloids and Surfaces. A, Physicochemical and Engineering Aspects*, *187*, 11–21. DOI: 10.1016/S0927-7757(01)00614-8

Ray, S. C. (2015). Application and Uses of Graphene Oxide and Reduced Graphene Oxide. *Applications of Graphene and Graphene-Oxide Based Nanomaterials*, 39–55. DOI: 10.1016/B978-0-323-37521-4.00002-9

Raza, M. A., Maqsood, M. F., Rehman, Z. U., Westwood, A., Inam, A., Sattar, M. M. S., Ghauri, F. A., & Ilyas, M. T. (2020). Thermally reduced graphene oxide-reinforced acrylonitrile butadiene styrene composites developed by combined solution and melt mixing method. *Arabian Journal for Science and Engineering*, *45*(11), 9559–9568. DOI: 10.1007/s13369-020-04845-4

Reghat, M., Mirabedini, A., Tan, A. M., Weizman, Y., Middendorf, P., Bjekovic, R., & Fuss, F. K. (2021). Graphene as a piezo-resistive coating to enable strain monitoring in glass fiber composites. *Composites Science and Technology*, *211*, 108842. DOI: 10.1016/j.compscitech.2021.108842

Regulation (EC) No 1272/2008 of the European Parliament and of the Council of 16 December 2008 on classification, labelling and packaging of substances and mixtures Official Journal of the European Union L 353/1

Regulation (EC) No 1907/2006 of the European Parliament and of the Council of 18 December 2006 concerning the registration, evaluation, authorisation and restriction of chemicals (REACH) Official Journal of the European Union L 396/1

Reich, S., and C. Thomsen. "Raman spectroscopy of graphite." Philosophical Transactions of the Royal Society of London. Series A: Mathematical, Physical and Engineering Sciences 362.1824 (2004): 2271-2288. https://doi.org/.DOI: 10.1098/rsta.2004.1454

Reina, A., Jia, X., Ho, J., Nezich, D., Son, H., Bulovic, V., Dresselhaus, M. S., & Kong, J. (2009). Large area, few-layer graphene films on arbitrary substrates by chemical vapor deposition. *Nano Letters*, *9*(1), 30–35. DOI: 10.1021/nl801827v

Reina, A., Son, H., Jiao, L., Fan, B., Dresselhaus, M. S., Liu, Z., & Kong, J. (2008). Transferring and identification of single-and few-layer graphene on arbitrary substrates. *The Journal of Physical Chemistry. C, Nanomaterials and Interfaces*, *112*(46), 17741–17744. DOI: 10.1021/jp807380s

Reis, J. P., de Moura, M., & Samborski, S. (2020). Thermoplastic Composites and Their Promising Applications in Joining and Repair Composites Structures: A Review. *Materials 2020, Vol. 13, Page 5832*, *13*(24), 5832. DOI: 10.3390/ma13245832

Rennhofer, H., et al. "Pore development during the carbonization process of lignin microparticles investigated by small angle X-ray scattering." Molecules 26.7 (2021): 2087. https://doi.org/.DOI: 10.3390/molecules26072087

Revaiah, R. G., Kotresh, T. M., & Kandasubramanian, B. (2020). Technical textiles for military applications. *Journal of the Textile Institute*, *111*(2), 273–308. DOI: 10.1080/00405000.2019.1627987

Riaz, A., Khan, S., & Arslan, T. (2023). Design and modelling of graphene-based flexible 5G antenna for next-generation wearable head imaging systems. *Micromachines*, *14*(3), 610. DOI: 10.3390/mi14030610 PMID: 36985015

Robertson, A. W., Montanari, B., He, K., Allen, C. S., Wu, Y. A., Harrison, N. M., Kirkland, A. I., & Warner, J. H. (2013). Structural reconstruction of the graphene monovacancy. *ACS Nano*, *7*(5), 4495–4502. DOI: 10.1021/nn401113r

Roche, J., & Anand, V. (2025). Creating value since 2015: A review of the Journal of Creating Value. *Journal of Creating Value*, *11*(1), 58–72. Advance online publication. DOI: 10.1177/23949643241304874

Rodriguez, J. N., Zhu, C., Duoss, E. B., Wilson, T. S., Spadaccini, C. M., & Lewicki, J. P. (2016). Shape-morphing composites with designed micro-architectures. *Scientific Reports*, *6*(1), 1–10. DOI: 10.1038/srep27933

Rong, C., Su, T., Chu, T., Zhu, M., Zhang, B., & Xuan, F. (2025). Tradeoff between mechanical strength and electrical conductivity of MXene films by nacre-inspired subtractive manufacturing. *Small*, *21*(12), 2411329. Advance online publication. DOI: 10.1002/smll.202411329 PMID: 39937436

Rudenko, R., Voitsihovska, O., Abakumov, A., Bychko, I., Selyshchev, O., Zahn, D. R. T., & Poroshin, V. (2022). The hopping Hall effect in reduced graphene oxide. *Materials Letters*, *326*, 132932. DOI: 10.1016/j.matlet.2022.132932

Ruidíaz-Martínez, M., Álvarez, M. A., López-Ramón, M. V., Cruz-Quesada, G., Rivera-Utrilla, J., & Sánchez-Polo, M. (2020). Hydrothermal synthesis of rGO-TiO2 composites as high-performance UV photocatalysts for ethylparaben degradation. *Catalysts*, *10*(5), 520. DOI:10.3390/catal10050520

Ryan, E. A., Seibers, Z. D., Reynolds, J. R., & Shofner, M. L. (2024). Electrically conducting polymers and composites for applications in space exploration. *Journal of Applied Polymer Science*, *141*(15), e55225. DOI: 10.1002/app.55225

Ryu, S., Kim, H., Park, S., Kwon, S., Kim, S., Sundaray, B., & Park, C. (2022). Millimeter-scale percolated polyethylene/graphene composites for 5G electromagnetic shielding. *ACS Applied Nano Materials*, *5*(6), 8429–8439. DOI: 10.1021/acsanm.2c01544

Sabato, A., Dabetwar, S., Kulkarni, N. N., & Fortino, G. (2023). Noncontact sensing techniques for AI-aided structural health monitoring: A systematic review. *IEEE Sensors Journal*, *23*(5), 4672–4684. DOI: 10.1109/JSEN.2023.3240092

Sabet, M. (2024). Unveiling advanced self-healing mechanisms in graphene polymer composites for next-generation applications in aerospace, automotive, and electronics. *Polymer-Plastics Technology and Materials*, *63*(15), 2032–2059. DOI: 10.1080/25740881.2024.2365282

Sachdev, H., Soni, M., Gupta, A., Chansoria, H., Patil, M., Sharma, B., Kar, K., & Sharma, R. (2025). Comparative evaluation of flexural strength of pmma reinforced with graphene (g-cam) and monolithic zirconia - an in vitro study. *Journal of Pharmacy & Bioallied Sciences*, *17*(Suppl 1), S238–S241. DOI: 10.4103/jpbs.jpbs_1715_24

Sadanandan, K. S., Bacon, A., Shin, D. W., Alkhalifa, S. F., Russo, S., Craciun, M. F., & Neves, A. I. (2020). Graphene coated fabrics by ultrasonic spray coating for wearable electronics and smart textiles. *JPhys Materials*, *4*(1), 014004. DOI: 10.1088/2515-7639/abc632

Sadeghi, A., Bellavista, P., Song, W., & Yazdani-Asrami, M. (2024). Digital twins for condition and fleet monitoring of aircraft: Towards more-intelligent electrified aviation systems. *IEEE Access : Practical Innovations, Open Solutions*, *12*, 99806–99832. DOI: 10.1109/ACCESS.2024.3371902

Sadigh, M. A. S., & Marami, G. (2018). Enhancing fatigue life in adhesively bonded joints using reduced graphene oxide additive: Experimental and numerical evaluation. *International Journal of Adhesion and Adhesives*, *84*, 283–290. DOI: 10.1016/j.ijadhadh.2018.04.001

Safian, M. T., Haron, U. S., & Ibrahim, M. M. (2020). A review on bio-based graphene derived from biomass wastes. *BioResources*, *15*(4), 9756–9785. DOI: 10.15376/biores.15.4.Safian

Saghafi, H., Ahmadi, I., Khamedi, R., Saghafi, H., Saeedifar, M., Brugo, T. M., Maccaferri, E., Ortolani, J., Mongioì, F., Mazzocchetti, L., & Zucchelli, A. (2024). The thickness effect of rubbery nanofibrous mat on modes I–II fracture mechanism of composite laminates. *Thin-walled Structures*, *196*, 111556. DOI: 10.1016/j.tws.2024.111556

Sahoo, P., Das, A., Das, S., & Mallik, A. (2025). Role of electrochemically exfoliated graphene in property enhancement of electrodeposited copper nanocomposites. *ChemistrySelect*, *10*(13), e202405666. Advance online publication. DOI: 10.1002/slct.202405666

Sakhaee-Pour, A. (2009). Elastic properties of single-layered graphene sheet. *Solid State Communications*, *149*(1-2), 91–95. DOI: 10.1016/j.ssc.2008.09.050

Saleh, H. M., & Hassan, A. I. (2023). Synthesis and characterization of nanomaterials for application in cost-effective electrochemical devices. *Sustainability (Basel)*, *15*(14), 10891. DOI: 10.3390/su151410891

Sales, A., Pereira, V., & Dias, A. (2025). Advances in nanomaterials for radiation protection in the aerospace industry: A systematic review. *Nanotechnology*, *36*(10), 102002. DOI: 10.1088/1361-6528/ada38f PMID: 39729645

Sam-Daliri, O., Farahani, M., Faller, L. M., & Zangl, H. (2020). Structural health monitoring of defective single lap adhesive joints using graphene nanoplatelets. *Journal of Manufacturing Processes*, *55*, 119–130. DOI: 10.1016/j.jmapro.2020.03.063

Sánchez-Trujillo, D. J., Osorio-Maldonado, L. V., & Prías-Barragán, J. J. (2023). Temperature dependence of electrical conductivity and variable hopping range mechanism on graphene oxide films. *Scientific Reports*, *13*(1), 4810. DOI: 10.1038/s41598-023-31778-3

Santos-Ceballos, J. C., Salehnia, F., Romero, A., & Vilanova, X. (2024). Application of digital twins for simulation based tailoring of laser induced graphene. *Scientific Reports*, *14*(1), 10363. DOI: 10.1038/s41598-024-61237-6 PMID: 38710895

Sarlioglu, B., & Morris, C. T. (2015). More electric aircraft: Review, challenges, and opportunities for commercial transport aircraft. *IEEE Transactions on Transportation Electrification*, *1*(1), 54–64. DOI: 10.1109/TTE.2015.2426499

Scalia, T., Bonventre, L., & Terranova, M. L. (2023). From protosolar space to space exploration: The role of graphene in space technology and economy. *Nanomaterials (Basel, Switzerland)*, *13*(4), 680. DOI: 10.3390/nano13040680 PMID: 36839048

Scheibe, B., Borowiak-Palen, E., & Kalenczuk, R. J. (2010). Oxidation and reduction of multiwalled carbon nanotubes—Preparation and characterization. *Materials Characterization*, *61*(2), 185–191. DOI: 10.1016/j.matchar.2009.11.008

Schwierz, F. (2010). Graphene transistors. *Nature Nanotechnology*, *5*(7), 487–496. DOI: 10.1038/nnano.2010.89 PMID: 20512128

Sedki, M., Mirabedini, P. S., Nakama, K., Stephens, G., Groves, M., Lee, I., Neupane, M. R., & Mulchandani, A. (2022). Synthesis of pristine graphene-like behaving rGO thin film: Insights into what really matters. *Carbon, 186*, 437–451. DOI: 10.1016/j.carbon.2021.10.011

Sethuraman, K., & Alagar, M. (2014). Thermo-mechanical and dielectric properties of graphene reinforced caprolactam cardanol based benzoxazine–epoxy nanocomposites. *RSC Advances*, *5*(13), 9607–9617. DOI: 10.1039/C4RA14383C

Seyed Pourmand, N., & Asgharzadeh, H. (2020). Aluminum Matrix Composites Reinforced with Graphene: A Review on Production, Microstructure, and Properties. *Critical Reviews in Solid State and Material Sciences*, *45*(4), 289–337. DOI: 10.1080/10408436.2019.1632792

Shaheen, S., Saeed, Z., Ahmad, A., Pervaiz, M., Younas, U., Khan, R. R. M., Luque, R., & Rajendran, S. (2022). Green synthesis of graphene-based metal nanocomposite for electro and photocatalytic activity; recent advancement and future perspective. *Chemosphere*, *311*, 136982. DOI: 10.1016/j.chemosphere.2022.136982 PMID: 36309056

Shahil, K. M., & Balandin, A. A. (2012). Thermal properties of graphene and multilayer graphene: Applications in thermal interface materials. *Solid State Communications*, *152*(15), 1331–1340. DOI: 10.1016/j.ssc.2012.04.034

Shalabi, O. R. Y. (2022). Manufacturing of smart composite aircraft skins using 2D material coated glass fiber reinforcements Khalifa. *Universitas Scientiarum.*

Shamsaei, E., de Souza, F. B., Yao, X., Benhelal, E., Akbari, A., & Duan, W. (2018). Graphene-based nanosheets for stronger and more durable concrete: A review. *Construction & Building Materials*, *183*, 642–660. DOI: 10.1016/j.conbuildmat.2018.06.201

Sharma, H., Kumar, A., Rana, S., & Guadagno, L. (2022). An Overview on Carbon Fiber-Reinforced Epoxy Composites: Effect of graphene oxide incorporation on Composites performance. *Polymers*, *14*(8), 1548. DOI: 10.3390/polym14081548 PMID: 35458296

Sharma, N., Tomar, S., Shkir, M., Kant Choubey, R., & Singh, A. (2021). Study of optical and electrical properties of graphene oxide. *Materials Today: Proceedings*, *36*, 730–735. DOI: 10.1016/j.matpr.2020.04.861

Sharon, M., Rodriguez, A. S. L., Sharon, C., & Gallardo, P. S. (Eds.). (2019). *Nanotechnology in the Defense Industry*. Wiley., DOI: 10.1002/9781119460503

Sheinerman, A. G. (2022). Plastic deformation and fracture processes in metal/graphene composites: A review. *Critical Reviews in Solid State and Material Sciences*, *47*(5), 708–735. DOI: 10.1080/10408436.2021.1965955

Sheinerman, A. G., & Gutkin, M. Y. (2020). Model of Enhanced Strength and Ductility of Metal/Graphene Composites with Bimodal Grain Size Distribution. *Metallurgical and Materials Transactions. A, Physical Metallurgy and Materials Science*, *51*(1), 189–199. DOI: 10.1007/s11661-019-05500-w

Shen, M. Y., Liao, W. Y., Wang, T. Q., & Lai, W. M. (2021). Characteristics and Mechanical Properties of Graphene Nanoplatelets-Reinforced Epoxy Nanocomposites: Comparison of Different Dispersal Mechanisms. *Sustainability 2021, Vol. 13, Page 1788, 13*(4), 1788. DOI: 10.3390/su13041788

Shen, L., Jiang, X., Ni, L., Luo, Y., Liang, M., Yang, Y., Zhou, S., & Zou, H. (2025). Improved infrared stealth and electromagnetic interference shielding performance of polyimide-derived carbon aerogels by constructing heterogeneous nanopores. *Polymer Engineering and Science*, *65*(2), 846–857. DOI: 10.1002/pen.27049

Shen, X. J., Meng, L. X., Yan, Z. Y., Sun, C. J., Ji, Y. H., Xiao, H. M., & Fu, S. Y. (2015). Improved cryogenic interlaminar shear strength of glass fabric/epoxy composites by graphene oxide. *Composites. Part B, Engineering*, *73*, 126–131. DOI: 10.1016/j.compositesb.2014.12.023

Shetty, R., Hegde, A., Shetty SV, U. K., Nayak, R., Naik, N., & Nayak, M. (2022). Processing and Mechanical Characterisation of Titanium Metal Matrix Composites: A Literature Review. *Journal of Composites Science 2022, Vol. 6, Page 388, 6*(12), 388. DOI: 10.3390/jcs6120388

Shiino, M. Y., Pelosi, S., Hilá, M. O., Cioffi, R., & Donadon, M. V. (2017, March). The Role of Stitch Yarn on the Delamination Resistance in Non-crimp Fabric: Chemical and Physical Interpretation. *Journal of Materials Engineering and Performance*, *26*(3), 978–986. Advance online publication. DOI: 10.1007/s11665-016-2460-2

Shi, J. L., Tian, G.-L., Zhang, Q., Zhao, M.-Q., & Wei, F. (2015). Customized casting of unstacked graphene with high surface area (>1300 m^2/g) and its application in oxygen reduction reaction. *Carbon*, *93*, 702–712. DOI: 10.1016/j.carbon.2015.06.007

Shi, L., Schröder, M., Meyer, H., Peláez, D., Wodtke, A., Golibrzuch, K., Schönemann, A.-M., Kandratsenka, A., & Gatti, F. (2025). Full quantum dynamics study for h atom scattering from graphene. *The Journal of Physical Chemistry A*, *129*(7), 1896–1907. DOI: 10.1021/acs.jpca.4c06712

Shin, M. K., Lee, B., Kim, S. H., Lee, J. A., Spinks, G. M., Gambhir, S., Wallace, G. G., Kozlov, M. E., Baughman, R. H., & Kim, S. J. (2012). Synergistic toughening of composite fibres by self-alignment of reduced graphene oxide and carbon nanotubes. *Nature Communications*, *3*(1), 1–8. DOI: 10.1038/ncomms1661

Shivakumar, K. N., Panduranga, R., & Sharpe, M. (2013, November). Interleaved Polymer Matrix Composites - A Review. *54th AIAA/ASME/ASCE/AHS/ASC Structures, Structural Dynamics, and Materials Conference*. DOI: 10.2514/6.2013-1903

Short, M. A., & Walker, P. L.Jr. (1963). Measurement of interlayer spacings and crystal sizes in turbostratic carbons. *Carbon*, *1*(1), 3–9. DOI: 10.1016/0008-6223(63)90003-4

Siburian, R., Sihotang, H., Raja, S. L., Supeno, M., & Simanjuntak, C. (2018). New route to synthesize of graphene nano sheets. *Oriental Journal of Chemistry*, *34*(1), 182. DOI:10.13005/ojc/340120

Sihver, L., Ploc, O., Puchalska, M., Ambrožová, I., Kubančák, J., Kyselová, D., & Shurshakov, V. (2015). Radiation environment at aviation altitudes and in space. *Radiation Protection Dosimetry*, *164*(4), 477–483. DOI: 10.1093/rpd/ncv330 PMID: 25979747

Silverman, B. A., & Kunkel, B. A. (1970). A comparison of the warm fog clearing capabilities of some hygroscopic materials. *Journal of Applied Meteorology*, *9*(4), 634–638. DOI: 10.1175/1520-0450(1970)009<0634:ACOTWF>2.0.CO;2

Simões, A., Battocchi, D., Tallman, D. E., & Bierwagen, G. P. (2008). Assessment of the corrosion protection of aluminium substrates by a Mg-rich primer: EIS, SVET and SECM study. *Progress in Organic Coatings*, *63*(3), 260–266. DOI: 10.1016/j.porgcoat.2008.02.007

Singh, T., & Sehgal, S. (2021). Structural Health Monitoring of Composite Materials. *Archives of Computational Methods in Engineering 2021 29:4, 29*(4), 1997–2017. DOI: 10.1007/s11831-021-09666-8

Singh, A., Lalotra, N., Shah, S. S., Rather, M. D., López-Maldonado, E. A., Pathania, K., Abaszade, R. G., Stetsenko, M., & Arya, S. (2025). A comprehensive review on the synthesis and properties of MXenes for supercapacitor applications. *ACS Applied Energy Materials*, *8*(8), 4884–4914. DOI: 10.1021/acsaem.4c03194

Singh, B., Kumar, I., Saxena, K. K., Mohammed, K. A., Ijaz Khan, M., Ben Moussa, S., & Shukhratovich Abdullaev, S. (2023). A future prospects and current scenario of aluminium metal matrix composites characteristics. *Alexandria Engineering Journal*, *76*, 1–17. DOI: 10.1016/j.aej.2023.06.028

Singh, S. K., Jain, R., Palaniappan, D., & Parmar, K.. (2022). *Spatiotemporal analysis and machine learning-based prediction of air quality in Indian urban areas.* Environmental Research and Technology., DOI: 10.35208/ert.1587308

Singh, S. K., Jain, R., Palaniappan, D., & Parmar, K.. (2023). Data-driven civil engineering: Applications of artificial intelligence, machine learning, and deep learning. *Turkish Journal of Engineering*, *9*(2), 354–377. DOI: 10.31127/tuje.1581564

Somanath, S., & Spaceport, K. Indian Space Research Organisation.

Song, P., Liu, B., Qiu, H., Shi, X., Cao, D., & Gu, J. (2021). MXenes for polymer matrix electromagnetic interference shielding composites: A review. *Composites Communications*, *24*, 100653. DOI: 10.1016/j.coco.2021.100653

Song, Y., Xu, J., Sun, Y., Xu, X., Tan, X., Du, M., Hu, J., Shi, Z., Wang, H., & Zhang, X. (2025). Structure and properties of au/n-g composites prepared based on different nitrogen sources. *Journal of Physics: Conference Series*, *2951*(1), 012108. DOI: 10.1088/1742-6596/2951/1/012108

Souza, D. F. d. (2024). Design and development of an electric vertical take off and landing aircraft system control using sliding mode control system Universidade Tecnológica Federal do Paraná].

Srinivasan, V., Kunjiappan, S., & Palanisamy, P. (2021). A brief review of carbon nanotube reinforced metal matrix composites for aerospace and defense applications. *International Nano Letters 2021 11:4*, *11*(4), 321–345. DOI: 10.1007/s40089-021-00328-y

Stan, G., & King, S. W. (2020). Atomic force microscopy for nanoscale mechanical property characterization. *Journal of Vacuum Science and Technology. B, Nanotechnology & Microelectronics : Materials, Processing, Measurement, & Phenomena : JVST B*, *38*(6), 6. DOI: 10.1116/6.0000544

Stankovich, S., Dikin, D. A., Dommett, G. H., Kohlhaas, K. M., Zimney, E. J., Stach, E. A.,…Ruoff, R. S. (2006). Graphene-based composite materials. nature, 442(7100), 282-286.

Steinke, K., Groo, L., & Sodano, H. A. (2021). Laser induced graphene for in-situ ballistic impact damage and delamination detection in aramid fiber reinforced composites. *Composites Science and Technology*, *202*, 108551. DOI: 10.1016/j.compscitech.2020.108551

Stobinski, L., Lesiak, B., Malolepszy, A., Mazurkiewicz, M., Mierzwa, B., Zemek, J., Jiricek, P., & Bieloshapka, I. (2014). Graphene oxide and reduced graphene oxide studied by the XRD, TEM and electron spectroscopy methods. *Journal of Electron Spectroscopy and Related Phenomena, 195*, 145–154. DOI: 10.1016/j.elspec.2014.07.003

Stoll, A. M., CmANTA, M. A., & PIEROALLINI, J. (1979). Thermal r Human Skin. *Aviation, Space, and Environmental Medicine*, •••, 779.

Stoller, M. D., Park, S., Zhu, Y., An, J., & Ruoff, R. S. (2008). Graphene-based ultracapacitors. *Nano Letters, 8*(10), 3498–3502. DOI: 10.1021/nl802558y

Sudan, K. (2024). Development and processing of shape memory polymer composites (SMPCS) for application in structural robotics and robotic sensors.

Suenaga, K., Sandré, E., Colliex, C., Pickard, C. J., Kataura, H., & Iijima, S. (2001). Electron energy-loss spectroscopy of electron states in isolated carbon nanostructures. *Physical Review. B, 63*(16), 165408. DOI: 10.1103/PhysRevB.63.165408

Suganya, B., Gopi, R., Kumar, A. R., & Singh, G. (2024). Dynamic task offloading edge-aware optimization framework for enhanced UAV operations on edge computing platform. *Scientific Reports, 14*(1), 16383. Advance online publication. DOI: 10.1038/s41598-024-67285-2 PMID: 39013972

Su, J., He, K., Li, Y., Tu, J., & Chen, X. (2025). Soft Materials and Devices Enabling Sensorimotor Functions in Soft Robots. *Chemical Reviews, 125*(12), 5848–5977. DOI: 10.1021/acs.chemrev.4c00906 PMID: 40163535

Sunderrajan, S., Miranda, L. R., & Pennathur, G. (2018). Improved stability and catalytic activity of graphene oxide/chitosan hybrid beads loaded with porcine liver esterase. *Preparative Biochemistry & Biotechnology, 48*(4), 343–351. DOI: 10.1080/10826068.2018.1446153

Sun, H., Hwang, U., Kim, S., Sung, J., Kim, T., Suhr, J., Park, I.-K., & Nam, J.-D. (2025). Reversible thixotropic rheological properties of graphene-incorporated epoxy inks for self-standing 3d printing. *ACS Macro Letters, 14*(2), 135–141. DOI: 10.1021/acsmacrolett.4c00765 PMID: 39801069

Sun, W., Geng, W., Qian, P., Wang, T., Jing, L., Feng, J., Wang, Y.-H., Wang, Y.-X., Zhou, H.-L., & Geng, H.-Z. (2025). Composite flexible films based on mwcnt and modified graphene for electric heating and emi shielding. *ACS Applied Polymer Materials, 7*(12), 7798–7810. Advance online publication. DOI: 10.1021/acsapm.5c00373

Sun, Y. W., Papageorgiou, D. G., Humphreys, C. J., Dunstan, D. J., Puech, P., Proctor, J. E., Bousige, C., Machon, D., & San-Miguel, A. (2021). Mechanical properties of graphene. *Applied Physics Reviews*, *8*(2), 021310. DOI: 10.1063/5.0040578

Sun, Y., & Zhang, J. (2021). Strategies for scalable gas-phase preparation of free-standing graphene. *CCS Chemistry*, *3*(4), 1058–1077. DOI: 10.31635/ccschem.020.202000289

Suomalainen, E. (2025). Growth over climate? Aviation and the Paris Agreement. https://doi.org/DOI: 10.21203/rs.3.rs-5986385/v1

Surekha, G., Krishnaiah, K. V., Ravi, N., & Suvarna, R. P. (2020, March). FTIR, Raman and XRD analysis of graphene oxide films prepared by modified Hummers method. []. IOP Publishing.]. *Journal of Physics: Conference Series*, *1495*(1), 012012. DOI: 10.1088/1742-6596/1495/1/012012

TabkhPaz, M., Park, D.-Y., Lee, P. C., Hugo, R., & Park, S. S.TabkhPaz. (2018). Development of nanocomposite coatings with improved mechanical, thermal, and corrosion protection properties. *Journal of Composite Materials*, *52*(8), 1045–1060. DOI: 10.1177/0021998317720001

Tafur, C. L., Camero, R. G., Rodríguez, D. A., Rincón, J. C. D., & Saenz, E. R. (2024). Applications of artificial intelligence in air operations: A systematic review. *Results in Engineering*, *25*, 103742. DOI: 10.1016/j.rineng.2024.103742

Taha, T. H., Elnouby, M. S., Abu-Saied, M., & Alamri, S. (2020). The green exfoliation of graphite waste and its suitability for biosensor applications. *RSC Advances*, *10*(16), 9347–9355. DOI: 10.1039/C9RA09602G PMID: 35497257

Tang, C., Wang, Y., Li, Y., Zeng, S., Kong, L., Li, L., Sun, J., Zhu, M., & Deng, T. (2023). A review of graphene-based temperature sensors. *Microelectronic Engineering*, *278*, 112015. DOI: 10.1016/j.mee.2023.112015

Tang, D., Wang, Q., Wang, Z., Liu, Q., Zhang, B., He, D., Wu, Z., & Mu, S. (2018). Highly sensitive wearable sensor based on a flexible multi-layer graphene film antenna. *Science Bulletin*, *63*(9), 574–579. DOI: 10.1016/j.scib.2018.03.014

Tang, H., Li, M., Chen, D., Tahir, M. H., Yin, L., Qian, K., Feng, Y., & Hu, Y. (2025). One-step production of isoalkane-rich sustainable aviation fuel fractions via catalytic pyrolysis of low-density polyethylene over y-type zeolite. *ACS Sustainable Chemistry & Engineering*, *13*(23), 8761–8772. DOI: 10.1021/acssuschemeng.5c02722

Tang, L. C., Wan, Y. J., Yan, D., Pei, Y. B., Zhao, L., Li, Y. B., Wu, L., Jiang, J.-X., & Lai, G.-Q. (2013). The effect of graphene dispersion on the mechanical properties of graphene/epoxy composites. *Carbon*, *60*, 16–27. DOI: 10.1016/j.carbon.2013.03.050

Tang, X., Debliquy, M., Lahem, D., Yan, Y., & Raskin, J. (2021). A review on functionalized graphene sensors for detection of ammonia. *Sensors (Basel), 21*(4), 1443. DOI: 10.3390/s21041443 PMID: 33669589

Tang, X., & Yan, X. (2017). Dip-coating for fibrous materials: Mechanism, methods and applications. *Journal of Sol-Gel Science and Technology, 81*(2), 378–404. DOI: 10.1007/s10971-016-4197-7

Tang, Y., Friec, A., Chen, M., & Sun, D. (2025). Designed biomaterial-enhanced cell transplantation for neural tissue engineering. *Aggregate, 6*(5), e70022. Advance online publication. DOI: 10.1002/agt2.70022

Tareq, M. S., Jony, B., Zainuddin, S., Al Ahsan, M., & Hosur, M. V. (2021). Fatigue analysis and fracture toughness of graphene reinforced carbon fibre polymer composites. *Fatigue & Fracture of Engineering Materials & Structures, 44*(2), 461–474. DOI: 10.1111/ffe.13371

Tomashov, N. D. (1966). *Theory of corrosion and protection of metals*. Macmillan.

Tong, C. (2022). *Advanced Materials for Printed Flexible Electronics*. Springer. DOI: 10.1007/978-3-030-79804-8

Trai, V., & Sobhy, M. (2025). Vibration and transient response analysis of bi-directional functionally graded graphene platelets reinforced plates resting on a Kerr foundation using a refined finite-element procedure. *Proceedings - Royal Society. Mathematical, Physical and Engineering Sciences, 481*(2314), 20250297. Advance online publication. DOI: 10.1098/rspa.2025.0297

Tran, D. T. (2020). rGO/persulfate metal-free catalytic system for the degradation of tetracycline: Effect of reaction parameters. *Materials Research Express, 7*(7), 075501. DOI: 10.1088/2053-1591/ab9e47

Tran, T. S., Balu, R., de Campo, L., Dutta, N. K., & Choudhury, N. R. (2023). Sulfonated polythiophene-interfaced graphene for water-redispersible graphene powder with high conductivity and electrocatalytic activity. *Energy Advances, 2*(3), 365–374. DOI: 10.1039/D2YA00298A

Trivedi, D. N., & Rachchh, N. V. (2022). Graphene and its application in thermoplastic polymers as nano-filler-A review. *Polymer, 240*, 124486. DOI: 10.1016/j.polymer.2021.124486

Tseghai, G. B., Malengier, B., Fante, K. A., Nigusse, A. B., & Van Langenhove, L. (2020). Development of a flex and stretchy conductive cotton fabric via flat screen printing of PEDOT: PSS/PDMS conductive polymer composite. *Sensors (Basel), 20*(6), 1742. DOI: 10.3390/s20061742 PMID: 32245034

Tucker, R. C.Jr. (1994). Thermal spray coatings. In. ASM Handbook: Vol. 5. *Surface Engineering* (pp. 497–509). ASM International. DOI: 10.31399/asm.hb.v05.a0001282

U.S. Environmental Protection Agency (EPA). (2019). Significant new use rules on certain chemical substances (FRL No. 9991–19; Docket ID No. EPA-HQ-OPPT-2017-0575). https://www.epa.gov/reviewing-new-chemicals-under-toxic-substances-control-act-tsca/significant-new-use-rules-certain

Ullah, S., et al. "Advances and trends in chemically doped graphene." Advanced Materials Interfaces 7.24 (2020): 2000999. https://doi.org/.DOI: 10.1002/admi.202000999

Uniyal, P., Gaur, P., Yadav, J., Bhalla, N., Khan, T., Junaedi, H., & Sebaey, T. A. (2025). A comprehensive review on the role of nanosilica as a toughening agent for enhanced epoxy composites for aerospace applications. *ACS Omega*, *10*(16), 15810–15839. DOI: 10.1021/acsomega.4c10073 PMID: 40321578

Vaidya, U. K., & Chawla, K. K. (2008). Processing of fibre reinforced thermoplastic composites. *International Materials Reviews*, *53*(4), 185–218. DOI: 10.1179/174328008X325223

Valorosi, F., De Meo, E., Blanco-Varela, T., Martorana, B., Veca, A., Pugno, N., & Bertocchi, F. (2020). Graphene and related materials in hierarchical fiber composites: Production techniques and key industrial benefits. *Composites Science and Technology*, *185*, 107848. DOI: 10.1016/j.compscitech.2019.107848

Vantrepotte, Q., Dubois, E., Camachon, C., & Causse, M. (2025). Effect of pilot expertise on visual scanning and precision of flight maneuvers in military aviation. *Aviation Psychology and Applied Human Factors*, *15*(1), 16–26. DOI: 10.1027/2192-0923/a000287

Varghese, S. S., Lonkar, S., Singh, K. K., Swaminathan, S., & Abdala, A. (2015). Recent advances in graphene based gas sensors. *Sensors and Actuators. B, Chemical*, *218*, 160–183. DOI: 10.1016/j.snb.2015.04.062

Vasiljević, J., Demšar, A., Leskovšek, M., Simončič, B., Čelan Korošin, N., Jerman, I., Šobak, M., Žitko, G., Van de Velde, N., & Čolović, M. (2020). Characterization of polyamide 6/multilayer graphene nanoplatelet composite textile filaments obtained via in situ polymerization and melt spinning. *Polymers*, *12*(8), 1787. DOI: 10.3390/polym12081787 PMID: 32785048

Vázquez-Sánchez, P., Rodríguez-Escudero, M. A., Burgos, F. J., Llorente, I., Caballero-Calero, O., González, M. M., Fernández, R., & García-Alonso, M. C. (2019). Synthesis of Cu/rGO composites by chemical and thermal reduction of graphene oxide. *Journal of Alloys and Compounds*, *800*, 379–391. DOI: 10.1016/j.jallcom.2019.06.008

Venkatesan, S., & Anthony Xavior, M. (2018). Tensile behavior of aluminum alloy (AA7050) metal matrix composite reinforced with graphene fabricated by stir and squeeze cast processes. *Science and Technology of Materials*, *30*(2), 74–85. DOI: 10.1016/j.stmat.2018.02.005

Verma, M., Verma, P., Dhawan, S., & Choudhary, V. (2015). Tailored graphene based polyurethane composites for efficient electrostatic dissipation and electromagnetic interference shielding applications. *RSC Advances*, *5*(118), 97349–97358. DOI: 10.1039/C5RA17276D

Verma, P., Goldar, D., Singh, S. K., & Md, D. (2024). Performance Evaluation and Enhancement of Geopolymer Concrete using Silica Fume. *International Journal of Civil Engineering*, *11*(11), 78–88. DOI: 10.14445/23488352/IJCE-V11I11P108

Verma, R., Yadav, S., Singh, D., & Singh, J. (2025). Fabrication of a mesoporous $CoFe_2O_4$/RGO nanohybrid and laccase interface biosensor for rapid detection of adrenaline for neurodegenerative disease diagnosis. *Materials Advances*, *6*(6), 1988–2001. DOI: 10.1039/D4MA01216J

Vieille, B., Casado, V. M., & Bouvet, C. (2014). Influence of matrix toughness and ductility on the compression-after-impact behavior of woven-ply thermoplastic- and thermosetting-composites: A comparative study. *Composite Structures*, *110*(1), 207–218. DOI: 10.1016/j.compstruct.2013.12.008

Vinci, G., Gobbi, L., Ruggieri, R., Ruggeri, M., & Tiradritti, M. (2023). Challenges and opportunities for the graphene industry. Sustainable approaches for the circular economy. In *Elsevier eBooks* (pp. 319–339). https://doi.org/DOI: 10.1016/B978-0-323-90914-3.00013-9

Visser, M., Gosens, I., Bard, D., Van Broekhuizen, P., Janer, G., Kuempel, E., Riediker, M., Vogel, U., & Dekkers, S. (2022). Towards health-based nano reference values (HNRVs) for occupational exposure: Recommendations from an expert panel. *NanoImpact*, *26*, 100396. DOI: 10.1016/j.impact.2022.100396 PMID: 35560294

Vlassiouk, I., Polizos, G., Cooper, R., Ivanov, I., Keum, J. K., Paulauskas, F., Datskos, P., & Smirnov, S. (2015). Strong and Electrically Conductive Graphene-Based Composite Fibers and Laminates. *ACS Applied Materials & Interfaces*, *7*(20), 10702–10709. DOI: 10.1021/acsami.5b01367

Voros, R. (2021). Generalizing Aspects of System Safety to Broaden Applicability. SAE International Journal of Advances and Current Practices in Mobility, 3(2021-01-0037), 1272-1284.

Wagner, C. D., Davis, L. E., Zeller, M. V., Taylor, J. A., Raymond, R. H., & Gale, L. H. (1981). Empirical atomic sensitivity factors for quantitative analysis by electron spectroscopy for chemical analysis. *Surface and Interface Analysis*, *3*(5), 211–225. DOI: 10.1002/sia.740030506

Wang, L. (2024). Achieving Electric Aviation with Advanced Battery Technology and Nanomaterials. *E3S Web of Conferences, 553*, 02015. DOI: 10.1051/e3sconf/202455302015

Wang, X., Zhao, J., Cui, E., Tian, X., & Sun, Z. (2021). Effect of Interfacial Structure on Mechanical Properties of Graphene Reinforced Al2O3–WC Matrix Ceramic Composite. *Nanomaterials 2021, Vol. 11, Page 1374, 11*(6), 1374. DOI: 10.3390/nano11061374

Wang, B., Duan, Y., Xin, Z., Yao, X., Abliz, D., & Ziegmann, G. (2018). Fabrication of an enriched graphene surface protection of carbon fiber/epoxy composites for lightning strike via a percolating-assisted resin film infusion method. *Composites Science and Technology*, *158*, 51–60. DOI: 10.1016/j.compscitech.2018.01.047

Wang, F., Ko, S. Y., Park, J. O., Park, S. H., & Kee, C. D. (2015). Electroactive polymer actuator based on PVDF and graphene through electrospinning. *Advanced Materials Research*, *1105*, 311–314. DOI: 10.4028/www.scientific.net/AMR.1105.311

Wang, G., Liu, X., Wang, M., Yu, M., Zhang, H., & He, L. (2024). Enhancing interlaminar fracture toughness of carbon fiber/epoxy composites toughened by polyethersulfone resin and hybrid carbon nanotubes/graphene oxide. *Polymer Composites*. Advance online publication. DOI: 10.1002/pc.29256

Wang, G., Yang, J., Park, J., Gou, X., Wang, B., Liu, H., & Yao, J. (2008). Facile synthesis and characterization of graphene nanosheets. *The Journal of Physical Chemistry. C, Nanomaterials and Interfaces*, *112*(22), 8192–8195. DOI: 10.1021/jp710931h

Wang, J., Chang, W., Islam, M. S., Huang, F., Wu, S., Rose, L. R. F., Zhang, J., & Wang, C. H. (2024). Toughening epoxy by nano-structured block copolymer to mitigate matrix microcracking of carbon fibre composites at cryogenic temperatures. *Composites Science and Technology*, *251*, 110548. DOI: 10.1016/j.compscitech.2024.110548

Wang, J., Liu, Y., Fan, Z., Wang, W., Wang, B., & Guo, Z. (2019). Ink-based 3D printing technologies for graphene-based materials: A review. *Advanced Composites and Hybrid Materials*, *2*(1), 1–33. DOI: 10.1007/s42114-018-0067-9

Wang, J., Li, Z., Fan, G., Pan, H., Chen, Z., & Zhang, D. (2012). Reinforcement with graphene nanosheets in aluminum matrix composites. *Scripta Materialia*, *66*(8), 594–597. DOI: 10.1016/j.scriptamat.2012.01.012

Wang, J., Mu, X., & Sun, M. (2019). The Thermal, Electrical, and Thermoelectric Properties of Graphene Nanomaterials. *Nanomaterials (Basel, Switzerland)*, *9*(2), 218. DOI: 10.3390/nano9020218 PMID: 30736378

Wang, J., Xing, Z., Chen, X., Cheng, Z., Li, X., & Liu, T. (2020). Recent progress in waveguide-integrated graphene photonic devices for sensing and communication applications. *Frontiers in Physics (Lausanne)*, *8*, 37. DOI: 10.3389/fphy.2020.00037

Wang, Q., Tian, Y., Duongthipthewa, A., Zhang, J., Liu, M., Su, Z., & Zhou, L. (2023). An embedded non-intrusive graphene/epoxy broadband nanocomposite sensor co-cured with GFRP for in situ structural health monitoring. *Composites Science and Technology*, *236*, 109995. DOI: 10.1016/j.compscitech.2023.109995

Wang, T., Huang, D., Yang, Z., Xu, S., He, G., Li, X., Hu, N., Yin, G., He, D., & Zhang, L. (2016). A review on graphene-based gas/vapor sensors with unique properties and potential applications. *Nano-Micro Letters*, *8*(2), 95–119. DOI: 10.1007/s40820-015-0073-1 PMID: 30460270

Wang, W., Cui, R., Zhang, P., Wan, C., & Pan, W. (2021). High electromagnetic interference shielding effectiveness in MgO composites reinforced by aligned graphene platelets. *Journal of the American Ceramic Society*, *104*(6), 2868–2878. DOI: 10.1111/jace.17667

Wang, Z., Soutis, C., & Gresil, M. (2021). Fracture Toughness of Hybrid Carbon Fibre/Epoxy Enhanced by Graphene and Carbon Nanotubes. *Applied Composite Materials*, *28*(4), 1111–1125. DOI: 10.1007/s10443-021-09906-x

Warren, A. S. (2004). Developments and Challenges for Aluminum - A Boeing Perspective. *MATERIALS FORUM*, *28*, 24–31.

Watanabe, K., Sakai, R., Tanaka, S., Nagata, M., Osaka, H., Kuwahara, Y., & Fujii, T. (2024). Electromagnetic interference with the mobile communication devices in unmanned aerial vehicles and its countermeasures. *IEEE Access : Practical Innovations, Open Solutions*, *12*, 11642–11652. DOI: 10.1109/ACCESS.2024.3351216

Weheabby, S., Liu, Z., Pašti, I., Rajić, V., Vidotti, M., & Kanoun, O. (2025). Enhanced electrochemical sensing of methyl parathion using agnps@il/go nanocomposites in aqueous matrices. *Nanoscale Advances*, *7*(8), 2195–2208. DOI: 10.1039/D4NA00919C

Weng, L., Zhang, L., Chen, Y. P., & Rokhinson, L. P. (2008). Atomic force microscope local oxidation nanolithography of graphene. *Applied Physics Letters*, *93*(9), 093107. DOI: 10.1063/1.2976429

Wen, S., Chen, K., Li, W., Zhou, Y., Wei, Q., & Shi, Y. (2019). Selective laser melting of reduced graphene oxide/S136 metal matrix composites with tailored microstructures and mechanical properties. *Materials & Design*, *175*, 107811. DOI: 10.1016/j.matdes.2019.107811

Wernet, G., Conradt, S., Isenring, H. P., Jiménez-González, C., & Hungerbühler, K. (2010). Life cycle assessment of fine chemical production: A case study of pharmaceutical synthesis. *The International Journal of Life Cycle Assessment*, *15*(3), 294–303. DOI: 10.1007/s11367-010-0151-z

White, S. R., Sottos, N. R., Geubelle, P. H., Moore, J. S., Kessler, M. R., Sriram, S. R., Brown, E. N., & Viswanathan, S. (2001). Autonomic healing of polymer composites. *Nature*, *409*(6822), 794–797. DOI: 10.1038/35057232 PMID: 11236987

Williams, D. B.. (2009). *The Transmission Electron Microscope*. Springer US., DOI: 10.1007/978-0-387-76501-3

Worsley, E. A., Margadonna, S., & Bertoncello, P. (2022). Application of Graphene Nanoplatelets in Supercapacitor Devices: A Review of Recent Developments. *Nanomaterials 2022, Vol. 12, Page 3600*, *12*(20), 3600. DOI: 10.3390/nano12203600

Wu, D. Y., Meure, S., & Solomon, D. (2008). Self-healing polymeric materials: A review of recent developments. *Progress in Polymer Science*, *33*(5), 479–522. DOI: 10.1016/j.progpolymsci.2008.02.001

Wu, K., Niu, Y., Zhang, Y., Yong, Z., & Li, Q. (2021). Continuous growth of carbon nanotube films: From controllable synthesis to real applications. *Composites. Part A, Applied Science and Manufacturing*, *144*, 106359. DOI: 10.1016/j.compositesa.2021.106359

Wu, T., Shen, J., Li, Z., Xing, F., Xin, W., Wang, Z., Liu, G., Han, X., Man, Z., & Fu, S. (2021). Microfluidic-integrated graphene optical sensors for real-time and ultra-low flow velocity detection. *Applied Surface Science*, *539*, 148232. DOI: 10.1016/j.apsusc.2020.148232

Wu, X., Ning, H., Liu, Y., Hu, N., Liu, F., Wang, S., Huang, K., Jiao, Y., Weng, S., Liu, Q., & Wu, L. (2020). Synergistic Delamination Toughening of Glass Fiber-Aluminum Laminates by Surface Treatment and Graphene Oxide Interleaf. *Nanoscale Research Letters*, *15*(1), 1–14. DOI: 10.1186/s11671-020-03306-z

Wu, Y., An, C., & Guo, Y. (2023). 3D printed graphene and graphene/polymer composites for multifunctional applications. *Materials (Basel)*, *16*(16), 5681. DOI: 10.3390/ma16165681 PMID: 37629973

Xie, X., Zhou, Y., & Huang, K. (2019). Advances in microwave-assisted production of reduced graphene oxide. *Frontiers in Chemistry*, *7*, 355. DOI: 10.3389/fchem.2019.00355 PMID: 31214562

Xiong, B., Liu, K., Xiong, W., Wu, X., & Sun, J. (2020). Strengthening effect induced by interfacial reaction in graphene nanoplatelets reinforced aluminum matrix composites. *Journal of Alloys and Compounds*, *845*, 156282. DOI: 10.1016/j.jallcom.2020.156282

Xu, S. (2024). Nanostructured Materials for Energy Storage in Future Electric Aircraft. *E3S Web of Conferences*, *553*, 02021. DOI: 10.1051/e3sconf/202455302021

Xu, F., Wu, X., Wang, H., Liu, H., & Ye, Z. (2022). Experimental research on moulding of graphene/PEKK composite powder by spark plasma sintering technology. *Applied Physics. A, Materials Science & Processing*, *128*(2), 122. DOI: 10.1007/s00339-022-05262-0

Xu, J., Li, R., Ji, S., Zhao, B., Cui, T., Tan, X., & Qiao, Y. (2021). Multifunctional graphene microstructures inspired by honeycomb for ultrahigh performance electromagnetic interference shielding and wearable applications. *ACS Nano*, *15*(5), 8907–8918. DOI: 10.1021/acsnano.1c01552 PMID: 33881822

Xu, L., Lu, H., Zhou, Y., Chi, Z., Li, Z., Md, Z. I., Dong, Y., Fu, Y., Zhu, Y., & Ni, Q. (2021). Ultrathin, ultralight, and anisotropic ordered reduced graphene oxide fiber electromagnetic interference shielding membrane. *Advanced Materials Technologies*, *6*(12), 2100531. DOI: 10.1002/admt.202100531

Xu, P., Loomis, J., Bradshaw, R. D., & Panchapakesan, B. (2012). Load transfer and mechanical properties of chemically reduced graphene reinforcements in polymer composites. *Nanotechnology*, *23*(50), 505713. DOI: 10.1088/0957-4484/23/50/505713 PMID: 23196792

Xu, W., Ren, C., Wang, Z., Lochab, B., Li, J., & Zhang, Y.. (2025). Fabrication of graphene oxide/silk protein core-sheath aerogel fibers for thermal management. *Journal of Materials Chemistry. A, Materials for Energy and Sustainability*, *13*(10), 7081–7090. DOI: 10.1039/D5TA00580A

Xu, X., Huang, J., Miao, G., Yan, B., Chen, Y., Zhou, Y., & Cai, W. (2025). Visualizing Thermal Reduction in Graphene Oxide. *Materials (Basel)*, *18*(10), 2222. DOI:10.3390/ma18102222

Yang, J., Luo, S., Zhou, X., Li, J., Fu, J., Yang, W., & Wei, D. (2019). Flexible, tunable, and ultrasensitive capacitive pressure sensor with microconformal graphene electrodes. *ACS Applied Materials & Interfaces*, *11*(16), 14997–15006. DOI: 10.1021/acsami.9b02049 PMID: 30869860

Yang, Q., Hu, J., Li, H., Du, Q., Feng, S., Yang, D., Zhang, Y., & Shen, J. (2025). All-optical modulation photodetectors based on the cds/graphene/ge sandwich structures for integrated sensing-computing. *Advanced Science (Weinheim, Baden-Wurttemberg, Germany)*, *12*(11), 2413662. Advance online publication. DOI: 10.1002/advs.202413662

Yang, S., Sha, S., Lu, H., Wu, J., Ma, J., Wang, D., Hou, C., & Sheng, Z. (2021). Graphene oxide and reduced graphene oxide coated cotton fabrics with opposite wettability for continuous oil/water separation. *Separation and Purification Technology*, *259*, 118095. DOI: 10.1016/j.seppur.2020.118095

Yang, Y., Li, B., Zhang, C., Wang, S., Liu, K., & Yang, B. (2015). Fabrication and properties of graphene reinforced silicon nitride composite materials. *Materials Science and Engineering A*, *644*, 90–95. DOI: 10.1016/j.msea.2015.07.062

Yang, Y., Liu, M., Zhou, S., Ren, W., Zhou, Q., & Zhang, W. (2021). Strengthening behaviour of continuous graphene network in metal matrix composites. *Carbon*, *182*, 825–836. DOI: 10.1016/j.carbon.2021.06.067

Yang, Y., Yin, Z., & Zhang, W. (2023). Graphene-based physical sensors for humidity, temperature and strain detections. *Microelectronic Engineering*, *279*, 112055. DOI: 10.1016/j.mee.2023.112055

Yang, Y., Yuan, H., Liu, M., Cheng, S., Li, W., Liang, F., Zheng, K., Liu, L., Yang, F., Liu, R., Su, Q., Qi, Y., & Liu, Z. (2025). Premelted-substrate-promoted selective etching strategy realizing cvd growth of high-quality graphene on dielectric substrates. *ACS Applied Materials & Interfaces*, *17*(4), 6825–6834. DOI: 10.1021/acsami.4c20313 PMID: 39809474

Yang, Y., Zhang, S., Huang, Z., Wu, H., Hu, D., Liu, H., & Bai, Y. (2025). Improvement of corrosion resistant of waterborne polyurethane coatings by addition of modified graphene oxide. *Surface Topography : Metrology and Properties*, *13*(2), 025008. DOI: 10.1088/2051-672X/add280

Yan, W., Liu, H. Y., & Mai, Y. W. (2004). Mode II delamination toughness of z-pinned laminates. *Composites Science and Technology*, *64*(13–14), 1937–1945. DOI: 10.1016/j.compscitech.2004.02.008

Yao, X., Kinloch, I. A., & Bissett, M. A. (2021). Fabrication and Mechanical Performance of Graphene Nanoplatelet/Glass Fiber Reinforced Polymer Hybrid Composites. *Frontiers in Materials*, *8*, 773343. DOI: 10.3389/fmats.2021.773343

Yao, Y., Zeng, X., Wang, F., Sun, R., Xu, J., & Wong, C. P. (2016). Significant Enhancement of Thermal Conductivity in Bioinspired Freestanding Boron Nitride Papers Filled with Graphene Oxide. *Chemistry of Materials*, *80*(4), 1357–1359. DOI: 10.1021/acs.chemmater.5b04187

Yavuz, M. T., & Ozkol, I. (2024). Thermal Structural Design Aspects of Military Aircraft. *Journal of Aeronautics and Space Technologies*, *17*(1), 124–158.

Yee, K., & Ghayesh, M. H. (2023a). A review on the mechanics of graphene nanoplatelets reinforced structures. *International Journal of Engineering Science*, *186*, 103831. DOI: 10.1016/j.ijengsci.2023.103831

Ye, J., Ji, X., Liu, Z., Liu, K., Li, J., Wang, R., Wang, J., & Lei, Q. (2024). Carbon fiber reinforced structural battery composites: Progress and challenges toward industrial application. *Composites. Part B, Engineering*, *277*, 111411. DOI: 10.1016/j.compositesb.2024.111411

Yi, M., & Shen, Z. (2015). A review on mechanical exfoliation for the scalable production of graphene. *Journal of Materials Chemistry. A, Materials for Energy and Sustainability*, *3*(22), 11700–11715. DOI: 10.1039/C5TA00252D

Yoon, D., & Cheong, H. (2012). Raman spectroscopy for characterization of graphene. In Raman Spectroscopy for Nanomaterials Characterization (pp. 191–214). Springer Berlin Heidelberg. DOI: 10.1007/978-3-642-20620-7_9

Yudhanto, A., Lubineau, G., Ventura, I. A., Watanabe, N., Iwahori, Y., & Hoshi, H. (2015). Damage characteristics in 3D stitched composites with various stitch parameters under in-plane tension. *Composites. Part A, Applied Science and Manufacturing*, *71*, 17–31. DOI: 10.1016/j.compositesa.2014.12.012

Yu, H., Chen, C., Sun, J., Zhang, H., Feng, Y., Qin, M., & Feng, W. (2022). Highly Thermally Conductive Polymer/Graphene Composites with Rapid Room-Temperature Self-Healing Capacity. *Nano-Micro Letters*, *14*(1), 135. DOI: 10.1007/s40820-022-00882-w PMID: 35704244

Yu, K., Wang, M., Qian, K., Lu, X., & Sun, J. (2016). The synergy effect of Graphene/SiO2 hybrid materials on reinforcing and toughening epoxy resin. *Fibers and Polymers*, *17*(3), 453–459. DOI: 10.1007/s12221-016-5862-8

Yuwen, C., Liu, B., Zhang, L., Guo, S., & Peng, J. (2019). Synthesis high-qulity graphene oxide and temperature-dependent dielectric properties of reduced graphene oxide. Materials Research Express, 6(9), 0950b4. DOI: 10.1088/2053-1591/ab149f

Zare, P., Aleemardani, M., Seifalian, A., Bagher, Z., & Seifalian, A. M. (2021). Graphene oxide: Opportunities and challenges in biomedicine. *Nanomaterials (Basel, Switzerland)*, *11*(5), 1083. DOI: 10.3390/nano11051083 PMID: 33922153

Zarepour, A., Ahmadi, S., Rabiee, N., Zarrabi, A., & Iravani, S. (2023). Self-Healing MXene- and Graphene-Based Composites: Properties and Applications. *Nano-Micro Letters*, *15*(1), 100. DOI: 10.1007/s40820-023-01074-w PMID: 37052734

Zeinedini, A. (2024). Fracture toughness of graphene/polymer nanocomposites: Well dispersion, agglomeration and toughening mechanisms. *Theoretical and Applied Fracture Mechanics*, *131*, 104449. DOI: 10.1016/j.tafmec.2024.104449

ZEROe: our hydrogen-powered aircraft | Airbus. Retrieved June 25, 2025, from https://www.airbus.com/en/innovation/energy-transition/hydrogen/zeroe-our-hydrogen-powered-aircraft

Zhang, S. L. (2015). Polypyrrole/graphene oxide films based microwave absorbing fabric through layer-by-layer assembly.

Zhang, B., Asmatulu, R., Soltani, S. A., Le, L. N., & Kumar, S. S. A. (2014). Mechanical and thermal properties of hierarchical composites enhanced by pristine graphene and graphene oxide nanoinclusions. *Journal of Applied Polymer Science*, *131*(19), app.40826. Advance online publication. DOI: 10.1002/app.40826

Zhang, H., Cao, J., Wu, W., Cao, Z., & Ma, H. (2016). Layer-by-layer assembly of graphene oxide on viscose fibers for the fabrication of flexible conductive devices. *Cellulose (London, England)*, *23*(6), 3761–3770. DOI: 10.1007/s10570-016-1088-6

Zhang, H., Gan, X., Lu, Z., Li, L., & Lu, L. (2025). Polycarboxylate superplasticizer-modified graphene oxide: Dispersion and performance enhancement in cement paste. *Nanomaterials (Basel, Switzerland)*, *15*(6), 419. DOI: 10.3390/nano15060419

Zhang, H., Huang, J., Wang, Y., Liu, R., Huai, X., Jiang, J., & Anfuso, C. (2018). Atomic force microscopy for two-dimensional materials: A tutorial review. *Optics Communications*, *406*, 3–17. DOI: 10.1016/j.optcom.2017.05.015

Zhang, H., Yang, Y., Ren, D., Wang, L., & He, X. (2021). Graphite as anode materials: Fundamental mechanism, recent progress and advances. *Energy Storage Materials*, *36*, 147–170. DOI: 10.1016/j.ensm.2020.12.027

Zhang, J., Wang, X., Chen, X., Xia, X., & Weng, G. J. (2022). Piezoelectricity enhancement in graphene/polyvinylidene fluoride composites due to graphene-induced α→ β crystal phase transition. *Energy Conversion and Management*, *269*, 116121. DOI: 10.1016/j.enconman.2022.116121

Zhang, R., Long, Z., Long, G., Wang, J., Wang, X., Zhang, X., & Jiang, Y. (2024). Mechanism of graphene oxide concrete macro-micro properties evolution under large temperature difference freeze-thaw action. *Construction & Building Materials*, *415*, 135019. DOI: 10.1016/j.conbuildmat.2024.135019

Zhang, X., Shi, C., Liu, E., He, F., Ma, L., Li, Q., Li, J., Bacsa, W., Zhao, N., & He, C. (2017). Achieving high strength and high ductility in metal matrix composites reinforced with a discontinuous three-dimensional graphene-like network. *Nanoscale*, *9*(33), 11929–11938. DOI: 10.1039/C6NR07335B

Zhao, X., Zhang, Q., Chen, D., & Lu, P. (2011). Enhanced mechanical properties of graphene-based poly (vinyl alcohol) composites. *Macromolecules*, *44*(7), 2392–2392. DOI: 10.1021/ma200335d

Zhong, Y. L., Tian, Z., Simon, G. P., & Li, D. (2015). Scalable production of graphene via wet chemistry: Progress and challenges. *Materials Today*, *18*(2), 73–78. DOI: 10.1016/j.mattod.2014.08.019

Zhou, C., Wu, S., Yang, B., Gao, Y., Wu, G., & Zhang, H. (2014). Toughening Polystyrene by Core-shell Rubber Particles: Analysis of the Internal Structure and Properties. *Polymers & Polymer Composites*, *23*(5), 317–324. DOI: 10.1177/096739111502300505

Zhou, L., Qiu, J., Wang, X., Wang, H., Wang, Z., Fang, D., & Li, Z. (2020). Mechanical and dielectric properties of reduced graphene oxide nanosheets/alumina composite ceramics. *Ceramics International*, *46*(12), 19731–19737. DOI: 10.1016/j.ceramint.2020.04.293

Zhu, D. (2018). *Aerospace Ceramic Materials: Thermal, Environmental Barrier Coatings and SiC/SiC Ceramic Matrix Composites for Turbine Engine Applications*. http://www.sti.nasa.gov

Zhu, C., Zhang, Y., & He, G. (2025). Graphene-Based Textile Sensors for Intelligent Structural Health Monitoring. *Polymers, 17*(11), 1484. DOI: 10.3390/polym17111484 PMID: 40508727

Zhu, S. E., Yuan, S., & Janssen, G. C. A. M. (2014). Optical transmittance of multilayer graphene. *Europhysics Letters, 108*(1), 17007. DOI: 10.1209/0295-5075/108/17007

Zhu, S., Wang, M., Qiang, Z., Song, J., Wang, Y., Fan, Y., & Ye, C. (2021). Multi-functional and highly conductive textiles with ultra-high durability through 'green' fabrication process. *Chemical Engineering Journal, 406*, 127140. DOI: 10.1016/j.cej.2020.127140

Zitoun, A., Fakis, D., Jayasree, N., Omairey, S., Oikonomidis, F., Stoeva, Z., & Kazilas, M. (2022). Graphene-based strain sensing in composites for structural and health monitoring applications. *SN Applied Sciences, 4*(2), 58. DOI: 10.1007/s42452-022-04940-1

Zotti, A., Zuppolini, S., Borriello, A., Vinti, V., Trinchillo, L., Borrelli, D., Caraviello, A., & Zarrelli, M. (2022). Effect of the Mixing Technique of Graphene Nanoplatelets and Graphene Nanofibers on Fracture Toughness of Epoxy Based Nanocomposites and Composites. *Polymers 2022, Vol. 14, Page 5105, 14*(23), 5105. DOI: 10.3390/polym14235105

Zuo, D., Li, G., Ling, Y., Cheng, S., Xu, J., & Zhang, H. (2021). Durable UV-blocking property of cotton fabrics with nanocomposite coating based on graphene oxide/ZnO quantum dot via water-based self-assembly. *Fibers and Polymers, 22*(7), 1837–1843. DOI: 10.1007/s12221-021-0750-2

About the Contributors

R.G. Abaszade, Ph.D., is an associate professor of Azerbaijan State Oil and Industry University (2022), Azerbaijan University of Architecture and Construction (2024). He is the author of 149 articles, 10 textbooks, 11 subject programs and 6 inventions, and papers in refereed journals with over 929 citations and h-index of 18, according to Google Scholar database. Dr. Abaszade received the degree of physical electronics (Bachelor) at Baku State University in 2003, Optoelectronics (Master) at Baku State University in 2006, and PhD in physics in 2018. In 2019, he created the "Innovation research" department at the International Academy of Ecoenergetics and continues to research the purchase, analysis and applications of nanomaterials. Our research works are carried out in the framework of joint cooperation with the scientific centers of the USA, Germany, Belarus, Kazakhstan, Turkey, Russia, Ukraine, Uzbekistan, Pakistan, Hungary, England, Mexico, Chili, China, İndia and Italy. Together with our research group, we are engaged in complex work related to the acquisition, addition, analysis and applications of graphene-based samples. At the same time, we create various nanoelements (negatrons, diodes, capacitors and transistors) based on graphene-based materials. In our department, my research group is continuing experimental work in the direction of obtaining various samples, studying them and preparing nanomatrix devices based on new smart technology for the implementation of smart city and alternative energy supply. He is the editor-in-chief of Ecoenergetics journal, and a leading reviewer of the editorial board of various journals. In 2019, he became a professor of the "International Academy of Ecoenergetics".

Saifullah Khalid stands as a beacon of innovation and excellence in Electrical Engineering and Aviation Technology. With a distinguished career that spans over two decades, his contributions have been instrumental in shaping the future of air traffic management and engineering research. His impressive credentials include serving as a Visiting Researcher at Nanyang Technological University,

Singapore, and holding a pivotal role as an Assistant General Manager (ATM) at SGRDJ International Airport Amritsar, in addition to his association with the Civil Aviation Research Organization (CARO) in India. Dr. Khalid's academic prowess is underlined by a Doctor of Philosophy in Electrical Engineering from Dr. APJ Abdul Kalam Technical University, India. His commitment to education and research is further evidenced by his role in nurturing the next generation of engineers and researchers and his mentorship of PhD students focusing on cutting-edge technologies in aviation. A testament to his inventive genius, Dr. Khalid has secured 72 patents, showcasing groundbreaking technologies that range from AI-enhanced fault detection systems to innovative solutions in renewable energy harvesting. His intellectual contributions also extend to over 127 journal articles and conference papers, alongside several authored and edited books that delve into the applications of artificial intelligence in aviation and electrical engineering. Dr. Khalid's accolades are numerous and noteworthy. He holds the world records for the most patents granted in a single day and has been recognized internationally with awards such as the Excellence in Innovation at International Research Awards and the Best Researcher Award across multiple platforms. These achievements highlight his unparalleled commitment to innovation and excellence in his field. Furthermore, Dr. Khalid has effectively bridged the gap between theoretical research and practical applications, significantly advancing technology in air traffic management and electrical engineering. His expertise is not just limited to research and development but also encompasses the planning and implementation of research facilities, showcasing his prowess in project management and strategic planning. In summary, Dr. Saifullah Khalid's curriculum vitae is a testament to a career built on innovation, leadership, and an unwavering commitment to advancing technology and engineering. His achievements not only underscore his personal dedication and talent but also his significant contributions to the fields of aviation and electrical engineering, making him an invaluable asset to the academic and professional communities he serves.

M.O. Stetsenko earned his Ph.D. in Solid State Physics from the V.E. Lashkaryov Institute of Semiconductor Physics, National Academy of Sciences of Ukraine, in 2016. He served as a Senior Researcher at the same institute from 2019 to 2022. In 2019, he joined Shenzhen University, China, as a Postdoctoral Fellow, and continued his postdoctoral research at Westlake University from 2021 to 2023. In 2024, he was appointed Research Assistant Professor at Westlake University, Hangzhou, China. His research interests include micro and nanofabrication, carbon nanomaterials, optical engineering, and nanophononics. Dr. Stetsenko has authored approximately 75 scientific publications, including 27 indexed in the Scopus database, along with three textbooks and three patents. His research contributions have garnered over 266 citations, with an h-index of 10.

***

Vugar Abdullayev is a doctor of technical sciences, Assos. Prof., Azerbaijan State Oil and Industry University. Baku, Azerbaijan. He has completed his Ph.D in Computer Science in the year 2005. He is currently as a Assos. Professor of the "Computer engineering" Department at the Azerbaijan State Oil and Industry University, Baku, Azerbaijan. He is author of 187 scientific papers.

Shyam Agasthya Civil Engineering Professional with an inclination towards Business Management, Financial Management and Business Strategy. MBA in Finance from State University of New York at Oswego. Five Years of Industry Experience handling over 40 construction projects at different stages of completion. Thirteen years of Academic Teaching Experience. Currently working at National University of Science and Technology, College of Engineering, Muscat, Sultanate of Oman as a Faculty in Department of Civil and Environmental Engineering.

Shashi Kant Gupta, Post-Doctoral Fellow and Researcher, Computer Science and Engineering, Eudoxia Research University, USA in collaboration with Eudoxia Research Centre, India. ORCID: 0000-0001-6587-5607. He is currently working as an Adjunct Research Faculty, Centre for Research Impact & Outcome, Chitkara University Institute of Engineering and Technology. Chitkara University, Rajpura, 140401, Punjab, India. He is working as Honorary Adjunct Faculty, School of Computing, Maryam Abacha American University of Nigeria (MAAUN), Nigeria. He is working as an Honorary Senior Research Fellow, Department of Scientific Research, Innovation and Training of Scientific and Pedagogical Staff, University of Economics and Pedagogy, Karshi City, Uzbekistan. He is working as a Research Collaborator and Invited Visiting Senior Scientist at the Research Institute of IoT and Cybersecurity, Department of Electronic Engineering, National Kaohsiung University of Science and Technology, Taiwan. He has completed his Ph.D. (CSE) from Integral University, Lucknow, UP, India, & Worked as Assistant Professor in the Department of Computer Science and Engineering, ITM, Lucknow, U.P., India & Worked as Assistant Professor in the Department of Computer Science and Engineering, PSIT, Kanpur, U.P., India, Worked as Associate Professor, School of Computer Applications, BBD University, Lucknow, UP, India, Worked as Assistant Professor, Department of Computer Science and Engineering, Ambalika Institute of Management and Technology, Lucknow, UP, India. & Also worked as Senior Lecturer, Department of IT, MCSCET, Lucknow, UP, India. He is currently working as Founder and CEO of CREP PVT. LTD., Lucknow, UP, India. He is a member of Spectrum IEEE & Potentials Magazine IEEE since 2019 and many more international organizations for research activities. He is an Editor-in-chief of International Journal of Data Informatics and Intelligent computing (IJDIIC) also Editor-in-chief of International Journal of Emerging Technologies in Computer and Communication (IJETCC). Senior Editor in Global Research Journal form London Organization of Skill Development, London, United Kingdom. He has published many research papers in reputed international journals, conferences & seminars with SCOPUS and SCI indexing. He has got many awards from many International and national organizations. He has organized various Faculty Development Programs, Seminars, Workshops, and Short-Term Courses at University level. His main research work focuses on Cloud computing, Big Data Analytics, IoT and Computational Intelligence-based Education. He is currently working as a reviewer and editorial in various international journals. He is currently editor in many edited books of different publishers like CRC Press, Taylor and Francis group Publication, Routledge Publisher, Bentham Science Publishers, Springer Nature Publisher, Wiley Publisher etc. He has published many patents like Indian, Germany, UK etc in the field of CSE and IT. He has more than 12 years of teaching experience, 2 years of Industrial Experience and more than 2.5 years as CEO and Founder of a firm.

M K Harikeerthan has an experience in teaching for more than 22 years. Currently working as an Associate Professor at Dayanand Sagar Academy of Technology and Management, Bangalore, India. Skilled in testing of Highway Materials and Concrete, AutoCAD. Educational professional with a PhD in Highway Engineering from B. M. S. College of Engineering and focused in teaching Highway Technology and Transportation subjects in the undergraduate course of Civil Engineering.

J. Mangaiyarkkarasi is currently working as an Associate Professor of Physics in Nadar Mahajana Sangam S. Vellaichamy Nadar College, Nagamalai, Madurai-625 019, Tamil Nadu, India. She has 25 years of teaching experience. Her major research interest includes synthesis of ceramic materials, powder X-ray analysis and electron density studies. She has published 11 research papers in the refereed International journals and 17 book chapters. She has prepared study materials (UG & PG Physics) for Madurai Kamaraj University Distance Education Course.

Saifullah Khalid obtained two Ph.D. degrees: one in Electronics and Communication Engineering SNU, India. His research interests are in the area of UAVs. He is currently post doctoral researcher at Lincoln University College, Malaysia.

Leevesh Kumar-Associate Professor Civil Engineering Dr. Leevesh Kumar is a distinguished academician and researcher with over 13 years of progressive experience in the field of civil and construction engineering across India, Ethiopia, the United Kingdom, and Bangladesh. Currently serving as an associate professor at the World University of Bangladesh and holding a concurrent part-time role with Liverpool John Moores University and the University of East London (UK), Dr. Kumar has built a diverse and global portfolio in higher education and construction technology research. He holds a Ph.D. in Civil Engineering from Career Point University (India), preceded by both M.Tech and B.Tech degrees in Civil Engineering from Bhagwant University. His educational foundation is complemented by a postgraduate diploma and several certifications from prestigious institutions like IIT Madras and NITTTR Chandigarh. Throughout his academic journey, Dr. Kumar has remained committed to advancing sustainable practices in the construction industry. Dr. Kumar's research focuses on sustainable construction materials, with specific interest in the mechanical and durability

Raji V. Nair is an Assistant Professor in the Department of Chemistry at Sri Sathya Sai Institute of Higher Learning, Anantapur Campus, Andhra Pradesh, India. She obtained her B.Sc. and M.Sc. degrees in Chemistry from the University of Kerala and completed her Ph.D. in Chemical Sciences at the CSIR–National Institute for Interdisciplinary Science and Technology (NIIST), Thiruvananthapuram, Kerala. Her research expertise centers on the development of advanced nanomaterials and biomaterials for environmental remediation, drug delivery, and strategic sensing applications. Dr. Nair has published extensively in reputed international journals and conferences in the fields of nanotechnology and materials chemistry.

Sushila Sahani holds a doctorate in Apparel and Textile Science. Her academic and professional journey blends expertise in textiles from material properties to sustainable innovations, with interests in fashion design, intellectual property and ethical consumerism. Her research and educational work explore the connections between textile knowledge and design thinking, highlighting the economic, cultural and sustainable impact of the fashion and textile industry.

Sitesh Kumar Singh is an accomplished academic and Chartered Engineer with a Ph.D. in Civil Engineering and over nine years of diverse experience in education, research, consultancy, and educational administration. Currently serving as an Assistant Professor in the Department of Civil and Environmental Engineering at the National University of Science and Technology, Muscat, Sultanate of Oman, he specializes in Civil and Transportation Engineering, Project Management, and Data Analytics. Dr. Singh holds multiple advanced degrees including M.Tech,

B.Tech, PGDIPR, PGDUPDL, and an PGD in Education. His career encompasses curriculum development, research supervision, quality assurance, and strategic academic planning. He has supervised over 40 postgraduate and 2 doctoral students, contributing to academic leadership through workshops, conferences, and editorial roles, and has published over 40 research articles, several patents, and authored books and book chapters.

Index

Symbols

A

C

E

F

G

I

L

M

N

P

R

S

T

U

www.ingramcontent.com/pod-product-compliance
Lightning Source LLC
LaVergne TN
LVHW062327300926
849161LV00007B/1598